MERCURY, MERCURIALS And MERCAPTANS

A Proceedings Publication of the
Rochester International Conferences
on Environmental Toxicity

Organized by the

Department of Radiation Biology and Biophysics
School of Medicine and Dentistry
The University of Rochester
Rochester, New York

during 17–19 June 1971

Conference Committee

Thomas W. Clarkson
Morton W. Miller
Jaroslav J. Vostal

Grateful acknowledgment is extended to the Division of Nuclear Education and Training of the United States Atomic Energy Commission and the New York State Science and Technology Foundation which granted funds in partial support of the conference.

MERCURY, MERCURIALS And MERCAPTANS

Edited by

MORTON W. MILLER

THOMAS W. CLARKSON

*Department of Radiation Biology
and Biophysics
School of Medicine and Dentistry
The University of Rochester
Rochester, New York*

CHARLES C THOMAS • PUBLISHER
Springfield • *Illinois* • *U.S.A.*

Published and Distributed Throughout the World by
CHARLES C THOMAS • PUBLISHER
BANNERSTONE HOUSE
301–327 East Lawrence Avenue, Springfield, Illinois, U.S.A.

This book is protected by copyright. No part of it may be reproduced in any manner without written permission from the publisher.

© *1973, by* CHARLES C THOMAS • PUBLISHER

ISBN 0-398-02600-9

Library of Congress Catalog Card Number: 72-81709

With THOMAS BOOKS *careful attention is given to all details of manufacturing and design. It is the Publisher's desire to present books that are satisfactory as to their physical qualities and artistic possibilities and appropriate for their particular use.* THOMAS BOOKS *will be true to those laws of quality that assure a good name and good will.*

Printed in the United States of America

BB-14

CONTRIBUTORS

JAMES BARBER
*Department of Botany
Imperial College of Science and Technology
London, England*

W. BEAUFORD
*Department of Botany
Imperial College of Science and Technology
London, England*

MATHS BERLIN
*Institute of Hygiene
University of Lund
Lund, Sweden*

WANDA BOLANOWSKA
*Department of Biochemistry
Institute of Occupational Medicine
Lodz, Poland*

THOMAS W. CLARKSON
*Department of Radiation Biology and Biophysics
School of Medicine and Dentistry
The University of Rochester
Rochester, New York*

S. C. FANG
*Department of Agricultural Chemistry
Oregon State University
Corvallis, Oregon*

ERNEST C. FOULKES
*Departments of Environmental Health and Physiology
University of Cincinnati
Cincinnati, Ohio*

LARS FRIBERG

*Department of Environmental Hygiene
The Karolinska Institute
Stockholm, Sweden*

JOHN C. GAGE

*Imperial Chemical Industries Limited
Industrial Hygiene Research Laboratories
Macclesfield, Cheshire, England*

LEONARD J. GOLDWATER

*Duke University Medical Center
Durham, North Carolina*

CRAWFORD A. GRANT

*Pathological Institution
Karolinska Hospital
Stockholm, Sweden*

J. HELLBERG

*Institute of Hygiene
University of Lund
Lund, Sweden*

GERALD H. HIRSCH

*Department of Pharmacology
Michigan State University
East Lansing, Michigan*

JERRY B. HOOK

*Department of Pharmacology
Michigan State University
East Lansing, Michigan*

ARNE JERNELÖV

*Swedish Water and Air Pollution Research Laboratory
Stockholm, Sweden*

Contributors

HIROSHI KASHIWAZAKI

*Department of Human Ecology
School of Health Sciences
University of Tokyo
Tokyo, Japan*

LEONARD T. KURLAND

*Department of Medical Statistics and Epidemiology
Mayo Clinic
Rochester, Minnesota*

LASZLO MAGOS

*Medical Research Council Laboratories
Toxicology Unit
Carshalton, Surrey, England*

TOMOYO MIYAMA

*Japan Women's College of Physical Education
Tokyo, Japan*

JORMA K. MIETTINEN

*Department of Radiochemistry
University of Helsinki
Helsinki, Finland*

BOHDAN R. NECHAY

*Department of Pharmacology and Toxicology
University of Texas Medical Branch
Galveston, Texas*

F. NIELSEN KUDSK

*Institute of Pharmacology
University of Aarhus
Aarhus, Denmark*

GUNNAR NORDBERG

*Department of Environmental Hygiene
The Karolinska Institute
Stockholm, Sweden*

TOR NORSETH

*Institute of Occupational Health
Oslo, Norway*

JERZY K. PIOTROWSKI

*Department of Biochemistry
Institute of Occupational Medicine
Lodz, Poland*

ASER ROTHSTEIN

*Department of Radiation Biology and Biophysics
School of Medicine and Dentistry
The University of Rochester
Rochester, New York*

Y. J. SHIEH

*Department of Botany
Imperial College of Science and Technology
London, England*

TSUGUYOSHI SUZUKI

*Department of Human Ecology
School of Health Sciences
University of Tokyo
Tokyo, Japan*

TAI-ICHIRO TAKEMOTO

*Department of Human Ecology
School of Health Sciences
University of Tokyo
Tokyo, Japan*

BARBARA TROJANOWSKA

*Department of Biochemistry
Institute of Occupational Medicine
Lodz, Poland*

JAROSLAV J. VOSTAL

*Department of Pharmacology and Toxicology
School of Medicine and Dentistry
The University of Rochester
Rochester, New York*

Contributors

J. HENRY WILLS

Institute of Experimental Pathology and Toxicology
The Albany Medical College of Union University
Albany, New York

JUSTYNA M. WISNIEWSKA-KNYPL

Department of Biochemistry
Institute of Occupational Medicine
Lodz, Poland

WELCOME

What makes this conference timely it seems to me is the fact that we need just what I hope will happen here. We need a group of people with special knowledge and special skills in the field of toxicity, and particularly in the fields of mercurials, sitting down together and really critically analyzing just what we do know, just what the gaps in our knowledge are, and what the failures in our skills happen to be. This way each one of us will go away from the conference with a clearer knowledge than that with which we arrived concerning what needs to be done both in gaining new knowledge and in developing new techniques and new skills in this important area of concern. If this comes out of the conference then I'm sure it will be a success for each and every one of us. It will certainly be a success for Rochester. And it will be a success for the ecology in direct proportion to your ability to analyze the problem and prepare for new directions of your endeavors.

So, I'll welcome you to Rochester to what I hope will be a stimulating and productive conference.

J. LOWELL ORBISON, M.D.
Dean, School of Medicine and Dentistry
The University of Rochester
Rochester, New York

INTRODUCTION

I hadn't seen the program until about five minutes ago, which of course speaks very well for the efficiency of the arrangements committee, and now I see that I'm here in the beginning and I'm here at the end so, this is the Alpha and Omega, with all that that implies.

I am reminded of what happened a few years ago and of what was probably the greatest mistake I ever made in my life. When somebody wrote to me from this department and said they were planning a first conference on environmental toxicity and would welcome any suggestions that I might make, I said I would be very, very happy to send my suggestions on one condition and that is that I'm not expected to attend the meeting. I think that it will be a long time before I get over the regrets I felt because it was that meeting which resulted in the production of a very excellent book which I'm happy to endorse, *Chemical Fallout: Current Research on Persistent Pesticides*. I use it very often.

Now, whether or not that had anything to do with the arrangements or lack of arrangement that got me here this time I don't know, but here I am and I'm very happy to be here.

Leonard J. Goldwater
Duke University Medical Center
Durham, North Carolina

CONTENTS

	Page
Contributors	v
Welcome—J. Lowell Orbison	xi
Introduction—Leonard J. Goldwater	xiii

SESSION I

THE GENERAL TOXICITY OF MERCURY AND ITS COMPOUNDS

Thomas W. Clarkson, *Chairman*

Chapter

1. INORGANIC MERCURY—A TOXICOLOGICAL AND EPIDEMIOLOGICAL APPRAISAL—*Lars Friberg and Gunnar Nordberg* 5
2. AN APPRAISAL OF THE EPIDEMIOLOGY AND TOXICOLOGY OF ALKYLMERCURY COMPOUNDS—*Leonard T. Kurland* 23
3. ARYL- AND ALKOXYALKYLMERCURIALS—*Leonard J. Goldwater* . 56
4. MERCAPTANS, THE BIOLOGICAL TARGETS FOR MERCURIALS—*Aser Rothstein* 68

SESSION II

ACTION OF MERCURY ON MEMBRANES AND TRANSPORT PROCESSES

David W. Fassett, *Chairman*

5. SITE OF THE FUNCTIONAL LESION RESPONSIBLE FOR AMINOACIDURIA AFTER ADMINISTRATION OF ORGANOMERCURIALS AND OTHER METAL COMPOUNDS—*E.C. Foulkes* 99
6. ACTION OF MERCURY ON RENAL SODIUM TRANSPORT AND ADENOSINETRIPHOSPHATASE ACTIVITY—*Bohdan R. Nechay* .. 111

7. EFFECT OF ORGANIC MERCURIAL COMPOUNDS ON RENAL ORGANIC ION TRANSPORT—*Jerry B. Hook and Gerald H. Hirsch* 124
8. MERCURIC ION, ORGANOMERCURIALS AND RENAL DIURESIS—*Jaroslav J. Vostal and Thomas W. Clarkson* 139

SESSION III

PHARMACOKINETICS OF MERCURY METABOLISM

J. Henry Wills, *Chairman*

9. INTRODUCTION TO SESSION III—*J. Henry Wills* 163
10. FACTORS AFFECTING THE UPTAKE AND RETENTION OF MERCURY BY KIDNEYS IN RATS—*Laszlo Magos* 167
11. THE UPTAKE AND DISTRIBUTION OF METHYLMERCURY IN THE BRAIN OF *Saimiri Sciureus* IN RELATION TO BEHAVIORAL AND MORPHOLOGICAL CHANGES—*M. Berlin, G. Nordberg and J. Hellberg* 187
12. METABOLIC FATE OF ETHYLMERCURY SALTS IN MAN AND ANIMAL—*Tsuguyoshi Suzuki, Tai-Ichiro Takemoto, Hiroshi Kashiwazaki and Tomoyo Miyama* 209
13. ABSORPTION AND ELIMINATION OF DIETARY MERCURY (Hg^{2+}) AND METHYLMERCURY IN MAN—*Jorma K. Miettinen* 233

SESSION IV

BIO-COMPLEXES AND CHELATES OF MERCURY

George Kazantzis, *Chairman*

14. FURTHER INVESTIGATIONS ON BINDING AND RELEASE OF MERCURY IN THE RAT—*Jerzy K. Piotrowski, Barbara Trojanowska, Justyna M. Wisniewska-Knypl and Wanda Bolanowska* 247
15. BILIARY COMPLEXES OF METHYLMERCURY: A POSSIBLE ROLE IN ORGAN DISTRIBUTION—*Tor Norseth* 264

16. THE *In Vivo* KINETICS OF MERCURY BINDING OF KIDNEY SOLUBLE PROTEINS FROM RATS RECEIVING VARIOUS MERCURIALS—S.C. Fang 277
17. PATHOLOGY OF EXPERIMENTAL METHYLMERCURY INTOXICATION: SOME PROBLEMS OF EXPOSURE AND RESPONSE—C.A. Grant 294

SESSION V

BIOTRANSFORMATION OF MERCURY AND ORGANOMERCURIALS

Robert J.M. Horton, *Chairman*

18. A NEW BIOCHEMICAL PATHWAY FOR THE METHYLATION OF MERCURY AND SOME ECOLOGICAL IMPLICATIONS—*Arne Jernelöv* 315
19. SOME ASPECTS OF MERCURY UPTAKE BY PLANT, ALGAL AND BACTERIAL SYSTEMS IN RELATION TO ITS BIOTRANSFORMATION AND VOLATILIZATION—*J. Barber, W. Beauford and Y.J. Shieh* 325
20. THE METABOLISM OF METHOXYETHYLMERCURY AND PHENYLMERCURY IN THE RAT—*J.C. Gage* 346
21. BIOLOGICAL OXIDATION OF ELEMENTAL MERCURY—*F. Nielsen Kudsk* 355

Author Index 373

Subject Index 376

MERCURY, MERCURIALS And MERCAPTANS

SESSION I
THE GENERAL TOXICITY OF MERCURY AND ITS COMPOUNDS

Thomas W. Clarkson, *Chairman*

Chapter 1

INORGANIC MERCURY—A TOXICOLOGICAL AND EPIDEMIOLOGICAL APPRAISAL

Lars Friberg and Gunnar Nordberg

ABSTRACT

A summary is given of the metabolism of inorganic mercury as well as dose-response relationships. Urine and blood analyses as indicators of exposure and toxic effects are dealt with. The necessity of treating metallic mercury and different inorganic mercury compounds separately is pointed out.

REPORT

In this presentation we shall define inorganic mercury as "mercury in the form of elemental mercury, mercurous and mercuric salts, as well as those complexes in which mercuric ions can form reversible bonds to such ligands as thiol groups on proteins." This definition was set forth in November 1968, at the Stockholm Symposium on Maximum Allowable Concentrations of Mercury Compounds.

Acute intoxication caused by inorganic mercury is seldom seen nowadays. The critical organ is usually the kidneys, even though pronounced symptoms from the gastrointestinal tract may occur after oral exposure. After inhalation of massive doses, symptoms of pulmonary irritation may be seen. Signs from the central nervous system, prominent after long-term exposure, are by comparison of secondary importance after acute exposure.

Chronic intoxication is a well-documented result of industrial mercury exposure. The classical symptoms include tremor and psychological disturbances from the central nervous system, often called *erethismus mercurialis*. Kidney damage does not seem to be a common sign of chronic intoxication, though proteinuria may occur and may develop into a nephrotic syndrome.

There are no reports of prenatal mercury intoxication following exposure to inorganic mercury, perhaps because the transplacental passage of mercuric mercury is very limited. All mercury compounds have some genetic activity, particularly with regard to the causing of c-mitosis; the toxicity compared with that of methylmercury is, however, insignificant (Ramel, to be published).

Absorption

Mercury may enter the body via inhalation, ingestion or skin peneration. For *elemental mercury* the respiratory route is by far the most important. The absorption via the gastrointestinal tract is negligible. Experiments on rats by Bornmann et al. (6) have shown an absorption of ingested elemental mercury of less than 0.01 percent.

Whether there is a direct transfer of mercury via the skin has been a matter of dispute since the beginning of this century. Some valuable animal studies were made by Juliusberg (14) and Schamberg et al. (29), who prevented the animals from inhaling the vapors from the mercury applied onto the skin and showed that an appreciable skin penetration occurred. No exact quantitative evaluation of the penetration rate can be made. Metallic mercury was unfortunately not included in Wahlberg's (11,30,39) extensive studies in the 1960's on skin penetration in guinea pigs.

The absorption via inhalation is high. The percentage of inhaled mercury taken up by the body in animal experiments has been estimated at 75 to 100 percent by Hayes and Rothstein (13) and Magos (18), while Gage (12) reported about 50 percent absorption. In human beings, Teisinger and Fiserova-Bergerova (35) and Nielsen-Kudsk (23) found a retention of about 75 percent at mercury concentrations between $50 \mu g/m^3$ to $350 \mu g/m^3$ of inhaled air. There seems to be a very rapid diffusion of mercury vapor through the alveolar membrane as shown by Magos (19) and Berlin et al. (4) In the latter report, only about 30 percent of the whole-body burden was left in the lungs immediately after a ten minute exposure.

Inorganic Mercury Compounds

There is every reason to believe that aerosols of mercury will follow general physical laws governing deposition in the respiratory system (34). Particles with a high probability of deposition in the upper airways should be cleared fast. For particles deposited in the pulmonary compartment, longer half-lives will be found, the duration of which will depend on the solubility and other chemical properties of the substance. No systematic studies have been made for aerosols of different mercury compounds. Morrow et al. (20) found in dogs that about 45 percent or an HgO aerosol with a mean diameter of 0.16μ was cleared in less than 24 hours, but that the remainder was cleared with a half-life of 33 days.

For water-soluble mercury compounds, the gastrointestinal absorption is probably considerable. Data from studies on rats by Prickett et al. (26) favor a 10 to 30 percent absorption. Mercurous compounds are much less soluble and the absorption is smaller. These compounds have been extensively used as laxatives.

In the studies by Wahlberg referred to earlier, about 5 percent of the mercury in a 2 percent water solution of mercuric chloride was absorbed via intact skin of a guinea pig over a five-hour period. If the penetration rate is the same for humans, the absorption may be considerable under unfavorable conditions.

Biotransformation, Transport and Distribution

A rapid biotransformation of mercury from elemental mercury to mercuric ion was shown in *in vitro* studies by Clarkson et al. (8). Several later studies [see e.g. Berlin et al. (2)] have shown that the uptake of mercury in the brain and the red cells is considerably higher after exposure to mercury vapor than after injection of mercuric salt, indicating that mercury exists in the blood in a different state after exposure to vapor than to mercuric salt. It is now evident (18,19) that during the very early period after exposure to mercury vapor, mercury is partly distributed in the blood as mercury vapor. This can explain the higher penetration and uptake of mercury in the brain after mercury vapor exposure.

Excretion

All available data show that practically all mercury is excreted via the urine and feces. After a single exposure to mercuric mercury via the parenteral route in rats, the fecal excretion during the first week or weeks exceeded the urinary excretion (28,33) while later the urinary excretion equalled or surpassed the fecal excretion. The higher the dose, the greater is the excretion via the urine (28). This might be true also after repeated exposure. Friberg (10) in rats given subcutaneous injections corresponding to 0.5 mg Hg/kg seven days a week found at equilibrium that 70 percent of the injected dose was in the urine and 30 percent in the feces. Ulfvarson (38), after giving a dose about one-tenth of that given by Friberg, found about 30 percent of the dose in the urine and 45 percent in the feces. Sooner or later, a steady state is reached and the excretion becomes equal to the exposure. Ashe et al. (1) exposed rabbits, rats and dogs to varying concentrations of mercury vapor for different periods of time. They studied the mercury excretion in urine. An equilibrium was reached after an initial increase. The individual variations were considerable, as were the variations from day to day. The data show a rapid decrease in excretion once the exposure has ceased.

In the mentioned studies by Friberg (10) an equilibrium was reached after about two weeks when the daily excretion via urine and feces equalled the daily intake. The amount of time required for such an equilibrium to be reached is dependent on the turnover of inorganic mercury to different body organs and its elimination from them displays a complicated pattern, rendering impossible the approximation to a one compartment system successfully made for methylmercury. By far the highest concentration of mercury, representing a considerable part of the whole-body content, is found in the kidneys. The brain is the critical organ in long-term exposure to mercury vapor. The entrance of mercury into that organ is facilitated by the existence of dissolved mercury vapor in the blood during exposure as mentioned above. Once it has entered the brain, mercury is oxidized to the mercuric form and the elimination from the organ is consequently slow after vapor exposure (2,19). The interpretation of brain levels of mercury is also complicated by the fact that mercury is unevenly distributed in the organ and accumulates specifically in certain nerve cells (7,24).

The detailed mechanism for excretion is still a matter of discussion and will not be dealt with here. Blood and urine concentrations tend to be correlated with each other, but not correlated with accumulation of mercury in kidney and other tissues (3,17). This finding is worth noting when blood and urine concentration data are to be evaluated.

Excretion is also via feces and urine in human beings. Sodee (32) after intravenous injection of $^{197}HgCl_2$ found that 72-hour urinary excretion was about 75 percent of the administered dose.

There are two careful studies on humans in which time-weighted long- term inhalation exposures have been calculated and urinary (in one case, also fecal) excretion has been studied. Tejning and Öhman (36) reported for 15 chloralkali workers exposed to 0.50 to 0.1 mg/m^3 a mean urinary excretion of 0.12 mg/liter and a mean fecal excretion of 0.09 mg/day. In ten persons with a mercury exposure of 0.1 to 0.2 mg/m^3 corresponding values were 0.19 and 0.14 mg/day. Smith et al. (31) in a comprehensive study of the United States chlorine industries found similar values for urinary excretion. Their results are seen in Figure 1-1.

Dose-response Relationships

In animals there are very few reliable data concerning dose-response relationships after long-term exposure. Those of Ashe et al. (1) are extensive, but cover only microscopically detectable injuries. They exposed groups of dogs, rabbits and rats to mercury vapor in concentrations ranging from about 30 mg/m^3 to 0.1 mg/m^3 for different periods of time, up to seven hours a day, five days a week for 83 weeks. Severe damage was found in kidneys and brain at concentrations of 0.9 mg/m^3 after about 12 weeks exposure. Even after a very long exposure time, there were no microscopically detectable injuries resulting from the 0.1 mg/m^3 concentration. At the last mentioned exposure level and time, the mercury values among rabbits rose to an average of about 5 ppm wet weight in the kidney and 0.2 ppm wet weight in the brain.

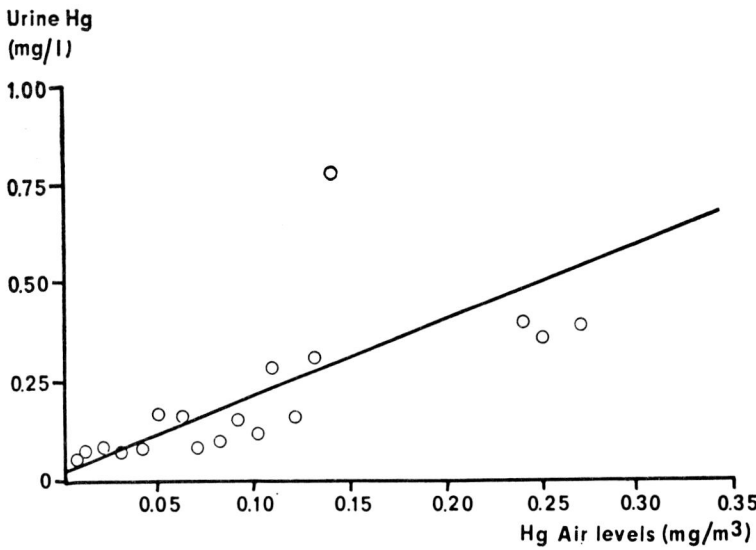

Figure 1-1. Concentrations of mercury in urine (uncorrected for specific gravity) in relation to time-weighted average exposure levels [From Smith et al. (31)].

A large individual variation was obvious. In the rabbits exposed to 0.9 mg/m^3 and in which microscopical evidence of intoxication was present, values of about 20 to 50 ppm were found in the kidneys and 1 to 2 ppm in the brain. Neither the behavior of the animals nor the renal function were studied in any of the experiments.

In the Russian literature, a number of animal studies have been reported during the 1950's and 1960's. Many experiments have included exposure to very low concentrations of mercury vapor for considerable periods of time. The data have been summarized in a monograph by Trachtenberg (37). The following descriptions are taken from this monograph and from personal discussions between one of us (Gunnar Nordberg) and scientists in the Soviet Union. Trachtenberg reported changes in the higher nervous activity in cats exposed to 0.085 to 0.2, 0.01 to 0.02, and 0.006 to 0.01 mg Hg/m^3. For three out of at least four cats in the first series a clear effect was seen on several parameters, including latency periods, already during the second week of exposure and the changes increased during further exposure. In the second series two cats out of four showed similar but less pronounced changes but the deviation from normal was not apparent until after 8 to 10 weeks of exposure. After termination of the exposure the parameters studied returned to normal within a few weeks.

Similar studies were carried out by Kournossov (16) who found changes in the higher nervous function in rats exposed to concentrations as low as 0.002 to 0.005 mg Hg/m^3. As this concentration of mercury is probably the lowest ever reported to have an effect on mammals, some details will be given. Kournossov exposed rate in four groups (five rats in each group) to different concentrations of mercury vapor: A: 0.02 to 0.03 mg Hg/m^3, B: 0.008 to 0.01 mg Hg/m^3, C: 0.002 to 0.005 mg Hg/m^3, D: 0.0000 to 0.0003 mg Hg/m^3. The exposure lasted 6.5 hours a day, six days a week. The mercury concentrations in the chambers were checked with a method according to Poleshajev (25).

According to this method, air to be analyzed is passed through a glass apparatus. Iodine vapor is mixed with the air. The mercury iodine mixture is absorbed in a solution of iodine and potassium iodine in water in an absorption flask. A solution of Na_2SO_3 and $CuCl_2$ is added, whereby a flocculation is formed which is precipitated by centrifugation. If mercury is present, red $Cu_2(HgI)_4$ is formed in amounts proportionate to the amount of mercury. Depending on the mercury concentration different shades of red are obtained in the precipitate. Evaluations of the sample are made in relation to a standard scale by subjective comparison.

The rats were placed in boxes where the door was connected by a mechanical device to a recorder which registered all movements of the door. Another device was connected to the floor of the box, permitting the registration of all of the movements of the rat inside the box. The rat was first trained to respond to a pattern of bell signals (food is placed outside the door, the door opens when the rat pushes it and the rat gets the food), light (food given, etc.), buzzer (no food given).

Changes in several parameters were observed when testing conditioned reflexes (Fig. 1–2).

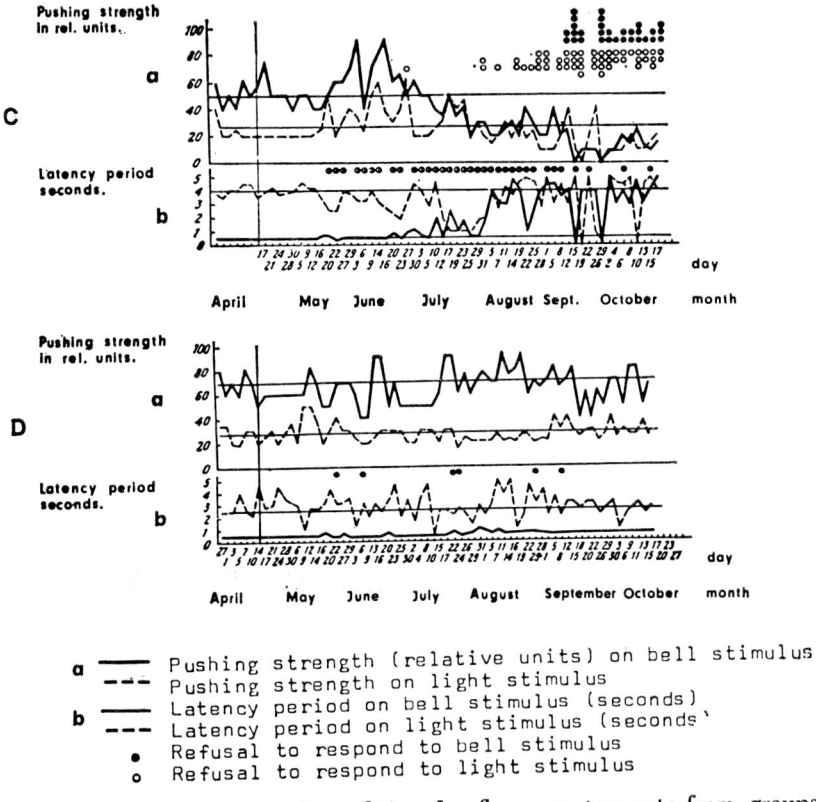

Figure 1-2. Registration of conditioned reflexes on two rats from groups exposed to C: 0.002–0.005 mg Hg/m³ and D: 0.0000–0.0003 mg Hg/m³ [From data by Kournossov (16)].

Trachtenberg (37) reported on studies on the uptake of radioactive iodine in the thyroid of rats chronically exposed to mercury vapor. In one series of 15 rats exposed to 0.01 to 0.03 mg Hg/m³ for 105 days, there was a difference compared to a control group and to preexposure values, indicating that mercury exposure gives rise to a hyperfunction of the thyroid. In contrast to Trachtenberg's observations there are unpublished data by Avetzkaja, in which a dose-related decrease in the uptake of radioactive iodine was observed with increasing mercury exposure.

In human beings a considerable number of studies relating

exposure and effects have been published. Most data are from exposure to mercury vapor but often it is not possible to decide to what extent exposure to aerosols of other forms of inorganic mercury is also involved. Very few studies are based on time-weighted averages which makes it difficult to evaluate dose-response relationships. Several studies [e.g. Neal et al. (21,22), Kesic and Haeusler (15), Bidstrup et al. (5), Friberg (9), Rentos and Seligman (27)] do, however, make it clear that classical mercury intoxication can occur at prolonged exposure to concentrations above 0.1 to 0.2 mg Hg/m^3.

A recent and comprehensive study by Smith et al.(31) tends to show that effects can be seen at considerably low concentrations. These authors examined 567 workers exposed to mercury (in general, more than 90 percent as mercury vapor) in the manufacture of chlorine and a control group of 382 persons. More than half of the all male study group had worked between six and 14 years in the industry and they came from 21 different plants. Every worker was examined once during a one-year period by plant physicians according to a predetermined procedure. The mercury concentrations in air were measured at different sampling places at least six times a year by means of ultraviolet meters. Time-weighted averages were calculated for each worker.

The prevalence of certain medical findings in relation to mercury exposure is illustrated in Figure 1–3. As can be seen from the figure, there is a clear dose-related response to mercury exposure for several parameters, including findings and symptoms from the nervous system expected in mercury poisoning. For diastolic blood pressure there was a negative correlation with mercury exposure. For some findings there is no evidence that even the lowest exposure (time-weighted average 0.01 to 0.05 mg/m^3) took place without effect.

Extensive studies performed by Trachtenberg and his collaborators in Kiev have been published in Trachtenberg's 1969 monograph. Mercury concentrations in air of the subjects' working places were determined by the colorimetric method of Poleshajev (25). In each work room a number of spot samples (usually more than 60) were made. For the exposed group

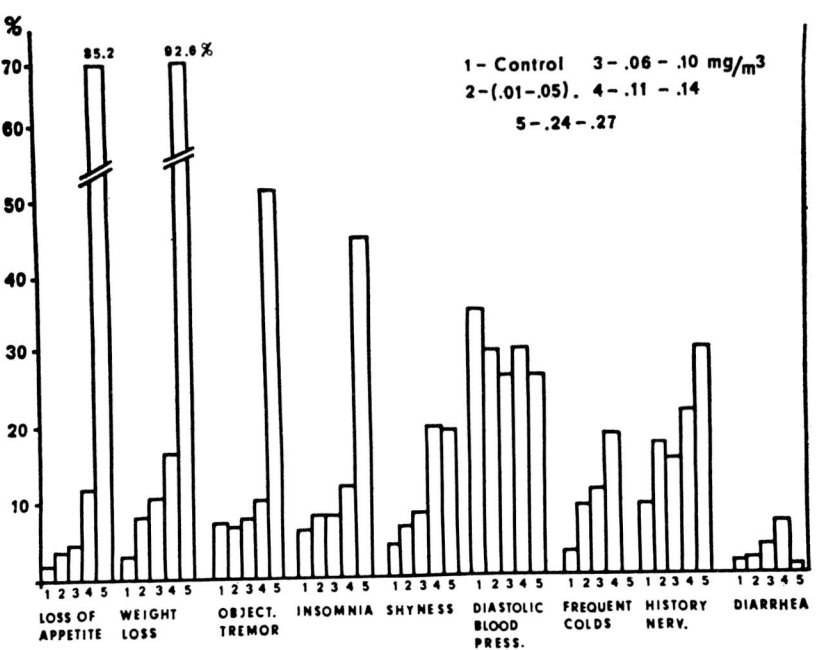

Figure 1-3. Percentage prevalence of certain signs and symptoms among workers exposed to mercury in relation to degree of exposure [From Smith et al. (31)].

(group A) which covered 506 persons the mercury concentrations on an average varied between 0.01 to 0.05 mg/m³ with only exceptional values above 0.1 mg/m³. The control group covered 68 persons and their exposures were less than 0.01 mg Hg/m³.

The medical examinations were made by plant physicians. Trachtenberg stated that an asthenic-vegetative syndrome was found in 51 percent of group A. Of these, he considered that 14 percent had an unspecific etiology while 38 percent could be traced to the mercury exposure.

Trachtenberg and his collaborators also studied the function of the thyroid by means of radioactive iodine. The results from an exposed and a control group are shown in Figure 1-4. The workers in the exposed group were among those workers already referred to. As can be seen, the differences in uptake of radio-

Uptake of radioactive iodine	Groups exposed to mercury			Control groups		
	Men (36 persons)	Women (31)	Total (67)	Men (26)	Women (19)	Total (45)
<25%	39	29	34	84	68	78
>25%	61	71	66	16	32	22

Figure 1–4. Relative number of workers (%) with uptake of radioactive iodine (after 24 hours) in the thyroid [From Trachtenberg (37)].

active iodine between exposed workers and controls are substantial, both for men and for women.

We shall not in this connection enter into a detailed discussion of either the study by Smith *et al.* (31) or the Russian studies. There is a possible bias in interpreting some minor subjective symptoms that should be mentioned, however. That is, there could be an "interviewer" effect. In all of the studies the medical examinations were made by the factory physicians. One has to presuppose that they had at the time of the examinations a knowledge of the exposure situations for different categories of patients which might have had some influence. On the other hand, it is by no means certain that there was a bias and one has to appreciate the well-standardized questionnaire employed by Smith *et al.* (31). The data presented so far tend to indicate that we would have to reexamine Threshold Limit Value-values for mercury exposure if we are looking for a no-effect level. There also seems to be need for studies on conditioned reflexes in countries other than the U.S.S.R.

Relation Between Mercury in Urine and Effects

Several studies have related mercury excretion via urine with symptoms of mercury poisoning. Knowledge of the metabolism of mercury makes it clear, however, that one should not expect to find a high correlation between mercury excretion and symptoms from the central nervous system.

Animal data of Ashe *et al.* (1) make it obvious that the individual variation is great, but also that the urinary excretion decreased late in the experiment when the damage to the central nervous system and the kidneys was evident. Our own studies on human beings (9) give about the same results. In an examination of 91 workers in a chlorine plant, seven cases of pronounced tremor were found. Four of the workers with pronounced tremor had been exposed to mercury vapor for about 25 years and had mercury levels in urine of only between 0.2 to 0.3 mg Hg/liter (Fig. 1–5). Several workers had a high mercury excretion without symptoms. The extensive study by Smith *et al.* (31) showed no good correlation between urinary mercury levels and symptoms.

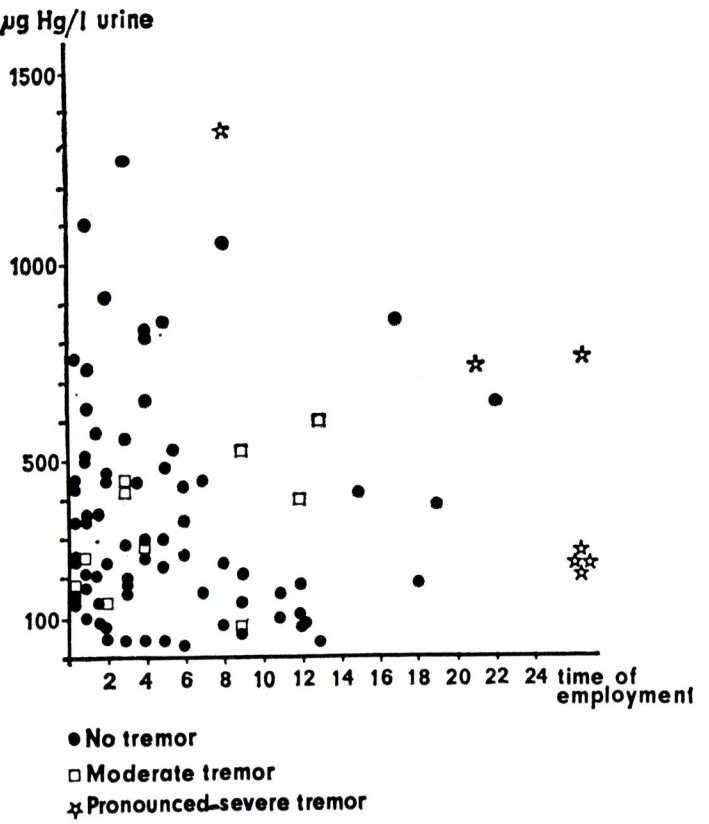

Figure 1-5. Prevalence of tremor in workers exposed to mercury in a chlorine plant [From Friberg (9)].

In summary, it can be noted that although on a group basis high mercury levels lead to a higher probability of mercury poisoning, in the individual case, high values of mercury can occur without symptoms, while symptoms can occur in association with low levels of mercury in the urine.

REFERENCES

1. Ashe, W., Largent, E., Dutra, F., Hubbard, D. and Blackstone, M.: Behavior of mercury in the animal organism following inhalation. *AMA Arch Ind Occup Med*, 7:19–43, 1953.
2. Berlin, M., Fazackerly, J. and Nordberg, G.: The uptake of mercury in the brains of mammals exposed to mercury vapour and to mercuric salts. *Arch Environ Health*, 18:719–729, 1969 (English).
3. Berlin, M. and Gibson, S.: Renal uptake, retention and excretion of mercury. I. A study in the rabbits during infusion of mercuric chloride. *Arch Environ Health*, 6:617–625, 1963 (English).
4. Berlin, M., Nordberg, G. and Serenius, F.: On the site and mechanism of mercury vapor resorption in the lung. *Arch Environ Health*, 18: 42–50, 1969.
5. Bidstrup, P., Bonnell, J., Harvey, D. and Locket, S.: Chronic mercury poisoning in men repairing direct-current meters. *Lancet*, 251: 856–861, 1951.
6. Bornmann, G., Henke, G., Alfes, H. and Möllmann, H.: Ueber die enterale resorption von metallischem quecksilber. *Arch Toxicol*, 26:203–209, 1970.
7. Cassano, G., Viola, P., Ghetti, B. and Amaducci, L.: The distribution of inhaled mercury (Hg^{203}) vapors in the brain of rats and mice. *J Neuropath Exp Neurol*, 28:308–320, 1969.
8. Clarkson, T., Gatzy, J. and Dalton, C.: Studies on the equilibration of mercury vapor with blood. Division of Radiation Chemistry and Toxicology, University of Rochester Atomic Energy Project, Rochester, New York, UR–582, 1961, pp. 64.
9. Friberg, L.: Aspects of chronic poisoning with inorganic mercury based on observed cases. *Nord Hyg Tidskr*, 32:240–249, 1951 (In Swedish with English summary).
10. Friberg, L.: Studies on the accumulation, metabolism and excretion of inorganic mercury (Hg^{203}) after prolonged subcutaneous administration to rats. *Acta Pharmacol Toxicol*, 12:411–427, 1956.
11. Friberg, L., Skog, E. and Wahlberg, J.: Reabsorption of mercuric chloride and methyl mercury dicyandiamide in guinea pigs through normal skin and through skin pre-treated with acetone, alkylarylsulphonate and soap. *Acta Derma Venerol*, 41:40–52, 1961 (English with German and Spanish summary).

12. Gage, J.: The trace determination of phenyl- and methylmercury salts in biological material. *Analyst,* 6:457–459, 1961 (English).
13. Hayes, A. and Rothstein, A.: The metabolism of inhaled mercury vapor in the rat studied by isotope techniques. *J Pharmacol, 138*:1–10, 1962.
14. Juliusberg, F.: Experimentelle untersuchungen über die quecksilberresorption bei der schmierkur. *Arch Derm Syph,* 56:5–88, 1901.
15. Kesic, B. and Haeusler, V.: Hematological investigation on workers exposed to mercury vapor. *Ind Med Surg,* 20:485–488, 1951.
16. Kournossov, V.: Materiali k gigienitsjekomej . . . *Gig Sanit,* 1:7–15, 1962 (In Russian).
17. Lundgren, K., Swensson, A. and Ulfvarson, U.: Studies in humans on the distribution of mercury in the blood and the excretion in urine after exposure to different mercury compounds. *Scand J Clin Lab Invest,* 20:164–166, 1967 (English).
18. Magos, L.: Mercury-blood interaction and mercury uptake by the brain after vapor exposure. *Environ Res,* 1:323–337, 1967.
19. Magos, L.: Uptake of mercury by the brain. *Br J Ind Med,* 25:315–318, 1968.
20. Morrow, P., Gibb, F. and Johnson, L.: Clearance of insoluble dust from the lower respiratory tract. *Health Phys, 10*:543–555, 1964.
21. Neal, P., Flinn, R., Edwards, T., Reinhart, W., Hough, J., Dallavalle, J., Goldman, F., Armstrong, D., Gray, A., Coleman, A. and Postman, B.: Mercurialism and its control in the felt hat industry. Public Health Bulletin No. 263, U.S. Public Health Service, Federal Security Agency, 1941, pp. 1–132.
22. Neal, P., Jones, R., Bloomfield, J., Dallavalle, J. and Edwards, T.: A study of chronic mercurialsim in the hatters fur-cutting industry, Public Health Bulletin No. 234, U.S. Treasury Department, Public Health Service, 1937, pp. 1–70.
23. Nielsen-Kudsk, F.: Absorption of mercury vapour from the respiratory tract in man, *Acta Pharmacol,* 23:250–262, 1965.
24. Nordberg, G. and Serenius, F.: Distribution of inorganic mercury in the guinea pig brain. *Acta Pharmacol Toxicol,* 27:269–283, 1969.
25. Poleshajev, N.: K metodike opredelenija rtoty v atmosfernom vozdoche. *Gig Sanit,* 6:74–76, 1956 (In Russian).
26. Prickett, C., Laug, E. and Kunze, F.: Distribution of mercury in rats following oral and intravenous administration of mercuric acetate and phenylmercuric acetate. *Proc Soc Exp Biol Med,* 73:585–588, 1950 (English).
27. Rentos, P. and Seligman, E.: Relationship between environmental exposure to mercury and clinical observation. *Arch Environ Health, 16*:794–800, 1968.
28. Rothstein, A. and Hayes, A.: The metabolism of mercury in the rat

29. Schamberg, J., Kolmer, J., Raiziss, G. and Gavron, J.: Experimental studies of the mode of absorption of mercury when applied by inunction. *JAMA*, 70:142–145, 1918.
30. Skog, E. and Wahlberg, J.: A comparative investigation of the percutaneous absorption of metal compounds in the guinea pig by means of the radioactive isotopes: ^{51}Cr, ^{58}Co, ^{65}Zn, ^{110m}Ag, ^{115m}Cd, ^{203}Hg. *J Invest Derm*, 43:187–192, 1964.
31. Smith, R., Vorwald, A., Patil, L. and Mooney, T.: Effects of exposure to mercury in the manufacture of chlorine. *Am Ind Hyg Assoc J*, 31:687–700, 1970.
32. Sodee, D.: A new scanning isotope, Mercury^{197}; A preliminary report. *J Nucl Med*, 4:335–344, 1963.
33. Takeda, Y., Kunugi, T., Hoshino, O. and Ukita, T.: Distribution of inorganic, aryl and alkylmercury compounds in rats. *Toxic Appl Pharmacol*, 13:156–164, 1968 (English).
34. Task Group on Lung Dynamics. Deposition and retention models for internal dosimetry of the human respiratory tract. *Health Phys*, 12:173–208, 1966.
35. Teisinger, J. and Fiserova-Bergerova, V.: Pulmonary retention and excretion of mercury vapors in man. *Ind Med Surg*, 34:580–584, 1965.
36. Tejning, S. and Ohman, H.: Uptake, excretion and retention of metallic mercury in chloralkali workers. Proc. 15th Int. Congr. Occup. Health, Wien, 1966 pp. 239–242, (English).
37. Trachtenberg, I.: *The Chronic Action of Mercury on the Organism, Current Aspects of the Problem of Micromercurialism and its Prophylaxis.* Kiev, ZDOROV'JA, 1965, p. 292.
38. Ulfvarson, U.: Distribution and excretion of some mercury compounds after long term exposure. *Int Arch Gewerbepath*, 19:412–422, 1962 (English).
39. Wahlberg, J.: Percutaneous toxicity of metal compounds. *AMA Arch Environ Health*, 11:201–204, 1965.

DISCUSSION

Goldwater: I'd like to comment about skin absorption. I found an x-ray picture of a hand of somebody who had been rubbing mercury into the skin which actually demonstrated little droplets of radio opaque material appearing in the hand. I think this shows that mercury can go through the skin.

Foulkes: What are the changes in the urine volume?

Friberg: It was not measured.

McDuffie: Is there any correlation of symptoms with urine cencentrations before the kidney is damaged?

Friberg: Smith et al* tried to correlate symptoms to the mercury level in urine but the association was very weak.

Weiss: How reliable would you consider the Soviet work to be in conditioned reflexes? Have you gone to the laboratories and seen their facilities and experimental examples, methods, etc?

Friberg: I have not been there myself but Dr. Gunnar Nordberg working in our institute has. His observation is that first of all they are not as sophisticated as in our countries; they make their own notations on paper when the rat pushes the door and so on. There are no electronic devices as there are here. Their analytical methods seem to be crude, being based on subjective evaluations of color shades which may be one thing that has to be considered.

But we have epidemiological data now from both the United States and the Soviet Union; these studies tend to show that there are some effects at lower concentrations than was thought of before. That doesn't necessarily mean of course that one should use these effects as a criteria for setting TLV values but there might be effects at very low concentrations.

Kazantzis: Studies of thyroid function in mercury-exposed workers would be worth repeating; radioactive iodine uptake can be misinterpreted because if the store of iodine in the thyroid gland is low the uptake of ^{131}Iodine will be high, and this need not necessarily be associated with an increased secretion of thyroxin. It is now possible to measure serum thyroxin directly and I feel this and other studies of thyroid function should be performed so that we can interpret the results that Dr. Friberg has shown us in relation to this organ.

Berlin: It's very difficult to expose to such low levels because there is a tendency to absorb on the walls of the exposure chamber and if very crude devices for checking the concentration inside it are used it may well be that the real concentration may have been much higher.

Clarkson: These studies were not with the radioactive mercury were they?

Berlin: No.

Fassett: Have you or others made studies of autopsy materials from workers who had been exposed for long periods to inorganic mercury in their occupation?

Friberg: We don't have any. There are some data on animals; Ashe et al,† found that when they observed symptoms it was at concentrations of 5 ppm wet weight in the kidneys and 0.2 ppm wet weight in the brain. They were exposed to 0.1 without symptoms, but in the rabbits exposed to 0.9 mg Hg/m^3, there was microscopical evidence of intoxication. Twenty to 50 ppm was found in the kidneys and 1 to 2 ppm in the brain.

* Smith et al.: Am Ind Hyg Assoc J, 31:687–700, 1970.
† Ashe et al: AMA Arch Hyg Occup Med, 7:19–43, 1953.

Berlin: There is no data available at this moment on the actual concentration of mercury in the brain. We are in the process of exposing squirrel monkeys to rather high levels for three or four months. We monitor the blood concentration during the exposure in the monkeys and no spectacular signs were seen the first month; then signs appeared including changes in the morphology of the brain.

Suzuki: In the 16th International Congress on Occupational Health, Tokyo, Watanabe* reported the mercury content in the brain of two workers who had been exposed to mercury vapor for many years in a mercury mine and died after about ten years' lapse from the end of mercury exposure. By the activation analysis, he got the values ranging from 2.69 to 33.56 ppm of wet brain tissue in one case and 2.07 to 18.0 ppm of wet brain tissue in the other. I feel these values of mercury content in the brain show the long persisting character of mercury in the case of inorganic mercury exposure.

Goldwater: With regard to Dr. Fassett's question about autopsy studies on people who had been exposed to inorganic mercury compounds, these people don't usually die of their exposure and so one rarely has a chance to do any autopsy studies. But, a few years ago just by chance I had an opportunity to examine histologically kidneys from a 75-year-old former hat maker who for 25 or 30 years had been exposed to mercury nitrate in Danbury, Connecticut. He died in the Presbyterian Hospital in New York of an unrelated cause and we did get his kidneys and they looked like the kidneys of a 20-year-old boy they were so beautiful.

Kazantzis: We have heard a little about mercury concentrations in blood, and mercury concentrations in brain. Have any studies been carried out on mercury concentrations in the cerebrospinal fluid? This is a body compartment which so far has not been brought into any model of mercury transfer to the brain; yet it is a body compartment which may play an important role.

Friberg: I do not know of any such studies.

Clarkson: I think perhaps one reason is the concentrations are very low in the cerebrospinal fluid.

Friberg: I should like to mention, concerning the excretion from the brain, some studies† we carried out several years ago where we first gave radioactive mercury to rats subcutaneously and then after a couple of months gave nonradioactive mercury. It was so obvious in these experiments that there was no exchange whatsoever in the brain while there was quite a lot of exchange within the kidneys and other organs so this tends also to support some of the things you mentioned.

Clarkson: Was this with inorganic salt?

Friberg: Yes, mercuric mercury.

* Watanabe, S.: Proceedings XVI International Congress Occupational Health. Tokyo, 1969, p. 553–554.
† Friberg, L.: *Acta Pharmacol Toxicol,* 12:411–427, 1956.

Norseth: We can load the rat kidneys with mercury by repeated injections and have very high levels of mercury in the kidney without kidney symptoms. I think the same thing is found regarding the release of inorganic mercury from phenylmercury where you can have a high level of inorganic mercury released from phenylmercury in the kidney without symptoms.

Chapter 2

AN APPRAISAL OF THE EPIDEMIOLOGY AND TOXICOLOGY OF ALKYLMERCURY COMPOUNDS

LEONARD T. KURLAND

ABSTRACT

The recent discovery that methylation of many forms of mercury occurs in the silt of waterways and enters the food chain of fish has aroused concern on the part of ecologists and public health officials. To date, only one case of suspected methylmercury intoxication from excessive consumption of fish has been reported in North America, and some contend that there is an unwarranted element of panic in the public pronouncements on this problem. But several hundred cases of methylmercury intoxication have resulted from eating contaminated fish in Japan, and others from eating contaminated pork in New Mexico, from misuse of fungicide-treated grain as food, and from industrial exposure. This review of recent studies attempts to place the methyl-mercury problem in a reasonable perspective.

Metallic mercury and the inorganic, alkoxyalkyl, phenyl, and other compounds of mercury are toxic to man and other animals; but the alkylmercuries and especially methylmercury pose a serious problem because of their ability to cross cell membranes. This results in their high absorption from the gastrointestinal tract, their propensity for the nervous system, their stability and retention within the body, their disregard for the placental barrier, and their affinity for fetal hemoglobin with subsequent deleterious effects on the developing nervous system and other tissues. The clinical features include paresthesia, ataxia, loss of hearing and peripheral vision, and intellectual deterioration. Cerebral palsy-like disease is said to occur in some children with transplacental exposure to methylmercury. Principal autopsy findings in the adult include atrophy of the granular-cell layer of the neocerebellum and areas of cortical atrophy, particularly the area striata. Peripheral-nerve degeneration is probably also present. The reasons these tissues are selectively affected are unknown.

Some of the factors which influence methylmercury absorption, excretion, distribution in various tissues, and dose-response relationships are reviewed here. The rate of metabolism of organic mercury compounds to inorganic mercury appears to determine tissue distribution and rates of excretion. The brain is the first organ to show functional disturbance from exposure to methylmercury, but it is largely inaccessible to biopsy, and no satisfactory means exist for detection of minimal effects while neuronal damage may yet be reversible. A critical, unanswered question is whether methylmercury in low doses can have a subclinical or delayed effect which emerges as retarded motor and intellectual development in an infant or child or as presenile dementia or motor-neuron disease in an adult.

Methylmercury is active genetically (c-mitotic and chromosome breakage effect). A mutagenic effect appears to be small. It is not yet possible to relate the magnitude of genetic risks to the degree of exposure. Uncertainty regarding genetic and teratogenic effects can be expected to influence toward conservatism the recommendations regarding methylmercury exposure of fertile women.

The basis for individual variation in sensitivity to methylmercury is yet to be determined. The relationship of genetic and nutritional factors and of symbiotic or antagonistic agents to this sensitivity is a problem needing study.

Experimental studies in animals and the observed effects in man (particularly in the developing fetus) have led to recommendations of an allowable daily intake of mercury which may serve as a sound basis for guidelines regarding amounts to be permitted in fish and other food.

A series of unresolved problems with respect to the site and mechanism of action and of the genetic effects of methylmercury is presented.

REPORT

Preface

In recent months several important volumes pertaining to the epidemiology and toxicology of the alkylmercuries have been published or circulated in manuscript form. These include the reports on the outbreak of mercury poisoning in the vicinity of Niigata, Japan (52, 54), which serve as a sequel to the volume on Minamata disease from Kumamoto University (44); the impressive collection of new information obtained by research workers in the past few years and published as *Methyl Mercury in Fish* by Professor Lars Friberg and his colleagues (8); and reports from several laboratories around the world, notably those of Professor Miettinen and his colleagues in Helsinki (27) and Dr. Clarkson here in Rochester, New York (5).

Having been privileged to work on the Minamata Bay outbreak years ago, I was invited to serve as a member of a committee of the Department of Health, Education, and Welfare to appraise the mercury problem and the recent advances of knowledge in Sweden and Finland. The report of our committee, recently published in *Environmental Research* (31), includes a review based on information made available to us by our generous hosts in Sweden and Finland. It contains sections on methods of chemical analysis; airborne mercury; the ecologic effects of methylmercury contamination; the sources, distribution, and control of mercury (in farming, food, and forestry); and its microbial transformation. New developments in some of these aspects will be brought to our attention at this session. Much that I will present is derived from the report which several members of our committee prepared on the medical implications of the ingestion of mercury (31).

My task of reporting on the alkylmercuries, and particularly methylmercury, is made easier by the availability of so much new data; and yet it is complicated by the need to present a current appraisal that is brief and comprehensive in regard to human experience with these compounds but selective and critical in regard to experimental studies bearing on their effects in man. I hope that this general presentation, though it cannot give credit to all those whose work deserves to be cited, will help set the stage for some of the later papers in this symposium by listing questions that must still be resolved on the nature and extent of the problem and reasonable means of controlling it.

The toxicologic characteristics of mercury and its compounds can be classified as follows:

Inorganic, as metallic (elemental) mercury or inorganic salts.
    Elemental mercury, which when vaporized and inhaled seems to produce its critical effect on the central nervous system.

Ionic forms, such as mercurous or mercuric salts, which tend to concentrate in and affect the kidney and liver before recognizable neurologic signs appear.

Organic, in which the mercury is bound covalently to at least one carbon atom.

Alkylmercury, in which the strong carbon-to-mercury bond does not readily dissociate. Its toxic effects have been attributed to the action of the intact molecule and compounds with shorter carbon chains (methyl, ethyl) which readily pass across membranes and have their major clinical effect on the nervous system.

Other organic compounds such as salts of aryls (for instance, phenyl mercuric acetate) and alkoxyalkyls (for instance, methoxymethyl mercury nitrate), which tend to be metabolized to mercuric ion compounds and thus affect the kidney and liver before neurologic disease develops.

In this report the effects of alkylmercury compounds, and especially methylmercury, will be stressed.

Development of Basic Knowledge

Greco (11), in 1930, opened his description on the effects of alkylmercury as follows:

> It can be said that the history of the alkyl compounds began with a tragedy; in 1865 two young chemists, while working on the preparation of diethyl mercury, inhaled the vapors and were poisoned by the very volatile product. One of them died in eleven days, and the other died after a year, during which he had been blind, deaf and almost completely demented.

In the decades that followed Edward's descriptions of these cases in 1865 and 1866 (7), it became clear that the alkylmercuries somehow could pass the blood-brain barrier and disrupt the nervous system itself. Although a few alkylmercury compounds were tried as therapeutic agents for various infections in the 1870's (35), their use was discouraged by reports such as that of Hepp (13) in 1887, which warned that diethylmercury compounds "proved to be highly toxic in a totally peculiar and unpredictable manner, and in every case the toxicity was so much greater than that of mercury alone that such dangerous substances absolutely should not be introduced into therapy." It was further noted that

> When the common mercurial compounds are used . . . we are immediately appraised of the danger of mercury poisoning (stomatitis, tenesmus, bloody stools), so that the administration can be stopped immediately, [but] we do not find these signs in poisoning by alkyl mercury, and when the first signs of poisoning appear, it is already too late . . . (13).

In the early part of this century it was recognized that the vascular supply to the central nervous system was unique in having a space between the blood vessels and the neuroglia which then forms a protective velum around the vessels even to the terminal capillaries. This is the morphologic basis of the blood-brain barrier concept. Since the neurons were affected in central-nervous-system syphilis, efforts were being made to discover a therapeutic agent that would pass the barrier and destroy the spirochete within the brain tissue itself. The inorganic mercurial compounds and most of the organic ones could not cross this barrier effectively. Cerletti (4), noting that the presence of an alkyl radical such as that in alcohol or ether could affect brain function, resorted to ethyl compounds which could take the mercury through the barrier. His studies in 1914 confirmed the toxicologic distinction of the short-carbon-chain alkylmercuries, namely their ability to cross membranes and the blood-brain barrier and affect the function of the nervous system. This property of the alkylmercurials decreases as the length of the carbon chain increases.

Although the toxicology of several of the alkylmercuries other than methylmercury has also been studied extensively, and these compounds have been used to varying degrees as fungicidal agents, we shall be concerned primarily with methylmercury—partly because these compounds have been used widely in agriculture, but mainly because this is the only alkylmercury that seems to be formed in nature and accounts for practically all mercury found in fish. And fish—in the absence of medicinal or inadvertent industrial exposure—is the major source of methylmercury in man.

Until recently it was thought that all alkylmercury compounds in the environment were man-made, directly, in laboratory or industrial production. In the past few years, however, it has been shown (19, 58) that various mercury compounds coming in contact with the bottom mud of waterways are converted to methylmercury. This compound is passed up the food chain and concentrated several thousand-fold in fish and shellfish. The mercury contamination in the sediments therefore constitutes a depot which this methylating process makes a continuing potential source of methylmercury.

Most of the other mercury compounds used in agriculture and industry (such as alkoxyalkyls and aryls) can be grouped, on the basis of their effects in man, with mercuric ion, to which such compounds appear to be metabolized, usually with resulting damage to the kidney before neuronal damage is recognized.

The covalent bond betwen C and Hg in methylmercury seems to provide considerable stability to the compound in tissues; and this—in conjunction with its ability to pass across the cell membrane—accounts for its unusual toxicologic characteristics, including its high absorption from the gastrointestinal tract, its concentration within the erythrocytes, its ready passage through the blood-brain barrier with resultant neuronal damage, its long retention within the body, its disregard for the placental barrier, and its affinity for fetal hemoglobin with subsequent deleterious effects on the developing nervous system and other tissues. These characteristics of methylmercury were manifested in the Japanese outbreaks (44, 54) where the toxin was transmitted through fish, and in the recent episode in New Mexico (23, 41–43) where it was transmitted through contaminated pork. Several other episodes of methyl and other alkylmercury poisoning, described in the Swedish report (8), have resulted from laboratory and industrial exposures, the use of ointments containing methylmercury for skin disorders, or direct consumption of mercury-treated cereal grains intended only for planting.

Principal Episodes

Minamata and Niigata

The two large outbreaks of methylmercury poisoning identified in Japan occurred in Minamata (44, 48) where 121 cases with 46 deaths were identified up to 1970, and in the villages adjacent to the Agano River in Niigata (53, 54) prefecture, where 47 cases with six deaths were observed through 1970. One of the Niigata cases and 22 of the 121 cases in Minamata were cerebral palsy-like disease of infants who had not consumed contaminated fish. Among the surviving patients—adults and children—there has been appreciable persistent disability. The Minamata and Niigata studies established the general nature of the disease, related it to the level of intake of fish, and defined the associated burdens of mercury in several tissues of the victims of the disease (55, 56).

It is likely that during the height of the Minamata epidemic, cases with only mild or atypical symptoms were not being diagnosed; this may account in part for the reported case-fatality ratio of 38 percent versus about 13 percent in Niigata. The conjecture is supported by the recent diagnosis of previously missed cases and the report of Takeuchi (47) in 1970 that autopsy upon a medical practitioner in a village near Minamata Bay who had had tremor and some intellectual symptoms provided clear-cut evidence of Minamata disease. It also appears that compensatory mechanisms of the nervous system can delay clinical recognition of the disease even though the patient has partial brain damage. No specific therapy for methylmercury poisoning was successful in the Japanese epidemics, although penicillamine, glutathione, and BAL were tried in Niigata and found to increase, transiently, the urinary excretion of mercury.

The sources of mercury contamination in Minamata Bay and the Agano River were discharges from factories using mercuric chloride catalysts in the manufacture of vinyl chloride and acetaldehyde. It was reported that methylmercury was preformed in the effluent from the factories, but the evidence for it is not substantial and aquatic biotransformation seems a more likely source of the methylmercury.

Alamogordo

In 1969 seven persons, including a pregnant woman, in a New Mexico family ate pork over a period of four months from animals which had been fed seed grain coated with methylmercury dicyandiamide. The concentration of mercury in the grain was 32 ppm, and in the pork it was 28 ppm (23, 41). Two children and one young adult in the family had severe brain damage of the type seen in Minamata disease. The infant, born eight weeks after the contaminated pork had been impounded, had a slight seizure but appeared well otherwise at birth, although urinary levels of mercury were high. He was not breast-fed but was given a formula presumably free of any toxic agent. At eight and ten months of age, however, hypertonia was noted; and he apparently is blind and severely retarded (41, 41A).

Clinical and Pathologic Features

General

Methylmercury may be absorbed through the skin, respiratory tract, and gastrointestinal tract. The reported clinical and pathologic effects of alkylmercury vary somewhat from case to case but generally agree with the description in the classic report by Hunter and associates (16) of four men exposed to methylmercury nitrate. Whether the exposure has been chronic or sudden seems not to affect the clinical manifestations, but the severity of the illness undoubtedly is influenced by the cumulative level of exposure. The clinical features may appear several weeks after exposure has been terminated and may progress in severity for several months even without fresh exposure. The four men described by Hunter and associates (16) developed symptoms about a month or more after exposure to methylmercury. The clinical features included numbness and tingling of the extremities, mouth, and lips; dysarthria; gross constriction of the visual fields; loss of hearing; unsteadiness of gait; loss of coordination; reflex changes, at times with extensor plantar responses; and progressive psychologic and intellectual deterioration. Although sensing of pain and light touch was usually normal, sensing of position and vibration and two-point discrimination were impaired. In the Japanese cases other features were noted, though not consistently: intention or resting tremor; loss of motor activity with hyporeflexia and flaccid paresis in some cases and with clonus, Babinski's sign, spastic paresis, and even contractures in other cases; extrapyramidal features such as hyperkinesia and rigidity; and autonomic symptoms such as sweating and salivation.

The report from Hunter and Russell (17) of an autopsy examination made on one of the patients cited above (16) who died years later of pneumonia provided a critical lead to the recognition of the agency of methylmercury in the Minamata catastrophe. The significant autopsy findings in this case, which revealed the same main pathologic features noted repeatedly in the Japanese cases, were cerebellar cortical atrophy, selectively involving the granular-cell layer of the neocerebellum, and areas of cortical atrophy, particularly severe in the area striata. The spinal cord was reported as normal.

Hunter and associates (16) had also published earlier their experimental finding that the dorsal tracts, the dorsal roots, and the sensory component of peripheral nerves were affected; and clinical features suggest such involvement in human sensory nerve and dorsal roots. Pathologic confirmation of such changes in man has been lacking, however, although Takeuchi (46, 47) is reported to have observed peripheral-nerve damage on review of earlier autopsy material. I am not aware of any report on nerve-conduction velocity in patients or experimental animals.

Recent animal studies have revealed that administration of methylmercury is followed by degeneration of peripheral nerve but not of the posterior-root ganglion (24). There is evidence that regeneration occurs much later (28), which may account for the clinical recovery occasionally reported and the apparent absence of peripheral-nerve changes, as in the study by Hunter and Russell (17).

Several of the Japanese patients had muscle fasciculations and atrophy, and marked neurogenic atrophy of muscle was noted in one case (53). Some microscopic changes have been described recently by Miyakawa and co-workers (28), but whether this effect is due to the direct action of the toxin on muscle fiber or due to degeneration of the nerve is unknown.

Brown (3) reported that several patients with amyotrophic lateral sclerosis (ALS) had told of exposure to ethylmercuric chloride (Ceresan), but studies by Mulder (30) give no indication that ALS occurs more frequently in patients with such exposure than in other patients or controls. Exposure to treated seed in Iraq was followed by several cases of a disease whose clinical features resembled those of ALS (20) but were not typical of it. At present one can say that the reports of mercury-induced ALS are yet to be confirmed.

Dahhan and Orfaly (6) in 1964 reported electrocardiographic abnormalities in patients who had used seeds treated with ethylmercury p-toluene sulphanilide (in Granosan-M) for the preparation of bread. However, Taylor and associates (49) in 1969 found no significant electrocardiographic abnormalities in employees of a seed-dressing plant who were exposed to low levels of various organomercurials including methylmercury.

Prenatal and Postnatal Exposure

It appears that methylmercury is more readily transferred across the placental barrier than are other compounds such as mercuric chloride or phenylmercuric acetate, as reported by Suzuki et al. (45) from studies in mice in 1967, and that it binds readily with fetal hemoglobin—even more readily than with adult hemoglobin (51). The toxicologic significance of these findings is still to be clarified, but this is probably the basis for the detrimental effect on developing brain tissue in animals which was described by Moriyama (29) in 1968 and presumably occurred in infants in Minamata. Since methylmercury also is passed through mother's milk, how to distinguish postnatal from prenatal effects of exposure remains a problem.

Congenital (Fetal) Disease

Prenatal poisoning in man has been reported in the Minamata outbreak, the Alamogordo episode, and several of the population studies where treated seed was misused, including a 1968 U.S.S.R. publication by Bakulina (1) that ten children had prenatal intoxication from ethylmercury.

Specific data on exposure of the mother and on concentration in the organs of the mother, fetus, or infant are almost entirely lacking in these episodes, and are limited even in animal experiments.

Twenty-two (6 percent) of 359 children born in the villages near Minamata Bay from 1954 through 1959 were reported to be affected with a disease resembling cerebral palsy. Its severity varied, some children having moderate to severe spasticity, ataxia, athetosis, intellectual retardation, seizures, or other evidence of brain damage. The method of case ascertainment that resulted in this rate of 6 percent is not presented in detail by Harada (12), but the rate is considerably higher than the cerebral palsy birth incidence of 0.1 to 0.6 percent among other populations.

Several features are particularly noteworthy in these cases: (1) The affected children had not eaten contaminated fish or shellfish; (2) The mothers apparently were not affected according to the diagnostic criteria which were applied, but most of the cases were from households where much fish was consumed and where other persons had been afflicted; and (3) Clinical symptoms were more difficult to elicit and more varied than in cases of Minamata disease among older children and adults. Possibly the intensive search for cases merely disclosed coincidental cerebral palsy in the population, since clinically and even pathologically no distinguishing feature is obvious. Furthermore, some of the infants and mothers in the Niigata area who apparently were unaffected had mercury concentrations in the hair exceeding some of those reported in the Minamata cases (53). However, it is likely that some of the Minamata congenital cases were due to mercury, since the incidence rate appears to be excessive, most of the infants and the mothers studied had more than the usual concentration of mercury in their hair, and those dying had more than normal concentrations of mercury in brain, liver, and kidney. Furthermore, the two reported autopsies showed depletion of granular cells in the cerebellum; damage to the cerebral cortex (but not selectively the visual cortex); and evidence of hypoplasia, disarrangement, and malformation of nerve cells. Although these pathologic changes may be observed in other forms of cerebral palsy too, the epidemiologic evidence seems to favor methylmercury toxicity during fetal development.

Since the brains of these infants continued developing after birth, and since there could have been severe exposure to mercury by way of mother's milk, it seems best to refer to such illness as "infantile Minamata disease." However, Harada (12) has maintained that the distribution of brain lesions in these cases differs from that in older children who presumably were exposed to contaminated fish. It would be helpful if comparisons with retarded and palsied infants from other localities had been made by the same investigative team.

Organ Concentrations as Indicators of Exposure

The Swedish report (8) contains an excellent assessment of the relationship of exposure to methylmercury, the resulting levels in various tissues, and their clinical effects. Valuable data from the Niigata episode not yet generally available have been carefully and critically reviewed in that report.

The central nervous system is, in a practical sense, the critical organ of methylmercury poisoning, since it is the first to show functional disturbances on exposure to toxic concentrations. But it is largely inaccessible to biopsy, and there are still no satisfactory means of detecting minimal central-nervous-system effects while the neuronal damage may yet be reversible. Also there is no biochemical or morphologic feature of the blood, cerebrospinal fluid, or other accessible fluids or tissues which currently can aid in the diagnosis of methylmercury poisoning—except the measured concentration of the mercury itself. Blood cells accumulate methylmercury in a concentration about ten times that in the plasma. The concentration in the blood, and especially in the blood cells, is the most reliable available index of exposure and of the level in the central nervous system. But methylmercury accumulates in the hair too, and in a fairly constant proportion to the concentration in the blood; so the hair concentration also may be used as a measure of exposure, of total body burden, and of the concentration in the central nervous system at the time of formation of the hair.

Brain Concentrations: Because methylmercury can cross the

blood-brain barrier more readily than can other mercurial compounds, the brain concentrations are much higher after administration of methylmercury than after corresponding doses of other agents. Even so, the concentration in the brain is slightly below the average for the total body.

In fatal cases of Minamata disease, the ratio of mercury in liver or in kidney to mercury in brain varied considerably, but generally it was between 4:1 and 2:1 (46). (This ratio contrasts with the distribution of inorganic mercury, of which there usually is 20 to 30 times more in the liver or kidney than in the brain.)

In the brain tissue of human beings who died after acute exposure to methylmercury, amounts of mercury ranging from 3 to $48 \mu g/gm$ of brain tissue have been reported (2). The lowest value may have been measured by relatively insensitive methods and may be only a fraction of the actual concentration. Data from patients in Minamata who died at various intervals after onset point toward a concentration of $5 \mu g/gm$, although it was higher than that in most cases.

Experiments have shown that the lowest toxic exposure varied widely among species of mammals, probably because of differences in the elimination rate. However, it seems that the brain concentration required for manifestation of clinical symptoms is fairly constant from species to species, namely about $8 \mu g/gm$ to $10 \mu g/gm$.

Concentrations in Blood and Hair: Data on the concentrations in blood and hair of patients and the concentrations in organs of those who died in Minamata, Niigata, and other episodes have not been collected in a uniform manner. The intervals from exposure to onset of symptoms, clinical suspicion of mercury intoxication, and collection and testing of specimens varied and often were unknown. The test procedures were often not standardized or adequately described. In some instances exposure continued even after symptoms developed; but in the Japanese experience repeated tests showed that in such cases most of the concentrations decreased as would be expected in persons whose exposure had ceased. Hair specimens usually were taken earlier than blood samples, and in several cases the hair concentration decreased in such a way that it appears that the blood levels were already declining. The relationship between the concentration in the blood or hair and the time that elapsed between the onset of the disease and the obtaining of the specimens makes it possible to estimate the concentration at onset; but in some cases the time is not entirely clear.

From the Minamata incident it has been calculated that severe intoxication was present when the hair concentration was equivalent to 700μg/gm (40). Blood samples do not appear to have been obtained at the time of intoxication, but concentrations higher than 4μg/ml of blood (equivalent to 8μg/gm in the erythrocytes) have been reported (2).

In the Niigata incident, it appears that the lowest concentrations at which symptoms were observed were of the order of 0.2μg/gm of blood and more than 200μg/gm of hair. Again, we do not know the degree of exposure or the interval between the exposure and onset of symptoms. However, the reported blood concentration may be low, since the usual ratio of hair to blood is about 300:1. According to that, 200μg/gm in hair would correspond to about 0.67μg/gm rather than the 0.2μg/gm which was reported.

In the Alamogordo investigation, serum concentrations were measured (rather than whole-blood), and these were about 3μg/gm, as was the concentration in one specimen of cerebrospinal fluid from an affected child. Concentrations of mercury in the hair were as high as 2,000μg/gm. It is likely that the brain concentrations equaled or exceeded those in Minamata patients who died, and that supportive treatment available to the Alamogordo children sustained them, although they have permanent brain damage. Urinary concentrations, considered undependable as indications of methylmercury intoxication, were reported as about 0.2 ppm in the three cases; but this level was noted also in urine from the father of the affected children, and he apparently was unaffected (15, 23).

Numerous values are available for Swedish individuals who ate much fish from mercury-contaminated lakes. One, without demonstrable clinical symptoms, had 0.65μg/gm of blood and 185μg/gm of hair, corresponding to an intake of 0.8 mg/day of methylmercury (50). The highest blood concentrations of mercury reported from a few of these exposed Swedish individuals exceeded the lowest concentrations reported from the diagnosed cases in Japan. Differences outlined herein in analytic and diagnostic methods, in the interval between symptoms and specimen collection, and in the degree of individual variation in sensitivity to methylmercury may account for the discrepancies noted.

It appears that in man a direct relationship exists between blood concentrations of mercury and the intake of mercury via fish containing mercurial residues. A reasonable approximation appears to be that for a man of 70 kg, 1 mg/day of methylmercury ($14\mu g$/kg body weight/day) gives rise to an incremental whole-blood concentration of approximately $0.8\mu g$/gm at equilibrium.

In Swedish individuals who are considered to have reached equilibrium between dietary intake and body burden of mercury, there is a simple direct relationship between concentrations of mercury in blood and hair, $60\mu g$ of mercury/gm of hair being equivalent to $0.2\mu g$/gm of whole blood or $0.4\mu g$/gm of red blood cells (18).

URGENT BASIC PROBLEMS

Physiologic: Absorption, Metabolism, Distribution, and Excretion.

Some of the factors that influence the absorption, distribution in various tissues, excretion, and dose-response relationships of methylmercury are known; but the mechanism of its action on the nervous system is still poorly understood.

Available data are based on studies of distribution and excretion following oral administration to human volunteers of single small doses of methylmercury labeled with mercury-203. Absorption from the gastrointestinal tract exceeded 90 percent. The labeled methylmercury was taken up rapidly by erythrocytes and was distributed fairly evenly among the tissues, except that the concentrations in kidneys and liver were higher than elsewhere and there was some delay in distribution to the brain. In adults, about 1 percent of the body burden is found in 1 liter of blood. Of the mercury in blood, the cells contain about 90 percent and the plasma about 10 percent. Mercury was excreted principally in the feces, at rates averaging about 2 percent per day. Only about 0.1 percent per day was lost in the urine.

Whereas methylmercury tends to concentrate in the hair and epidermis, in relation to the body burden the amount thus excreted is small in man (though large in feathered and furred animals). The elimination in man (and other animal species) may be reasonably described as following an exponential course.

The biologic half-life of methylmercury-203, according to total-body measurement, was about 70 to 90 days—somewhat shorter in the blood and perhaps longer in the brain (26). From the results of these studies, it was calculated that at equilibrium a weekly unit dose of methylmercury-203 would maintain the whole-body burden at 15.2 units of mercury (9). The slow elimination in man delays achievement of a steady state between uptake and loss of methylmercury until about a year after the beginning of low-level exposure. The rate of metabolism of organic mercury compounds to inorganic mercury appears to determine tissue distribution and rates of excretion. Because of the stability of methylmercury this conversion is slow and elimination (largely of the intact methylmercury) is likewise slow. Norseth's studies (32) in rats showed a relatively high rate of conversion (about 40 percent) to inorganic mercury, apparently in the cecum; but most fecal mercury is derived from the enterohepatic circulation, pancreatic secretion, and shedded intestinal epithelium. The turnover of intestinal epithelial cells, which is several times greater in the rat than in man, may be important in controlling the rate of excretion of methylmercury (2).

The observations of Östlund (34) on the metabolism of dimethylmercury in the mammal show that it behaves differently from methylmercury. In mice exposed to dimethylmercury, either through ingestion or inhalation, most of the administered dose was rapidly exhaled as intact dimethylmercury. Smaller amounts of the compound were metabolized to methylmercury.

Methylmercury is known to react with SH groups in amino

acids and proteins, and it is believed bound in other ways in biologic systems, such as to carboxy and amide groups. The effect at the molecular level has not been elucidated, however; nor do we know which enzyme systems are affected and to what extent; nor has the relation between cellular lesions and symptoms of poisoning been clarified. Enzymes containing SH groups are known to be inhibited and the permeability of cell membranes is known to be affected. It has been suggested that the main effect of methylmercury in the brain is a disturbance in "protein synthesis," which "could then explain the latency period between exposure and onset of symptoms" (8).

Methylmercury *in vitro* is said to inhibit transaminase activity in tissue slices from rat brain, the lowest effective concentration being about $10 \mu g$ Hg/gm (22).

Genetic: Embryotoxic and Teratogenic Effects

Experimentally, methylmercury may disturb cell division in test systems of plants through inactivation of the mitotic spindle (or c-mitosis) (37). Although disturbances in cell division have been produced in *Drosophila* and in tissue culture (10), they have not been demonstrated in test mammals.

In a study with *Drosophila melanogaster* selected so that the c-mitotic effect could be traced on the gametes, treatment of female larvae and adults with methylmercury significantly increased the number of exceptional daughter offspring (XXY) showing irregularities of meiotic chromosomal disjunction (36, 38), and other chromosomal effects and developmental disturbances (mutant gene and outstretched and bent wing) were noted also.

In earlier experiments, administration of organic mercurial compounds to inbred CBA mice did not produce dominant lethal mutations (36). Khera (21) is said to have administered methylmercury to male rats for seven days and then mated them in sequential periods of five days with untreated females. Although the number of impregnated females was the same in all groups, the average litter size was reduced, the most marked reduction occurring 15 to 20 days after the exposure to methylmercury was discontinued. The results were positive at dose levels as low as 1 mg/kg body weight per day (39).

In other studies, wherein larger doses of methylmercury were given to pregnant rodents, embryotoxic and teratogenic effects were manifested as reduced weight, osseous abnormalities, and cleft palate (21,29,33).

There is continuing interest regarding the development of the brain and the behavior patterns of offspring whose pregnant mothers received low doses of methylmercury. A number of variations in biochemical determinations have been noted, but the significance of these effects in relation to possible behavioral changes is not known, nor is it known whether the observed changes are specific for mercury compounds (39).

Although no increase of abortions or congenital anomalies was recognized in conjunction with the Minamata tragedy, there was an increased incidence of nonspecific "cerebral palsy." The influence of methylmercury in this, and particularly its mechanism of action, are not understood. Certainly methylmercury is active genetically (c-mitotic and chromosome breakage effect), and this implies the possibility of damage to fetal cells. Thus far, however, the mutagenic effect appears to be small; and though methylmercury exposure may bring genetic risks, it is not possible to relate their magnitude to different degrees of exposure. Nevertheless, there is a need to clarify the effect of small and moderate doses of methylmercury on the germ cells as well as on the tissues of the developing embryo and fetus. In the meantime, the element of uncertainty here may have a conservative influence on recommendations regarding methylmercury exposure of fertile women.

The significance of the statistical increase in chromosomal breaks in blood cells of human consumers of large amounts of fish containing methylmercury reported by Skerfving et al. (40) remains to be clarified. No significant polyploidy or aneuploidy was found with it.

Practical: The Allowable Daily Intake

The Swedish Commission on Evaluating the Toxicity of Mercury in Fish has released its report (8) with recommendations on an "allowable daily intake" (ADI) of methylmercury. This concept amounts to a warning to all persons to restrict their intake of mercury-contaminated foods so as to keep their ADI within the recommended level.

Several methods of determination are available. The one which is perhaps the most conservative is based on the lowest toxic concentration rather than the lowest fatal concentration. The lowest whole-blood concentration associated with toxic symptoms was taken as $0.2\mu g/gm$; and the commission equated this with $60\mu g/gm$ in hair and with a daily intake of 0.3 mg of methylmercury for a 70-kg man. They then applied a safety factor of 10 and obtained the following hypothetically safe standards: for whole blood, $0.02\mu g/gm$; for hair, $6\mu g/gm$; and for ADI, 0.03 mg for a 70-kg man.

With these standards, the restriction upon fish consumption is severe. Thus the allowable weekly intake (7×0.03 mg $= 0.21$ mg/wk) would be supplied by 210 gm of fish containing 1 ppm of methylmercury. If fish containing 0.5 ppm of methylmercury were eaten daily, the limit (0.03 mg) would be reached by daily consumption of 60 gm.

It is of interest that the average blood concentrations observed by McDuffie (25) among members of Weight-Watchers, Inc. who ate much fish were as high as $0.01\mu g/gm$, or half of the maximum recommended above. Although this is appreciably less than the blood concentrations (up to $0.65\mu g/gm$) noted among some apparently unaffected eaters of much freshwater fish in Sweden, it is also appreciably higher than the level of $0.002\mu g/gm$ reported for others who ate little or no fish.

COMMENT AND RECOMMENDATIONS

It has been noted recently that the methylmercury concentration in about 90 percent of commercial swordfish tested was greater than the 0.5 ppm guideline of the US Food and Drug Administration, and in about 50 percent it exceeded 1.0 ppm. This has led to removal of swordfish from the marketplace in the United States; and as expected, there has been criticism (57) of this action because no case of toxicity from swordfish had even been identified. Herdman (14) has just described the case of a woman in New York who during a nine-month period of dieting had consumed about 300 gm of swordfish daily (average estimated methylmercury, 0.3 mg per day) and developed symptoms suggestive of methylmercury poisoning. Though a wide-based gait was noted, the illness was initially diagnosed and treated as psychoneurosis. The symptoms included lethargy, difficulty in "focusing" the eyes, some loss of memory and of reading comprehension, tremor of hands and tongue, and mispronunciation of words. Mercury in her hair during this illness was subsequently estimated as 42μg/gm, and whole-blood level of mercury measured several months after discontinuance of the fish diet was 0.06 μg/gm.

This may be the first recognized case in North America or Europe that was due to ingestion of fish. It has been argued that the patient's symptoms may have been a reflection of her neurotic and faddistic tendencies or were influenced by other dietary aids, medication, or nutritional deficiencies. The mercury concentrations mentioned are abnormal, however; and it seems reasonable to relate the symptoms to that substance. If this is a singular case, perhaps the loss of swordfish as a source of protein and the consequent substitution of other foods containing more cholesterol may be responsible, in the long run, for more disease than the removal of swordfish will prevent.

This dilemma leads us to reconsider a few of the serious problems to which I have already referred. There is a need for systematic studies of presumably exposed populations. Whether subtle harms result from low-dosage exposure remains to be seen; certainly all reasonable means to reduce exposure to mercury should be instituted as soon as possible.

Some of the unsolved problems follow:

1. The mechanism of action of methylmercury is not understood. Why are the granular cells of the cerebellum, dorsal roots, and some cerebral cortical cells selectively affected? What are the biochemical features that these cells have in common?

2. To produce its detrimental effect on the nervous system, must the methylmercury exceed a critical concentration, or does almost any exposure cause a corresponding neuronal loss? Are

tissues other than the nervous system affected to any significant degree before or after recognizable neuronal damage occurs?

3. What is the basis of individual variation in sensitivity? Is this sensitivity influenced by genetic and nutritional factors and by symbiotic or antagonistic effects of other toxic agents? Are children more sensitive to methylmercury than are adults?

4. Whereas diagnosis is based on neurologic evaluation and this, in its earliest phase, is a reflection of moderately severe damage (though much of that may still be reversible), how can neurologic involvement be detected earlier and how can treatment for early involvement be improved?

5. Are there subclinical or mild forms of the methylmercury intoxication that may interfere with inherent intellectual or motor capabilities? Could forms of cerebral palsy or presenile dementia be due to slight exposure to methylmercury?

6. Our data on genetic effects and prenatal intoxication are limited. Since the mothers of the Minamata cerebral palsy patients seem to have been well, can we assume that the fetus is more sensitive than the adult?

The question of genetic and teratogenic effects is perhaps one of the major reasons for the conservative stand which has been taken regarding consumption of saltwater predatory fish (swordfish, tuna), which appear naturally to have a relatively high level of methylmercury.

When we match the potential danger against the known economic, and perhaps nutritional, loss from the restrictive measures that have been imposed, we can realize the gravity of the problem. Although decision-making in this situation is not within the responsibility of many of us here, I know that all progress in our understanding of the methylmercury problem through further epidemiologic and laboratory studies will be appreciated and will result in the prompt institution of the most reasonable safeguards and adequate methods of control.

REFERENCES

1. Bakulina, A.V.: Effect of subacute Granosan poisoning on progeny. *Sov Med*, 31:60, 1968.

2. Berglund, F. and Berlin, M.; Risk of methylmercury cumulation in man and mammals and the relation between body burden of methylmercury and toxic effects. In M. W. Miller and G. G. Berg (Eds.): *Chemical Fallout: Current Research on Persistant Pesticides.* Publisher, Springfield, Thomas, 1969, pp. 258-273.
3. Brown, I. A.; Chronic mercurialism: A cause of the clinical syndrome of amyotrophic lateral sclerosis. *Arch Neurol Psychiat, 72:*674, 1954.
4. Cerletti, U.; Antiluetici neurotropici. *Riv Sper Freniat, 42:*374, 1917.
5. Clarkson, T. W.; Epidemiological and experimental aspects of lead and mercury contamination of food. *Food Cosmet Toxicol,* 9:1, 1971.
6. Dahhan, S. S. and Orfaly, H.: Electrocardiographic changes in mercury poisoning. *Am J Cardiol, 14:* 178, 1964.
7. Edwards, G. N.; Cited by Expert Group (8).
8. Expert Group; Methyl mercury in fish: a toxicologic-epidemiologic evaluation of risks. *Nord Hyg Tidskr,* (suppl 4) 1971.
9. Falk, R., Snihs, J. O., Ekman, L., Greitz, U. and Åberg, B.: Whole-body measurements on the distinction of mercury-203 in humans after oral intake of methylradiomercury nitrate. *Acta Radiol,* 9:55, 1970.
10. Fiskesjö, G.; The effect of two organic mercury compounds on human leucocytes *in vitro. Hereditas, 64:*142, 1970.
11. Greco, A. R.; Elective action of some mercurial compounds on the nervous system. I. Determination of the mercury in the blood and cerebrospinal fluid of animals treated with diethyl mercury and with the common mercurial compounds. *Riv Neurol,* 3:515, 1930.
12. Harada, Y.: Congenital (or fetal) Minamata disease. Minamata Disease (Study Group of Minamata Disease), Kumamoto University, Japan, 1968, pp. 93-117.
13. Hepp, H.; Cited by Greco (11).
14. Herdman, R. C.; Statement to the Subcommittee on the Environment of the Committee on Commerce of the United States Senate, May 20, 1971.
15. Hinman, A. N.: Personal communication.
16. Hunter, D., Bomford, R. R. and Russell, D. S.: Poisoning by methyl mercury compounds. *Q J Med,* 9:193, 1940.
17. Hunter, D. and Russell, D. S.; Focal cerebral and cerebellar atrophy in a human subject due to organic mercury compounds. *J Neurol Neurosurg Psychiat, 17:*235, 1954.
18. International Committee; Maximum allowable concentrations of mercury compounds. *Arch Environ Health, 19:*891, 1969.
19. Jensen, S. and Jernelöv, A.: Biological methylation of mercury in aquatic organisms. *Nature (Lond), 223:*753, 1969.
20. Kantarjian, A. D.; A syndrome clinically resembling amyotrophic lateral sclerosis following chronic mercurialism. *Neurology, 11:*639, 1961.
21. Khera, K. S.; Reported by D. Clegg, at Symposium on Mercury in

Human Environment, Royal Society of Canada, Ottawa, February 1971.
22. Kuwahara, S.; Cited by Expert Group (8).
23. Likosky, W. H., Pierce, P. E., Hinman, A. R. and Nickey, L.; Organic mercury poisoning, New Mexico. Read at the meeting of the American Academy of Neurology, Bal Harbour, Florida, April 27 to 30, 1970.
24. Matsumoto, H., Sinkei, L., Tomco, N. and Kameda, T.; Pathological study of toxic polyneuropathy. I. Experimental induction of polyneuropathy by methylmercury. *Adv Neurol Sci*, 13:660, 1969.
25. McDuffie, B. R.: Weight watchers, beware! *Newsweek* (no. 91) January 25, 1971.
26. Miettinen, J. K.: Organic mercurials as food chain problems. Report to the FAO/IAEA Panel on the use of isotope and radiation techniques in studies on the fate of pesticide residues in the food chain, IAEA, Vienna, 1969.
27. Miettinen, J. K., Rahola, T., Hattula, T., Rissanen, K. and Tillander, M.: Retention and excretion of ^{203}Hg-labelled methylmercury in man after oral administration of CH_3 ^{203}Hg biologically incorporated into fish muscle protein—preliminary results. Fifth RIS (Radioactivity in Scandinavia) Symposium. Department of Radiochemistry, University of Helsinki, Finland, 1969.
28. Miyakawa, T., Deshimaru, M., Sumiyoshi, S., Teraoka, A., Udo, N., Hattori, E. and Tatesu, S.; Experimental organic mercury poisoning —pathological changes in peripheral nerves. *Acta Neuropathol*, 15:45, 1970.
29. Moriyama, H.; Cited by Expert Group (8).
30. Mulder, D. W.; Personal communication.
31. Nelson, N., Byerly, T. C., Kolbye, A. C., Jr., Kurland, L. T., Shapiro, R. E., Shibko, W. H., Thompson, J. E., Van Den Berg, L. A. and Weissler, A.; Hazards of mercury: special report to the secretary's pesticide advisory committee, Department of Health, Education, and Welfare. *Environ Res, 4:* 1, 1971.
32. Norseth, T.; *Studies on the biotransformation of methylmercury salts in the rat*, Ph.D thesis, University of Rochester, New York, 1969.
33. Oharazawa, H.; Effect of ethyl mercuric phosphate in the pregnant mouse on chromosome abnormalities and fetal malformation. *J Jap Obstet Gynecol Soc*, 20:1479, 1968.
34. Östlund, K.: Studies on the metabolism of methyl mercury and dimethyl mercury in mice. *Acta Pharmacol Toxicol* 27 (suppl 1): 5, 1969.
35. Prumers, V.: Cited by Greco (11).
36. Ramel, C.: Genetic effects of organic mercury compounds. *Hereditas*, 57:445, 1967.

37. Ramel, C.: Genetic effects of organic mercury compounds. I. Cytological investigations on allium roots. *Hereditas, 61*:208, 1969.
38. Ramel, C. and Magnusson, J.; Genetic effects of organic mercury compounds. II. Chromosome segregation in Drosophila melanogaster. *Hereditas, 61*:231, 1969.
39. Shibko, S.; Personal communication.
40. Skerfving, S., Hansson, K. and Lindsten, J.; Chromosome breakage in humans exposed to methyl mercury through fish consumption: preliminary communication. *Arch Environ Health 21*:133, 1970.
41. Snyder, R.D.; Congenital mercury poisoning. *New Eng J Med 284*:1014, 1971.
41A. Synder, R.D.; The involuntary movements of chronic mercury poisoning. *Arch Neurol, 26*:379, 1972.
42. Storrs, B., Thomson, J., Fair, G., Dickerson, M. S., Nickey, L., Barthell, W. and Spaulding, J.E.; Organic mercury poisoning: Alamogordo, New Mexico *Morbidity Mortality, 19*:25, 1970.
43. Storrs, B., Thompson, J., Nickey, L., Barthel, W. and Spaulding, J. E.; Follow-up organic mercury poisoning: New Mexico. *Morbidity Mortality, 19*:169, 1970.
44. Study Group of Minamata Disease, Minamata Disease, Kumamoto University, Japan, 1968.
45. Suzuki, T., Matsumoto, N., Miyama, T. and Katsunuma, H.; Cited by Expert Group (8).
46. Takeuchi, T.; Pathology of Minamata Disease, Kumamoto University, Japan, 1968.
47. Takeuchi, T.: Unpublished data.
48. Takeuchi, T., et al., Pathological studies on Minamata disease with special reference to the histopathological findings in the central nervous system. *Kumamoto Igkz, 31* (suppl 2):262, 1957.
49. Taylor, W., Guirgis, H. A. and Stewart, W. K.; Investigation of a population exposed to organomercurial seed dressings. *Arch Environ Health 19*:505, 1969.
50. Tejning, S.; Mercury content in the blood system, blood plasma and in heavy fish eaters from different regions and the connections between mercury content in fish together with a proposal on international food hygienic limits on mercury applicable to fish and fish products. Report 670731, Clinic of Occupational Medicine, University Hospital, Lund, Sweden, 1967.
51. Tejning, S.; Personal communication.
52. Tsubaki, T.; Personal communication.
53. Tsubaki, T.; Unpublished data.
54. Tsubaki, T., Sato, T., Kondo, K., Shirakawa, K., Kambayashi, K., Hirota, K., Yamada, K. and Murone, I.; Outbreak of intoxication by organic mercury compound in Niigata perfecture: an epidemiological and clinical study. *Jap J Med, 6*:132, 1967.

55. Uchida, M. and Hirakama, K.; Biochemical studies on Minamata disease. III. Relationships between the causal agent of the disease and the mercury compound in the shellfish with reference to their chemical behaviors. *Kumamoto Med J, 14*:171, 1961.
56. Uchida, M., Hirakawa, K. and Indue, T.: Biochemical studies on Minamata disease. IV. Isolation and chemical identification of the mercury compound in the toxic shellfish with special reference to the causal agent of the disease. *Kumamoto Med J, 14*:181, 1961.
57. Widener, A.; Consumers: resist unscientific claims of ecology faddists. *Rochester Post Bulletin*, June 15, 1971.
58. Wood, J. M., Kennedy, F. S. and Rosen, C. G.; Synthesis of methylmercury compounds by extracts of a methanogenic bacterium. *Nature (Lond), 220*:173, 1968.

DISCUSSION

Suzuki: I'd like to comment on the safety level which Dr. Kurland has just presented. It will be interesting and important in many senses to compare the Swedish safety level to the actual Japanese level.

We have the data from women living in Tokyo without any obvious toxicological conditions. Mercury content in red cells and plasma were correlated to the average daily intake of various food items, such as rice, meat, fish and others, which was obtained from the two-week-record of diet by a substantial technique of the step-wise regression analysis. Only a significant regression coefficient was found on the dietary fish intake. Their diet consisted primarily of marine fish such as tuna, horse mackerel and salmon.

On the simple regression equation of mercury content in red cells (Y) based on the intake of fish (X), including that of processed fish flesh ($Y = 0.451 X + 12.40$), the already published Swedish safety level of 40 $\mu g/gm$ in red cells is understood to correspond the value of about 60 gm of daily fish intake in the women studied.* Our national average consumption of "fish" (not only fish but shellfish and mollusca are included) in 1968 was estimated at 86.2 gm per day. The proportion of shellfish and mollusca in quantity was about one fourth of the total "fish" consumption, then the true national average of fish consumption is considered to be about 65 gm per day.

If we accept the Swedish safety level, and consider the average mercury content in fish as methylmercury being 0.5 pg/gm, our average daily intake of methylmercury would be higher than the safety level.

Friberg: Dr. Kurland mentioned that the lowest level in hair for obtaining symptoms in Niigata was 200 ppm and the blood level was 0.2; but in fact the lowest level in hair was 50 and so the hair level corresponded with that

---

* Suzuki *et al.*: Industrial Health, 1971, in press.

level as found in our studies in Sweden where we have compared hair levels with blood levels. We have data from healthy people where we have studied the concentrations in hair and blood at the same time; it's from these studies we have drawn our conclusions concerning the relation between hair and blood levels. The lowest level of intoxication in Japan was 50 in hair and 0.2 gives a hair-blood ratio corresponding to what we have found in normal Swedish people.

Gage: At the Stockholm meeting (November 1968) the limit for occupational exposure to methylmercury was set at 10 $\mu g/100$ ml of blood which approximates 0.1 $\mu g/gm$. The suggested limit applies not only to workmen occupationally exposed, but also to very young and very old people. Is this five-fold difference due to the different populations involved or have ideas on the toxicity of methylmercury been revised?

Friberg: Our committee did not try to revise the TLV values for methylmercury in industry. On the other hand, if we had had good data at the time of the Stockholm meeting I think it would have been pretty unnecessary to write the present report on methylmercury in fish. We felt, however, that the scientific evidence wasn't very good and that is the reason why we have tried particularly to analyze the Japanese data.

Clarkson: With reference to Dr. Gage's question and the safety factor of ten, did you include possible teratological effects?

Friberg: Yes. But it is very difficult to decide whether a safety factor should be, for example, ten or eight.

The basic philosophy when we had our discussion was that we started out from the lowest level where we had found any evidence whatsoever of a toxic effect. That was one reason why we thought that the factor of ten was enough. Then during the whole discussions of Japanese and different animal data when we had to choose between a low level and a high level in evaluating something we always took a conservative approach. So the safety factor of ten might well be considerably higher than ten. I want to stress, however, that it is based on the evidence we had at the time. It might be necessary for us to reevaluate our recommendations a year or two from now based on data that will be available at that time.

Herman: There have been various populations in Sweden and Japan with elevated hair and blood levels that have been labeled "asymptomatic" populations. How thoroughly have these "asymptomatic" subjects been studied in terms of complete neurological examinations, in terms of visual field testing, peripheral nerve conductions, EEG's, audiography or any other objective tests? It seems the "no effect" dose of methylmercury in humans has not been determined precisely.

Friberg: On some a fairly extensive study was carried out but not on all.

Berlin: We can't exclude according to present techniques that there may have been some changes, particularly in view of the experimental evidence we have now that there are induced morphological changes. With the Japanese studies there were a number of patients especially in Niigata

An Appraisal of Alkylmercury Compounds

which were very thoroughly studied from your local point of view. These data are all published in the Niigata report and our special publications too. Dr. Tsubaki, who is a neurologist at the University of Niigata, has done the majority of these studies.

Suzuki: Dr. Kurland reported that the mercury content in hair can be used as an indicator of the total body burden of methylmercury. But, I think we should be careful when using the mercury content in hair as an indicator of body burden. If there is a constant intake of methylmercury, in other words, in a dynamic equilibrium between uptake and excretion of methylmercury, it would be a useful indicator. But in a cross-sectional study, a single and partial sampling of hair cannot provide the data covering the entire feature of mercury exposure. On the relationship between mercury content in red cells and in hair, we used to get different relationships due to the changing exposure and the hair sampling.

Friberg: I think that we could agree that one shouldn't emphasize too much the hair values and make calculations back to the red cell values but under certain circumstances hair values can be useful.

Foulkes: I wonder whether we are not oversimplifying this whole problem. Surely the distribution within the body, and therefore also the toxic effects within the body, are going to be influenced not only by the intake but by any number of other dietary, genetic or other factors. Could this be a part of the apparent discrepancy between your results?

Clarkson: The question has been raised by some people that, as in the case of the Itai Itai disease, other factors may have somehow contributed to the effects of methylmercury in Minamata and Niigata.

I understand, however, that it's the opinion of most people who have reviewed this situation that it is a straightforward toxic effect of methylmercury. One does not need to invoke the presence of other environmental agents and so on. I wonder if anyone has any comments on that because I know it's been discussed at some length, particularly in view of the fact that certain people in Sweden have had much higher blood levels without any symptoms.

Goldwater: I want to comment briefly on a case of a premature infant who, during cardiac catheterization, accidentally got some mercury injected into her heart and had been running blood levels of 60 μg to 80 μg per hundred ml of blood now for several months without any detectable adverse effects. This infant is being studied as vigorously and meticulously as any case can possibly be studied.

I was going to ask Dr. Kurland this same question about the possibility of related or subsidiary factors particularly since he had mentioned there were differences in the mortality figures between Niigata and Minamata. This, of course, is a very crude thing because it may have been just a dose effect: the Minamata people got more than the Niigata people. Dr. Suzuki mentioned that there were certain differences in the clinical manifestations between the Niigata and the Minamata people.

So perhaps we should be looking at other factors since we know that in Minamata there was arsenic and thallium and selenium and manganese and aldehydes that may have come into the picture.

Kurland: The Minamata experience prepared neurologists and others elsewhere for the outbreak which unfortunately was to come to Niigata several years later. Once the disease was recognized, they began to search for relatively mild cases. Therefore, earlier and milder cases were apparently better diagnosed in Niigata than in Minamata. I presented the case fatality ratios which were higher in Minamata; this may be taken as an indication that milder cases comprised a greater proportion of the total in Niigata than in Minamata.

Klein: Has there been any epidemiological study of individuals in a population at risk of methylmercury exposure by virtue of food intake or other environmental exposure who have blood levels just above the acceptable level and just below levels where clinical symptoms become manifest? Is there evidence of teratogenic or genetic effects in this population?

Grant: One way we tried was to look at material from patients who had died with a diagnosis of some neurological disease with which methylmercury intoxication could conceivably be confused, i.e. the spongy encephalopathies.

We've managed to find nine cases from two large Stockholm hospitals covering a period of about 11 years. Eight of the nine had very low mercury values in their brain tissues. The ninth is mentioned in the book *Methyl Mercury in Fish. A Toxicologic-Epidemiologic Evaluation of Risks*, Nordisk Hygienisk Tidskrift, suppl 2, 1971, p. 153, and had 19 μg Hg/gm. Very little of that was actually methylmercury.

McDuffie: I want to comment on Dr. Kurland's remarks about the studies we did in Binghamton, New York, on the tuna-swordfish eaters.° Table 2–I shows that in the tuna-swordfish diet group the average concentrations of

TABLE 2–I
MERCURY LEVELS IN TUNA—SWORDFISH DIET GROUP AND IN CONTROL GROUP*

Materials Analyzed	Average Concentration of Total Mercury Found†		
	Diet Group	Control Group	Ratio
Blood, whole, 10 ml.	1.0μg/100 ml (42)	0.2μg/100 ml (18)	5/1
Hair‡, unwashed, 0.1 to 0.2 g	8.8μg/g (41)	3.1μg/g (19)	3/1
Urine, 25 ml of spot specimen	6.4μg/liter (42)	1.3μg/liter (11)	5/1

* Estimated daily intake of mercury by diet group is shown in Figure 2–1.
† Averages based on number of participants shown in parentheses.
‡ Some hair samples cut several inches from scalp, thus reflecting previous body concentrations.

° Testimony of B. McDuffie, Hearings of the U.S. Senate Commerce Committee Sub-committee on the Environment, Washington, D.C., May 20, 1971; prepared for submission to *Science*.

An Appraisal of Alkylmercury Compounds

mercury found where about 0.010 ppm for whole blood, 8.8 ppm for hair, and 6.4 µg/liter for urine. The ratio to the control group concentrations was 5 to 1 for blood, 3 to 1 for hair, and 5 to 1 for urine. (The hair was in some cases cut too far out from the scalp to be an indication of the most recent growth.)

Figure 2-1 gives a plot of the blood concentrations found for the diet group *versus* daily intake rates, estimated from the amount of tuna and swordfish the people had been eating, adjusted to a 150-pound-person, and also adjusted for the number of months the person had been on the diet to allow for degree of approach to the steady state, assuming a 70-day body half-life.

One can see a rough correlation between the intake rate and the blood concentration, the correlation coefficient being 0.65 for this data. The hair data did not correlate quite as well (a correlation coefficient of 0.43) and

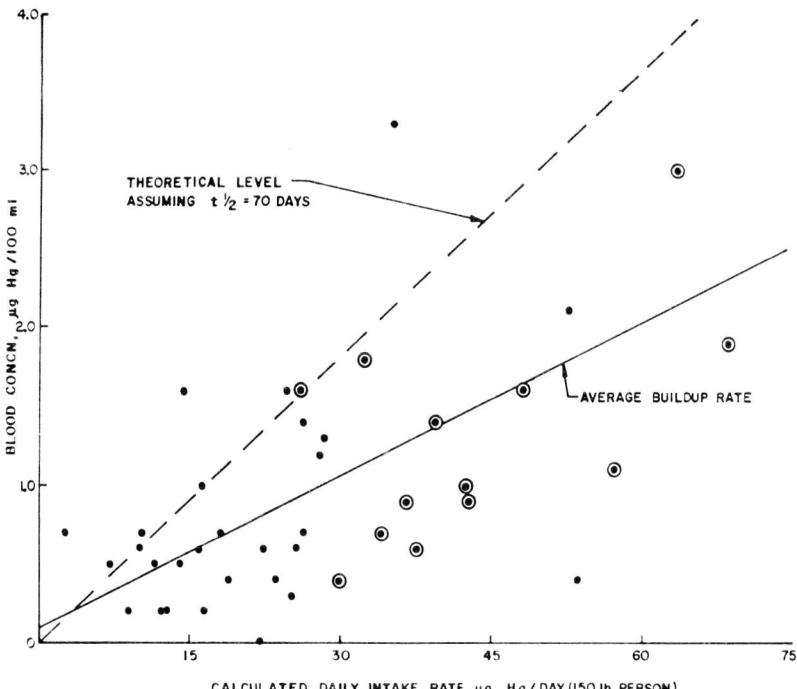

Figure 2-1. Buildup of mercury levels in human consumers. Daily intake rate calculated from diet information supplied by each person, assuming tuna @ 0.25 ppm, swordfish @ 1.00 ppm, per 150 lb. body weight, adjusted for degree of exponential buildup if on diet less than 16 months. Circled points for persons who ate significant amounts of swordfish.

no correlation was found for the urine data, where individual variations are known to be pronounced.

The dash line shows what we would have expected for a body half-life of 70 days. Our slope is only about half that, but our blood determinations are probably a little low. (We've been doing some recovery studies which indicate that our results represent around 70 or 80 percent for the total mercury in blood.) Also, the estimates of intake are based on what the people told us, figuring 0.25 ppm mercury for tuna and 1 ppm for swordfish, which may be a little bit inflated.

Table 2–II gives the quartile rankings of the top 40 dieters. There isn't much of a jump in average blood or hair level until the third or fourth quartiles where the intake rates are ≥ 27 μg/day; here one sees a distinct increase in the average blood and hair concentrations, a doubling and a tripling in these upper quartiles.

TABLE 2–II
QUARTILE RANKINGS OF TOP FORTY DIETERS

	Est. Dosage Range	Av. Blood Concn.	Av. Hair Concn.
	μg/day/150 lb	μg/100 ml	μg/g
1st Quartile	9 to 16	0.60	5.3
2nd Quartile	17 to 26	0.64	4.9
3rd Quartile	27 to 38	1.20	9.4
4th Quartile	40 to 75	1.73	14.4

Figure 2–2 shows data obtained in a follow-up study with five dieters who had above average blood levels. It also shows the symptom level, 0.20 ppm, based on the Swedish Commission findings and the recommended "safe level," 0.020 ppm. The average of the control group who hadn't eaten any tuna or swordfish in the last year was about 0.002 ppm according to our study, and the average dieter was at 0.010 ppm. After a rise on the part of two dieters, the concentrations started down, with a slope corresponding to about a 50-day half-life, certainly less that the 70-day whole-body half-life. (Miettinen has reported a whole-body half-life averaging 76 days, but a half-life for methylmercury in red blood cells of about 50 days, in agreement with the decay curves of Figure 2–2.)

All five were supposed to cut out tuna and swordfish, but I learned later that three of the four had continued eating some tuna; "A" had been eating 12 oz. of swordfish and 50 oz. of tuna a week, cut out the swordfish and kept on with the tuna, then in February cut back on tuna considerably. "B" switched from all swordfish to halibut and turbot; "D" increased tuna intake from 12 to 30 oz/week in February, then cut back to 12 oz/week in March.

The study shows that a number of people (10 percent or more of our diet group) were up between the "safe" and symptom level. (These persons have not been checked for any evidence of neurological symptoms.)

An Appraisal of Alkylmercury Compounds

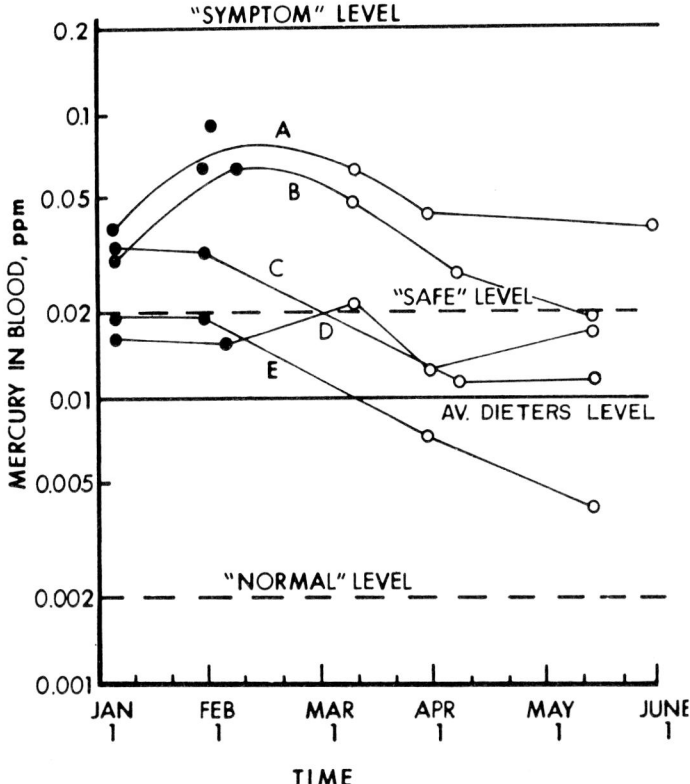

Figure 2–2. Logarithmic plot of blood mercury levels vs. time for five above average dieters. Adjusted estimated dosage rates, in μg/day/150 lb., prior to the first sampling in January (1971) were: 75 for A; 64 for B; 36 for C; 15 (plus mercury in 7 oz. cod/day) for D; 68 for E. Solid circles—single analyses; open circles—averages of duplicate analyses.

Miettinen: Did you determine how much of your tuna mercury was methylmercury and how much inorganic?

McDuffie: I did not but I think others have.

Smith: We're fairly satisfied that the bulk of the mercury that we've examined in the tuna and swordfish is methylmercury.

Miettinen: The percentage?

Smith: Ninety percent would be conservative.

Miettinen: This is a little unexpected this short half-life time both in accumulation where the levels were half of those expected and in elimination because they correspond to inorganic mercury which has a biological half-time of 42 days rather than methylmercury.

Nechay: Does anybody know what the mercury concentration is in shrimp, crab, lobsters, oysters, etc?

Kolbye: According to the information I have, very large lobster tails may be running in excess of the guideline, but whether they are commercially important in terms of human ingestion remains to be determined. We have no present problem with crabs or other forms of shellfish.

Kurland: Shrimp is apparently below 0.06 per million in total mercury content.

Berlin: I'd like to comment on Dr. Kurland's comment about our 200 ng or 0.2 ppm in blood as a toxic level. One should not regard that value as derived from one patient. It is very important to realize that it is the result of evaluating a population of values. We concluded that the lower part of the population should be around 200.

Kolbye: I think you have to differentiate, when you're talking about shellfish areas, as to whether they're open for commercial fishing or closed. Some of the areas in and around Galveston Bay, for example, are closed for a variety of pollution reasons. I'm not aware of any shrimp levels from approved beds that contain any appreciable methylmercury residue.

Herdman: I can mention two things. Firstly, we have performed a number of market basket surveys where we analyzed shrimp, lobster, oysters, and other shellfish and also studied a fair number of the edible shellfish along Long Island and other parts of the state. We agree with Dr. Kolbye that none of these were above FDA guidelines. Secondly, one seafood we do find which is markedly above the guideline and we haven't mentioned it so far is whale meat, which goes up to 2 ppm.

McDuffie: A few samples of shellfish we have tested have run around 0.1 to 0.2 ppm. Also, I understand that in Canada large halibut might be over the guideline.

Kolbye: Halibut over 150 pounds in size tend to run over the guideline, but again with respect to the United States the situation is well under control because the halibut fishermen have agreed to a uniform policy with respect to large fish. So we don't expect any halibut above the guideline to be reaching the market.

Kazantzis: Dr. Kurland told us about methylmercury poisoning occurring in the fetus, which I take to be a condition somewhat similar to adult methylmercury poisoning affecting the child at a stage when organ development is almost complete. We know very little about the effects of methylmercury at an earlier stage of development. The only report of a teratogenic effect that I know of is by Murakami in 1969.* Has this report been substantiated and on what has it been based?

Berlin: Murakami's report is based on clinical data from the Minamata report. He hasn't himself examined these patients.

Kazantzis: This is a very important claim but one which as far as I can see we have very little evidence for at present.

* *J Japan Med Assoc,* 61:1059, 1969.

Joselow: You mentioned that in infantile Minamata disease, mercury could have entered the infant either through transfer in mammary milk or actual transfer across the palcenta during gestation. Then you also mentioned that in the New Mexico episode, the infant was fed a carefully controlled diet. Did this rule out mammary transfer?

Kurland: I discussed with Dr. Russell Snyder* the case he reported and he assured me that the child was not nursed by the mother and presumably was affected prenatally.

* N Engl J Med, 284:1014, 1971.

Chapter 3

ARYL-AND ALKOXYALKYLMERCURIALS

LEONARD J. GOLDWATER

ABSTRACT

Hundreds of aryl-and alkoxyalkyl mercury compounds have been synthesized. Of the aryls, phenylmercuric acetate (PMA) is best known and most widely used. Extensive observations have been made on humans exposed occupationally to PMA, consistently showing a relatively low degree of toxicity. Available evidence suggests that other phenylmercurials have toxicological characteristics similar to those of PMA. Animal experiments are, for the most part, in consonance with the human experience. Longterm feeding experiments with PMA equivalent to 0.5 ppm of mercury in the diet have yielded inconsistent results which, nevertheless, suggest the possibility of reversible kidney injury in rats.

Very little toxicological information is available on the alkoxyalkylmercurials except for those related to the diuretics; many members of this group are practically nontoxic. Experimentally, methoxyethylmercury chloride administered parenterally to rats behaves like the arylmercurials, but there is some question as to similar behavior in humans.

REPORT

Introduction

Whitmore's classic monograph *Organic Compounds of Mercury* was published just 50 years ago (33). While written primarily for chemists, the work was stimulated in part "... because of the need for some non-ionized mercury compound for use with organic arsenicals in the treatment of syphilis." In his preface, the author points out that the true organic mercury compounds, in which mercury is attached directly to a carbon atom, have practically none of the properties of mercuric salts, while compounds of the types O-Hg, N-Hg and S-Hg for the most part have properties which are those of inorganic mercury salts. At least in part, the differences can be explained on the basis of the stability of the respective compounds but the complexities of the reactions are great, as illustrated by Wright in 1957 (36).

Hundreds of aryl-(8) and alkoxyalkyl (12) mercury compounds have been synthesized (Tables 3-I to 3-IV). Most of the existing pharmacological and toxicological knowledge about the latter has been derived in connection with their applications in medicine as diuretics. Arylmercurials have found their greatest use as biocides. For both classes there are substantial bodies of literature dealing with both human and animal observations (1, 32).

Strictly speaking, the arylmercurials embrace all mercury derivatives of aromatic hydrocarbons. This includes such groups as benzyls, tolyls, naphthyls and many others, including their halogenated and other derivatives. For purposes of the present review, consideration will be limited almost entirely to the phenyls since they are the most widely used and since very little is known of the toxicology of the other classes of aryls. Phenylmercuric acetate is the most important and best known in its class, there being some forty trademarked brands on the American market and at least a dozen more in which it is present in combination with other chemicals.

TABLE 3-I
SOME PHENYLMERCURIC SALTS OF INORGANIC ACIDS

Borate	Chloride	Iodide
Bromate	Cyanide	Nitrate
Bromide	Fluoride	Perchlorate
Carbonate	Hydroxide	Thiocyanate
Chlorate	Iodate	

TABLE 3-II
SOME PHENYLMERCURIC SALTS OF AROMATIC ACIDS

Atropate	Coumarate	Naphthenate
Benzoate	Hydrocinnamate	Phthalate
Benzoylacrylate	Mandelate	Salicylate
Cinnamate	Methacrylate	Tropate

TABLE 3-III
SOME PHENYLMERCURIC SALTS OF ALIPHATIC ACIDS

Acetate	Formate	Lactate
Ammonium acetate	Gluconate	Myristate
Acrylate	Glycolate	Oleate
Butyrate	Hydroxybutyrate	Propionate
Citrate	Hexanoate	Stearate
Formamide		

TABLE 3-IV
SOME ALKOXY- AND MISCELLANEOUS MERCURIALS

Methoxyethylmercury chloride
Methoxyethylmercury acetate
Methoxyethylmercury silicate
Chlormethoxypropyl mercuric acetate
Hydroxymercurichlorophenol
Mercurypentanedion

Phenylmercury 2-ethylhexyl maleate
PM Monoethanolammonium acetate
PM Monoethanolammonium lactate
PM Triethanolammonium lactate
PM Dimethyldithiocarbamate
Lactoxymercuriphenylammonium lactate
Sodium salt of hydroxymercurymethoxypropylcarbamylphenoxy acetic acid

Phenylmercurials

During recent years a number of reviews of the toxicology of mercury compounds have been published (3, 4, 6, 15, 29). Of special relevance to the present discussion is one appearing in 1964 which deals specifically with phenylmercurials (20). This article not only contains an extensive review of the literature but also describes some previously unreported studies on human subjects who had been occupationally exposed to several phenylmercuric compounds. The salient points brought out through the literature review in this article and from the original observations, include the following:

1. For human exposures there appears to be no significant difference in the toxicity of the various phenylmercurials which have been studied (20).

2. Phenylmercurials may be absorbed through the intact skin or mucous membranes, but relatively high concentrations must be applied before measurable amounts will be absorbed (15).

3. Solutions of phenylmercurials in concentrations in the percent range will regularly cause chemical second degree burns of the skin (17), while concentrations of a fraction of a percent are non-irritating. Skin sensation rarely occurs (15).

4. Numerous observations in a number of research centers have shown that persons occupationally exposed to phenylmercurials even in concentrations many times the accepted Threshold Limit Value (TLV, MAC) of 0.1 or 0.05 mg/m^3 in air and who for many years excrete mercury in the urine in the milligram

range daily, rarely develop any detectable physical or physiological abnormalities.

5. On a group basis there is some correlation between the levels of phenylmercurials to which workers are exposed and the concentration of mercury in their urine and blood. The relationship does not always hold for individuals (15, 18).

6. The occurrence of poisoning due to phenylmercurials, at least up to 1963, was extremely rare. Chronic occupational poisoning had not been reported.

7. Animal experiments with various phenylmercurials using different routes of administration have given results which show considerable variation in degrees of toxicity. With the exception of phenylmercuric propionate, the LD_{50} figures which have been published fall within a fairly narrow range (Table 3-V) but other types of toxicological experiments have produced a wide range of results (13, 14). Most observers have found that the phenyl-

TABLE 3-V
ACUTE TOXICITY OF PHENYLMERCURIALS FOR EXPERIMENTAL ANIMALS

Compound	Animal	Route	*LD_{50} Dose	Reference
Acetate	rat	I.V.	20	10
	mouse	I.V.	20	10
	rabbit	I.V.	5	10
	rat	oral	40	29
	chick	oral	60	25
	mouse	I.V.	15	22
	rat	oral	35	28
	rat	oral	60	27
	mouse	oral	13	27
Propionate	rat	oral	2.86 g	21
Propionate	rat	oral	90	28
Nitrate	rat	S.C.	63	23
Dinaphthylmethane disulfonate	guinea pig	oral	70	13
8-hydroxyquinolinate	rat	oral	50	28
Triethanolamine lactate	rat	oral	30 MLD†	28
Nitrate	rabbit	I.M.	10	5
	rabbit	I.P.	10	5
	rabbit	oral	30	5
Silver borate	mouse	S.C.	130	2

* All values are mg/kg of body weight.
† *M*inimal *L*ethal *D*ose.
I.V. = intravenous
S.C. = subcutaneous
I.M. = intramuscular
I.P. = intraperitoneal

mercurials were of a relatively low order of toxicity in animals, this being in consonance with human experience.

Several articles dealing with the toxicology of phenylmercurials have been published since 1964, for the most part confirming earlier observations. Piechocka (27), for example, found the LD_{50} dose for rats and mice to be in the same general range as that of previous investigations. She observed that PMA accumulated principally in kidney, liver and spleen with relatively little in blood and brain. PMA led to greater accumulation of mercury than did equivalent amounts of $HgCl_2$. Rats fed for six months on diets containing 8 ppm of mercury in PMA showed decreased fertility, but no such change was found when the mercury in the diet was 1, 2 or 4 ppm. Rabbits receiving 30 mg/kg of PMA exhibited a reversed albumin/globulin ratio with no decrease in total serum proteins. A comparative study by Petrovic and Soldatovic (26) confirmed the fact that PMA is less toxic than ethylmercury chloride but more toxic than the diuretic Salyrgan,® the latter being an alkoxyalkyl compound.

Since 1969 much attention has been paid to methods of analysis for alkylmercurials but very little to the determination of aryls. Yamaguchi *et al.* (37), however, has found that thin-layer chromatography may not be satisfactory in providing a clear separation between phenylmercuric acetate and methylmercury chloride. This is mentioned here because of the obvious importance of analytical techniques in toxicological studies. The relatively meagre attention paid to the arylmercurials may reflect their low order of toxicity when compared with the alkyls.

Lord Lister has been credited with the saying: "Next to the promulgation of truth, the best thing a man can do is the recantation of error" (19). He may have intended this to apply to one's own errors but by extension it may be applied to the errors of others. I refer here not so much to a particular article by Fitzhugh *et al.* (11) of FDA but rather to the many persons who have cited it in support of their views. The article in question merits scrutiny since it has been used for many years as a basis of condemnation of certain uses of phenylmercurials.

The only literature reference cited by Fitzhugh is an article by Cotter (7) published in 1947 ". . . who stated that the use of compounds of the type of phenylmercuric salts, as fungicides, germicides, etc., was entirely without danger to the public." The FDA group was not satisfied that sufficient evidence existed to support Cotter's conclusion and so they embarked on long-term feeding experiments on rats, using mercuric acetate and phenylmercuric acetate at various dosage levels: 0.1, 0.5, 2.5, 10, 40 and 160 ppm of mercury. The summary of the studies quoted *in toto* is as follows:

> Phenylmercuric acetate in the diet is chronically more toxic to the rat than is mercuric acetate. Phenylmercuric acetate in the diet causes 10 to 20 times as much storage of mercury in liver and kidney tissue as of mercuric acetate.
>
> Detailed histopathologic examination of the animals was done. The most important lesion that may be attributed to mercury was renal damage, consisting principally of an intensification in time and degree of the same type of damage as that occurring in the controls, and eventually leading to large, fibrous, granular kidneys. This effect was seen from even as little as 0.5 p.p.m. mercury given in the form of phenylmercuric acetate.

A careful reading of the entire article reveals a number of important points which do not appear in the summary. For one thing, it was only the female rats which showed the alleged damage, and not the males. The number of animals showing abnormalities is not stated but in one table and conclusions are drawn from groups averaging five and as few as one. In addition to the questionable significance of such small numbers, there was the anomalous finding of enlargement of the kidneys in rats fed at the 0.5 ppm level but not in those receiving 0.1 ppm, 2.5 ppm, or 10 ppm. This implies some critical factor operating at 0.5 ppm but not at lower or higher levels, an unlikely explanation. Another unexplained finding was that of approximately equal storage of mercury in male and female kidneys in the face of the statement that "Increased storage of mercury in the kidney can be related in a general way to increased injury in this organ." Similar injury in both sexes would have been expected.

In the 1964 report cited above (20) it is stated that while phenylmercurials did not appear to have a high degree of toxicity for humans, additional observations were needed before definite conclusions could be drawn. Six years and some 600 man-years later it can still be said that these compounds show no severe toxicity in man, at least when inhalation is the principal route of absorption (16). Fitzhugh's concern over the possible threat to the health of rats is brought further in question by extensive, but unpublished, observations made in the laboratories of the Wisconsin Alumni Research Foundation (WARF) between 1964 and 1966 (35).

Full details of the WARF studies are far too lengthy to be given, but summaries can be quoted:

> In a nine month study, the organic mercurials, phenylmercuric acetate (PMA) and phenylmercuric propionate (PMP) . . . produced no apparent toxic effects when applied in aqueous dilutions topically to skin of rabbits. At 90 days and eight months increased mercury levels were noted in the kidney tissues of animals receiving application of PMA (20 ppm) and PMP (5 ppm and 20 ppm) when compared to comparable negative control tissues. Less definite mercury levels were noted in other test groups receiving PMA (0.5 and 5 ppm) and PMP (0.5 ppm). At eight months there is no indication of accumulation of mercury in tissues when compared with mercury levels found at 90 days. At eight months liver tissue from animals receiving 20 ppm PMA and PMP do show an increased mercury content over comparable controls. No significant body weight, food consumption, clinical or pathological effects were noted in animals receiving applications of PMA or PMP solutions.
>
> In a 90-day feeding study, organic mercurials, phenylmercuric acetate and phenylmercuric propionate . . . were fed to rats at levels providing 0.05 ppm, 0.5 ppm and 5.0 ppm mercury in the diet. At these levels of feeding higher levels of mercury in the kidney were noted in test animals when compared to the negative control animals. At the highest levels of administration (5.0 ppm) a slight increase in kidney weight was observed. Mercury content of the liver was higher in some test groups where mercury was added to the diet. At 90 days minimal alternations of the kidneys of some rats from test groups 4 (5.0 ppm PMA), 5 (0.05 ppm PMP), 6 (0.5 PMP) and 7 (5.0 ppm PMP) were observed. At eight months no definite variation between kidney tissue of negative control and test animals was observed.
>
> In a one year feeding study, organic mercurials, phenylmercuric acetate (PMA) and phenylmercuric propionate (PMP) . . . were fed to rats at levels providing 0.05 ppm, 0.5 ppm and 5.0 ppm mercury in the diet. At these levels of feeding no weight gain or feed consumption effect was noted in females in any group while males receiving 0.5 and 5.0 ppm PMA and PMP showed a slight

decrease in rate of gain and a slightly lower feed intake. At the highest level of supplementation with PMA, 5.0 ppm, there was an increase in weight for heart, liver and kidney. These weight differences were not evident after the one month withdrawal period.

Histologic changes in the liver were minimal and were not related to the administration of PMA. Kidney alternations which appeared to be limited to typical senile changes, were more severe in animals receiving 5.0 ppm PMA but showed regression during the 1 month withdrawal period.

The WARF results differ from those of Fitzhugh in that such abnormalities as were found involved the male rats rather than the females. By continuing their observations for a month after removing the mercurials from the diet, regression in the kidney changes were found. The inconsistencies between the two studies are disquieting.

Feeding studies on piglets (31), using phenylmercuric chloride (PMC) at several dosage levels, showed that daily administration equivalent to 0.76 mg/kg of mercury depressed growth rate but produced no overt pathological lesions. At levels of 2.28 and 4.56 mg/kg there was irritation of the intestinal tract leading to diarrhea, and kidney injury in the form of nephrosis. Regenerative activity was noted in the damaged kidneys. The authors of this report point out the similarity of their findings to those of poisoning due to mercuric chloride.

Alkoxyalkylmercurials

Compared with the alkyl and aryl compounds of mercury, relatively little information is available on the alkoxyalkyls. The toxicological position of the alkys and aryls is fairly well established; that of the alkoxyalkys is not. This may be because all members of the family do not behave alike.

One of the early reports on the toxicity of an alkoxyalkylmercurial is misleading both as to the title of the article and in its conclusions (34). The compound described is methoxyethylmercury silicate, a pesticide known commercially as Abavit. Abnormalities found in six of 12 exposed workers were gingivitis, erethism and retention of urea, strongly suggesting that inorganic mercury was the cause of the difficulties. This suspicion is confirmed by the authors' description of the workplace, there being gross contamination with free metallic mercury on the floors. All of the victims recovered completely when they were removed from exposure. The role of the alkoxyalkyl compound is certainly not clear.

Those alkoxyalkylmercury compounds which are closely related to the mercurial diuretics might be expected to have a low order of toxicity due to the ready liberation and rapid excretion of their mercury (1). One such compound, chlormethoxypropylmercuric acetate, has been found to have an acute oral LD_{50} in rats of 224 mg/kg and in limited experiments to have no teratogenicity in mice (30). Workers engaged in formulating and packaging this material as well as phenylmercurials have shown no higher blood urine mercury ratios than those of men exposed to inorganic and phenylmercury compounds (16). Only the most guarded conclusions may be drawn from incomplete and limited observations of this nature.

As mentioned elsewhere in this symposium (see chap. 2) one of the characteristics of the alkylmercurials is their affinity for and long residence time in blood. Fragmentary evidence suggests that this may also be true in humans for methoxyethylmercury chloride (15). Seven workers who had spent about six weeks in preparing and packaging a batch of this material were found four weeks after completing this task to have whole blood mercury levels of 34 μg/100 ml to 109μg/100 ml, with an average of 65μg/100 ml. Here again conclusions must be guarded since these men had returned to their usual jobs making phenylmercuric acetate. It is of some interest, however, that at no previous time did any of these workers have a blood mercury level above 24μg/100 ml, and on subsequent examination the levels returned to the customary baselines. At no time did they exhibit any toxic effects.

Similarity in the metabolism of methoxyethylmercuric chloride and of phenylmercuric acetate was demonstrated by Merville et al. in 1967 (24). He injected these substances intraperitoneally into rats in doses of 1 to 3 mg/kg and found similar patterns of elimination and of accumulation of mercury in the kidneys.

In a brief review published in 1969 Daniel and Gage (9) commented on the neurotoxic effects of alkylmercury salts and their absence from aryls and that the alkoxyalkylmercury salts appear to resemble the aryl more than the alkyl compounds. After injecting 2-methoxy ^{14}C ethylmercury chloride subcutaneously in rats they found that about one-half of the labeled carbon could be recovered in the expired air within 48 hours and that after that time practically all of the mercury found in the urine was inorganic. Mercury stored in the kidneys had no radioactivity. This metabolic behavior resembles that of the aryls rather than that of the alkylmercury compounds. Quite obviously there is need for additional studies on the toxicology of the alkoxyalkylmercurials.

REFERENCES

1. Baer, J.E. and Beyer, K.H.: Renal pharmacology. *Ann Rev. Pharmacol,* 6:261, 1966.
2. Baer, M.: Quecksilberhaltige Desinfizientia in der Otorhino-laryngologischen Lokaltherapie. *Monatschr F Ohrenheilk,* 73:751, 1939.
3. Battigelli, M.C.: Mercury toxicity from industrial exposure, *J Occup Med,* 2:337, 394, 1960.
4. Bidstrup, P.L.: *Toxicity of Mercury and Its Compounds.* New York, Elsevier, 1964.
5. Birkhaug, J.E.: Phenyl-mercuric-nitrate. *J Infect Dis,* 53:250, 1933.
6. Brown, J.R. and Kulkarni, M.V.: A review of the toxicity and metabolism of mercury and its compounds. *Med Services J Can,* 23:786, 1967.
7. Cotter, L.H.: Hazards of phenylmercuric salts, *J Occup Med,* 4:305, 1947.
8. Cross, J.L. and Hale, R.W.: A technical survey of new and expanded applications for mercury, Battelle Memorial Institute, Columbus, Ohio (1959), Processed.
9. Daniel, J.W. and Gage, J.C.: The metabolism of 2−^{14}C methoxyethylmercury chloride. *Biochem J, 111*:20, 1969.
10. Eastman, N.J. and Scott, A.B.: Phenylmercuric acetate as a contraceptive. *Human Fertility* 9:33, 1944.
11. Fitzhugh, O.G. and Nelson, A.A.: Laug, E.P., Kunze, F.M.: Chronic oral toxicities of mercuri-phenyl and mercuric salts. *Arch Ind Hyg Occup Med,* 2 (1950) 433.
12. Friedman, H.L.: Relationship between chemical structure and biological activity in mercurial compounds. *Ann NY Acad Sci,* 65:461, 1957.
13. Goldberg, A.A. and Shapero, M.: Toxicological hazards of mercurial paints, *J Pharm Pharmacol,* 9:469, 1957.
14. Goldberg, A.A., Shapero, M. and Wilder, E.: Penetration of phenyl-

mercuric dinaphthylmethane disulfonate into skin and muscle tissue. *J Pharm Pharmacol,* 2:20, 1950. 89.
15. Goldwater, L.J.: Occupational exposure to mercury: The Harben Lectures 1964. *J R Inst Pub Health Hyg,* 27:279, 1964.
16. Goldwater, L.J. and Hoover, A.W.: Unpublished observations.
17. Goldwater, L.J., Ladd, A.C., Berkhout, P.G. and Jacobs, M.B.: Acute exposure to phenylmercuric acetate. *J Occup Med,* 6:227, 1964.
18. Goldwater, L.J., Ladd, A.C. and Jacobs, M.B.: Absorption and excretion of mercury in man, VII. Significance of mercury in blood. *Arch Environ Health,* 9:735, 1964.
19. Guthrie, D.: *Janus in the Doorway.* London, Pittman, 1963.
20. Ladd, A.C., Goldwater, L.J. and Jacobs, M.B.: Absorption and excretion of mercury in man: V. Toxicity of phenylmercurials. *Arch Environ Health,* 9:43, 1964.
21. Linfield, W.M., Sherrill, J.C., Casely, R.E., Noel, D.R. and Davis, G.A.: Studies on the development of antibacterial surfactants, I. Institutional use of antibacterial fabric softeners. *J Am Oil Chemists Soc,* 37:248, 1960.
22. Lundgren, K.D. and Swensson, A.: Phenylmercuric compounds as problem of industrial hygiene. *Nord Hyg Tidskr.* 31:207, 1950.
23. *THE MERCK INDEX,* 8th edition, Stecher, P.G. (Ed.): Rahway, N.J., Merck & Co., 1968.
24. Merville, R., Dequidt, J. and Lelong-Corteel, M.L.: Experimental intoxication by certain organomercurial derivatives. *Bull Soc Pharm Lille,* 2:95, 1967.
25. Miller, V.L.: Klavano, P.A., Jerstad, A.C. and Csonka, E.: Absorption, distribution and excretion of ethyl mercuric chloride. *Toxicol Appl Pharmacol,* 3:459, 1961.
26. Petrović, C. and Soldatović, D.: Comparative toxicity of some mercury compounds. *Chem Abstr* 71:28806g, 1969. (Original in Croat).
27. Piechocka, J.: Chemical and toxicological studies of the fungicide phenylmercuric acetate. *Chem Abstr,* 68:94834f, 1968, 70:19197m, 19198n 1969; 71:2464u, 1969. (Original articles in Polish).
28. Ramp, J.A. and Grier, N.: Industrial applications of phenylmercurials as anti-microbials, Proc. 47th Mid-Year Meeting Chem. Specialties Manufacturing Assn.
29. Stahl, Q.R.: Preliminary air pollution survey of mercury and its compounds, National Air Pollution Control Administration Pub. No. APTD 69–40, 1969.
30. Troy Chemical Company: Unpublished Observations, 1966.
31. Tryphonas, L. and Nielsen, N.O.: The pathology of arylmercurial poisoning in swine. *Can J Comp Med,* 34:181, 1970.
32. Voress, H.E. and Smelcer, N.K.: Mercury toxicity: A bibliography of published literature. USAEC Rep. TID–3067, 1957.

33. Whitmore, F.C.: *Organic Compounds of Mercury.* New York Chemical Catalog Co., 1921.
34. Wilkening, H. and Litzner, S.: Ueber Erkrankungen insbensordere der Niere durch Algylquecksilberverbindungen. *Deutsch Med Wschr,* 77:432, 1952.
35. Wisconsin Alumni Research Foundation: Reports to Great Lakes Biochemical Co., Inc., Unpublished Observations, 1964–6.
36. Wright, G.F.: Oxymercuration of alkenes. *Ann NY Acad Sci,* 65:436, 1957.
37. Yamaguchi, S., Matsumoto, H., Hoshide, M. and Akitake, K.: Microdetermination of organic mercurials by thin-layer chromatography. *Kurume Med. J* 16:53, 1969.

Chapter 4

MERCAPTANS, THE BIOLOGICAL TARGETS FOR MERCURIALS

Aser Rothstein

ABSTRACT

The role of blood in mercury toxicity is examined from two aspects. In the first, the blood is considered as the vehicle of transport of mercury from the sites of absorption to the tissues. The important factors are the fraction diffusible mercury, which is very small, and the rates of redistribution from protein-bound reservoirs in the plasma and within the red blood cells, the permeability of the red cell membrane being a primary factor. In the second, the movement of mercurials into red blood cells and the consequent disturbances of function are examined as a model of the behavior of mercurials with respect to other cells. Lipid soluble mercurials (non-ionic forms) penetrate rapidly and their primary effects will be on internal systems. Ionic forms penetrate slowly, and their initial effects will tend to be disturbances of membrane function.

REPORT

Mercaptans, or sulfhydryl groups as they are more commonly called, have an exceedingly high affinity for mercurials. Indeed, mercurials are generally used as highly specific reagents for titrating sulfhydryls or for disturbing biological functions associated with sulfhydryls. Conversely, almost every toxic action of mercurials is to some extent automatically attributed to an interaction with sulfhydryl groups to form a complex known as a mercaptide.

In biological systems, sulfhydryl groups are found in a few

This paper is based on work performed under contract with the United States Atomic Energy Commission at the University of Rochester Atomic Energy Project and has been assigned Report No. UR–49–1470.

diffusible low molecular weight substances such as cysteine, reduced glutathione, CoA, lipoate, and thioglycolate, but the predominant sulfhydryl constituents are the proteins. The primary targets for interaction and for consequent toxicological effects of the mercurials are therefore the proteins. For this reason it is not surprising that the mercurials are especially popular agents for studying the role of sulfhydryl groups in enzyme activity. Webb (29), for example, points out that until 1966 the total number of papers dealing with this topic was about 1,350, with 50 percent appearing between 1960 and 1965.

Despite the prolific output of papers concerned with the nature of the sulfhydryl-mercury interaction, and with the role of sulfhydryl groups in protein function, such knowledge has not been particularly useful in advancing our understanding of mercury toxicity. The gap between the mercury-sulfhydryl reactions in a chemical sense, the consequent disturbance of protein function on the one hand, and manifestations in animals or man, on the other hand, has been too large. For example, the complex pattern of disturbances of the central nervous system induced by methylmercury cannot yet be related to specific functional components of the brain, and certainly cannot yet be directly related to specific molecular events at the level of the sulfhydryl-mercury interaction. Even in the case of the kidney, another target organ for mercurials, in which the toxic manifestations can be attributed not only to specific segments of the tubule, but to specific transport functions, the responsible proteins and sulfhydryl groups that are altered by the metal cannot yet be identified with any certainty (18).

The gap in knowledge between the initial chemical event and the ultimate toxicological manifestation is in part due to lack of information concerning the molecular basis of functional disturbances, but in addition, in studying the mechanism of action of mercurials, we are faced with a severe logistical problem. Although mercurials are highly specific for sulfhydryl groups, they are *highly unspecific* in terms of proteins. Almost all proteins contain certain sulfhydryl groups that are metal-reactive. Furthermore, because sulfhydryl groups are important in most protein functions, mercurials can disturb almost all functions in which proteins are involved. Thus almost every protein in the body is a potential target.

Paradoxically, despite the plethora of potential sites of action of mercurials, the toxicological effects are highly specific, with unique patterns for different forms of the metal. In seeking explanations for the paradox it has become clear that the particular target of particular mercurials is determined not only by chemical specificities but also by unique distributional factors (16,18). Of the many potential protein targets, only those that are accessible to the mercurial will react and be modified. It is irrelevant from a toxicological point of view that perhaps hundreds of other proteins are also chemically capable of reacting.

The factors that determine accessibility may have little to do with the chemical specificities of sulfhydryl groups for mercury. Two such parameters are of primary importance, firstly the location of the target protein, and secondly the ability of the mercurial to reach the target. The location is fixed, depending on the geometrical arrangements of proteins within cells, of cells within organs, and of organs within the pathways of circulation. The ability of the mercurial to reach the target, on the other hand, depends on a variety of factors outlined in Figure 4–1, including circulatory dynamics, distributions of mercurials in blood between cells and plasma, distribution of mercurials between diffusible and non-diffusible forms, filtration of diffusible mercury into tissue spaces, and the ability of the mercurials to penetrate into the cells

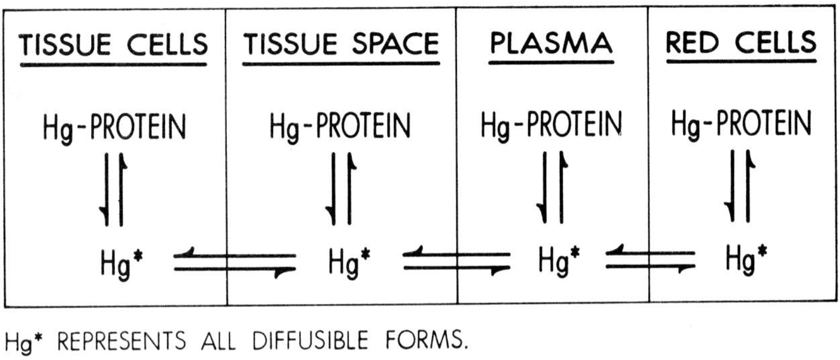

TISSUE CELLS	TISSUE SPACE	PLASMA	RED CELLS
Hg-PROTEIN	Hg-PROTEIN	Hg-PROTEIN	Hg-PROTEIN
$\updownarrow$	$\updownarrow$	$\updownarrow$	$\updownarrow$
Hg* $\rightleftharpoons$	Hg* $\rightleftharpoons$	Hg* $\rightleftharpoons$	Hg*

Hg* REPRESENTS ALL DIFFUSIBLE FORMS.

Figure 4–1. Compartmentation and equilibria of mercury in blood and tissues.

to reach particular functional proteins. Specific information concerning the kinetic parameters of the sulfhydryl-mercurial interactions, the kinetic parameters of the diffusion steps, or a precise description of the chemical equilibria are sparse. Nevertheless even the limited data allow certain tentative conclusions, and more importantly they point toward the kinds of information that will be needed for an adequate understanding of the nature of toxic responses to different compounds of mercury.

The present discussion will be restricted largely to interactions of mercurials with sulfhydryl groups of blood, with two distinct aspects in mind. Firstly, the blood is the delivery system that carries the mercurials to the tissues in which the toxicological insult will occur. Knowledge of the distribution of mercurials between blood components and of the kinetics of flow between those components can allow some understanding of tissue distributions and of excretion rates. Secondly, the behavior of mercurials with respect to red blood cells can serve as a model for understanding effects on other cellular systems. The red blood cell is an excellent model particularly for understanding the factors that control the penetration of mercurials through the cell membrane and for ascertaining the nature of actions of mercurials on the plasma membrane, the primary cellular target (18).

Mercaptides

The formation of mercaptides and the functional consequences in proteins is reviewed in some detail elsewhere (29). The essential information will be briefly summarized here. The sulfhydryl groups of small molecules such as cysteine have an extremely high affinity for mercurials, forming mercaptides. The reactions are reversible with dissociation constants reported to be as high as 10^{-27}, but strongly influenced by the pH and by anions (29). The effect of pH can be attributed to competition of protons for the sulfur moiety, whereas that of anions is due to formation of a complex with the mercury moiety. Nominally mercuric mercury in aqueous solution is present as a bivalent cation, Hg^{++}, and organic mercurials as RHg^+, but in physiological salt solutions the mercury may be neutral or negative due to formation of complexes with anions. With chloride, for example, Hg^{++} forms a series of complexes with as many as six Cl^- ions. Anions such as phosphate may form more stable complexes than Cl^-. In any case, the apparent affinity of the mercurials for sulfhydryls is considerably reduced because of the competitive effects of the anions. Unfortunately systematic data on affinities of different mercurials for protein sulfhydryls as a function of pH and anion composition (especially physiological saline) are not available.

Sulfhydryls in proteins show a wide array of behavior with respect to mercurials. Depending on the mercurial and protein used, some sulfhydryls are as reactive as those in simple compounds, but others are relatively unreactive (or "masked") unless the protein is first unfolded by denaturation. In addition, complex time dependencies may be present due to the fact that reaction of a mercurial with the accessible sulfhydryl groups may lead to a slow unfolding of the protein followed by a reaction of previously inaccessible ligands. Indeed, any conditions that influence protein configuration are likely to alter the reactions of protein sulfhydryl groups with mercurials.

The important factors in mercaptide formation are known in general, but are not sufficiently understood to allow prediction in particular cases. In addition to the factors discussed above, the following must be considered:

1. The reaction can be influenced by neighboring groups; for example, charged groups near the sulfhydryl may have a large attractive or repulsive effect on ionic forms of mercurials.

2. The mercurials because of the size or arrangement of its organic portion, may be excluded from reaction because of steric factors.

3. The sulfhydryls may be "hidden" because of the configuration of the protein or because of internal bonding to other protein ligands.

The consequences of mercaptide formation on proteins is best understood in terms of changes of enzyme activity. Some sulfhydryl groups are located in the active center involved in substrate binding. Their interaction with mercurials usually results in a relatively simple stoichiometric, competitive inhibition. Other sulfhydryls may be adjacent to the active center. Their interaction may lead to noncompetitive effects and partial inhibitions. Interactions of sulfhydryls at some distance from the active center may result in little inhibitory effect or in a pronounced but complicated effect associated with changes in the tertiary structure of the protein.

Distributions of Mercurials Between Blood Components and Between Blood and Tissues

Few systematic studies of the factors that determine the distributions of mercurials have been carried out. Yet such factors may be prime determinants of the patterns of toxicity. From the scattered sets of data and from knowledge of behavior of mercurials in other systems, some generalizations can be made. A generalized and simplified scheme is presented in Figure 4–1. In each fluid compartment mercury can be represented in two forms, protein-bound and diffusible, with only the latter component able to cross the membrane barriers. The important factors are the relative concentrations of free and diffusible sulfhydryl groups, the affinities of the components for mercury, the rates of movement of diffusible mercury across the membrance boundaries, and the rates of association and dissociation of the mercury protein complexes. The information concerning the above factors is minimal.

In blood, the distribution of sulfhydryl groups is roughly known. The amount in serum is relatively low, 0.5 mM/liter* (30) compared with that in the red blood cells, reported to be 12 mM/liter (4). Calculations from more recent data indicate that the values for cells may be higher, 20 mM/liter (6). The latter value agrees with that calculated from the hemoglobin content based on eight sulfhydryl groups per molecule. On the basis 95 percent of the sulfhydryl groups of blood are within the cells, with hemo-

* The total cysteine content of the plasma proteins is equivalent to about 3.5 mM/liter of blood calculated from compiled data (5). A large proportion is contributed by serum albumin which constitutes about 70 percent of the plasma proteins. In serum albumin most of the sulfur moieties are in the form of disulfides with only 1 molecule of free sulfhydryl per molecule of protein and only 70 percent of the molecules capable of mercaptide formation (6). Calculated from these data the free sulfhydryl concentration contributed by albumin is 0.4 mM/liter of blood, consistent with the cited value of 0.5 mM/liter (4).

globin accounting for over 90 percent, the membrane accounting for 4 percent, and reduced glutathione for most of the remainder (28). Serum albumin is the major sulfhydryl component of the plasma, with other proteins contributing minor fractions and with low molecular weight-diffusible components such as cysteine present in only trace amounts (5).

The interaction of hemoglobin with several mercurials has been determined. Under appropriate conditions, all eight sulfhydryl groups of hemoglobin are capable of reacting with inorganic mercury but only variable numbers with organic mercurials, six with the diuretic, chlormerodrin, and only two with parachloromercuribenzoate (PCMB) or parachloromercuribenzenesulfonate (PCMBS) (28). No comparative assessment has been made of the interactions of different mercurials with plasma proteins. In the case of inorganic mercury, serum albumin is the predominant factor (8).

If mercurials were equally distributed among the sulfhydryl groups in blood, then the preponderance of the agents (over 95%) should be found associated with the red blood cells. Even assuming that only 25 percent of the sulfhydryls of hemoglobin are reactive for a given agent and that all of the plasma protein sulfhydryls are reactive, the predicted distribution would be 85 percent in cells and 15 percent in plasma. In fact, large differences in distribution are found between different mercurials and between different species (Table 4-I). Unfortunately the data are

TABLE 4-I
REPORTED DISTRIBUTIONS OF MERCURIALS IN BLOOD

Compound	Species	% in cells	Reference
Methyl mercury	human	94–96	14
	human	66	24
	rabbit	80	25, 1
	rat	99	15
	dog	80	25
Phenyl mercury	rabbit	90	1
Inorganic	human	50	27
	monkey	25	2
	rabbit	25–31	2
	rabbit	35	3
Mercury vapor	human	63	23
	human	55–70	14
	monkey	67	2
	rabbit	82–84	24

not strictly comparable. Almost all represent analyses of blood from animals or humans exposed to or injected with different doses and after different times, with no assurance that they represent an equilibrium or steady state situation. Furthermore, the distribution may be dose dependent. With this cautionary note in mind, it can still be concluded that methylmercury is carried largely in the cells, whereas 50 percent or more inorganic mercury is carried in the plasma. The latter finding is unexpected since it has already been pointed out that 95 percent of the total of blood sulfhydryl groups is within the red blood cells. It is also clear that the distribution for methylmercury is considerably different in rats compared to rabbits and man. Such differences may play an important role in terms of species differences in response to mercurials. This point will be discussed in further detail in the paper by Vostal (Chapter 8) later in the program.

The differences in distribution between compounds and between species suggest that all reactive sulfhydryls do not have equal affinities for different mercurials. Unfortunately, data on relative affinities of sulfhydryls of blood proteins for various mercurials at physiological pH and in the presence of physiological anions, are not available. Furthermore, the amounts of mercurials in blood even at toxicological levels are very small relative to the numbers of sulfhydryls. For example, with inorganic mercury (19) or mercury vapor (19), only a few percent of the mercury in the body is in the blood even shortly after injection or exposure. At a toxic level (LD_{50}) of 3 mg of metal per kilogram of body weight (26) the maximal amount of mercury in blood would be .05 mM/liter (assuming that 5% of the body burden is in the blood). This concentration is sufficient to saturate only 1 in 400 of the sulfhydryl groups of blood. With more toxic compounds and with sublethal levels, the concentration of mercury in blood would be sufficient to combine with considerably less than 1/1000 of the sulfhydryl groups. On this basis, the behavior of the average protein sulfhydryl group may be irrelevant. A very small population with especially high affinities may be the determining factor in distribution of mercurials.

Another important component of blood is the diffusible mercury. Although reported to be exceedingly low (1, 3, 5, 15), it is the only form of mercury that can move between plasma and cells, and plasma and tissues, and it thereby is one of the primary determinants of rates of distribution and redistribution. The amount of diffusible mercury will depend on the relative concentrations of diffusible and nondiffusible binding ligands and on their relative affinities, but data on amounts and forms of diffusible mercury-complexing materials are not available. Cysteine is undoubtedly one such substance.

Concentrations and affinities determine equilibrium distributions but the important factors in tissue distribution may be the amount of diffusible mercury and the rate at which mercury can move from one compartment to another. During the passage of blood through a tissue, some fraction of the diffusible mercury will move across the capillary endothelium into tissue space, and a reequilibration of the metal between the serum protein red cells and diffusible components will tend to occur. The latter may change in amount in certain tissues. If the rate of equilibration is slow relative to the circulation time through the tissue, then the large reservoir of the mercury associated with the serum proteins and red cells at the time that the blood entered that tissue will not be deposited. During a single passage of blood, only a small fraction, determined by the size of the diffusible component, and the rate of reequilibration, will be left in a given tissue (Fig. 4–1). Unfortunately almost no data on rates of equilibration of mercury between cells and plasma, plasma protein and diffusible components are available.

The equilibration time between red blood cells and medium in *in vitro* experiments depends on the compound used. For example, inorganic mercury moves rapidly into red blood cells, whereas organic mercurials diffuse slowly and at rates that may differ considerably (Fig. 4–2). In this particular experiment, however, the cells were suspended in saline so that all of the added mercury was diffusible. In the presence of plasma proteins the diffusible mercury would be exceedingly low and the equilibration rates would have been much slower. A few sets of data on equilibration times between plasma and red blood cells *in vivo* indicate that such is the case. For example, with inorganic mercury (3), mercury vapor (5), and methylmercury (1), a

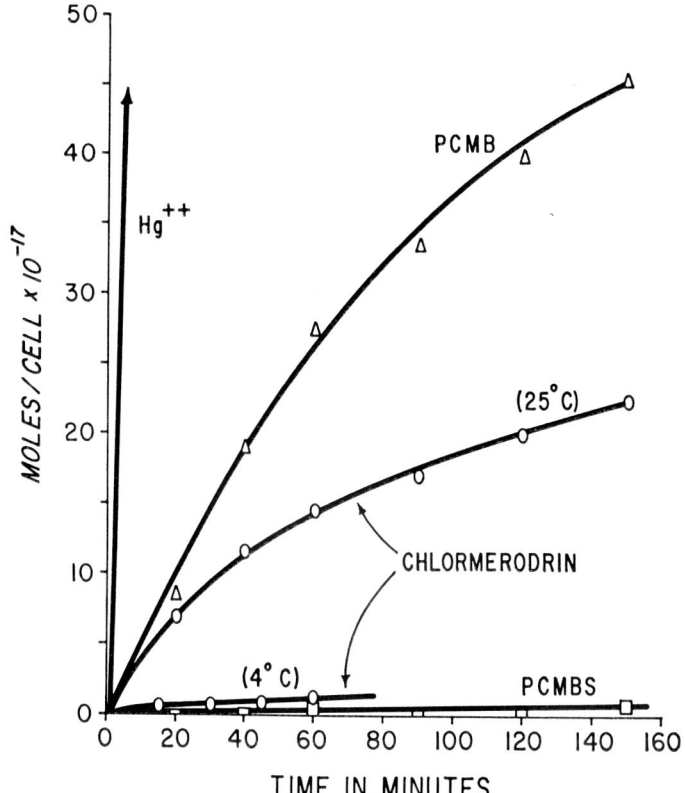

Figure 4–2. The time course of uptake of mercurials by human red blood cells (From reference 28).

constant distribution ratio between cells and plasma is not attained until an hour or more after exposure. Such information, though scanty, suggests that the rate of equilibration of mercury between blood compartments is relatively slow and that therefore the amount of diffusible mercury and the kinetics of redistribution may be the key factors in tissues distribution.

Organ Specificities

Almost every functional protein in the body is a potential target for mercurials, yet the toxicological actions of the mercurials are highly specific and to a degree unique for each particular mercurial. The central nervous system is the primary target of alkylmercurials such as methylmercury, whereas the kidney is the first target for certain organic compounds, the mercurial diuretics. Inorganic mercury is primarily a nephrotoxic agent, whereas the brain is the most sensitive target for mercury vapor, even though the latter is rapidly oxidized to inorganic mercury in the blood (5). These highly organ-specific actions cannot, at the present time, be explained on the basis of specificities for reaction with particular mercaptans. They may have more to do with the compartmentation in blood and the kinetics of movement from one compartment to the other. For example, in the case of mercury vapor, most of its oxidation to Hg^{++} occurs within the red blood cell (5). Because of slow redistributions, however, the partition between cells and plasma is quite different for mercury vapor than for Hg^{++} (Table 4-I).

Even with respect to a given organ, the kidney, the actions of mercurials are relatively unpredictable (4, 10). The organic mercurial, chlormerodrin, produces a diuresis whereas PCMB does not. Nevertheless intravenous PCMB can prevent the diuresis produced by chlormerodrin in a competitive manner as though the two agents react with a common site involved in a step preceding that which produces diuresis. If, on the other hand, PCMB is introduced into the renal tubule by retrograde injection, then it too produces a diuresis. These differences in action of mercurials on the kidney cannot be attributed to chemical specificities for particular sulfhydryl groups, but rather to the geographical arrangement of the functional sulfhydryls and on the ability of the various mercurials to reach them (16, 18). Such questions of arrangement and accessibility at the cellular level have been examined in the red blood cell and the findings described in the next section.

Uptake and Effects of Mercurials at the Cellular Level

The red blood cell is an excellent model for determining the general parameters involved in the action of mercurials at the cellular level. Its geography in terms of mercaptans is relatively simple. About 4 percent is in the proteins of the membrane and the remainder is within the cell with 6 percent in the form of reduced glutathione, 90 percent in hemoglobin, and only traces in other proteins that may have enymic functions (28).

Certain obvious generalizations can be made, applicable to the action of any sulfhydryl reactive chemical agent (18).

1) It is an obvious but important fact of geography that a chemical agent must pass into the cell from the outside toward the inside. Its first encounter with the cell is therefore at the outer face of the membrane. The agent passes through the membrane by whatever channels are available, driven by whatever forces act upon it (usually its electrochemical gradient).

2) As the agent contacts the membrane and passes through, it reacts with any accessible reactive ligands, according to the kinetics of the reaction and local conditions in the membrane, such as concentration of agents and ligands, pH, and so on. If any of the ligands are functionally important, disturbances of function appear in a sequence determined by accessibility, the most peripheral usually being the first to be affected.

3) If sulfhydryl agents are added in small concentrations relative to the potential number of targets in the cell, agents that form 'irreversible' bonds tend to be used up in the membrane, whereas agents that are bound reversibly tend eventually to accumulate in the interior of the cell where the bulk of the sulfhydryl groups are located. Thus the reversibly bound reagents give transient effects, whereas irreversibly bound agents give long-lasting effects. With larger concentrations of agents, the effects appear more rapidly and all "sensitive" systems tend to be affected.

4) The membrane is the primary target of the agents because it is the first part of the cell to be exposed and because it is exposed to the maximum concentration of agent. In the case of slowly penetrating agents, and in cases of agents added in small amounts relative to the number of cells, it may be the only part of the cell that is affected.

5) For agents that can penetrate only slowly, the membrane protects the cytoplasmic systems. Furthermore, the sensitive cytoplasmic systems are also protected by the presence of large amounts of functionally inert sulfhydryls that "soak up" the agent. Thus internal effects develop more slowly than membrane effects and usually require larger concentrations of agent.

6) Because of the presence of permeability barriers, and because of the presence of large amounts of chemically reactive but functionally inactive ligands, temporal relationships and dosage-affected relationships may be exceedingly complicated, obscuring a sometime simple relationship between chemical interaction and related effect at a sensitive site.

According to the above analysis the specific effects of particular mercurials will depend not only on the chemical specificities of their interactions with sulfhydryl groups of particular functional proteins, but also on the geographical location of particular populations of sulfhydryl groups as determined by cellular structure, and also on the rates at which the mercurials reach those populations. The latter factor is determined by the permeation properties of the mercurials such as size, charge, shape, and lipid solubility rather than their affinities for sulfhydryl groups. The specificity dictated by location and accessibility of sulfhydryl groups has been given the name "geographical specificity" in contrast to "chemical specificity" (18).

The role of the membrane in restricting access of certain mercurials to cellular sites is illustrated in Figure 4-2. Organic mercurials are taken up much more slowly than inorganic mercury. The membrane as a diffusion barrier accounts for their slow uptake and for the large differences between them (28). For example, interaction of PCMB, PCMBS, and chlormerodrin with isolated membranes and with hemoglobin is very rapid, completed in minutes, whereas interaction with the intact cell continues for hours. Indeed the uptake is diffusion-limited with uptake curves following first order kinetics. Differences in rates of permeation rather than rates of reaction also account for the differences between PCMB and PCMBS and between chlormerodrin at 25°C and 4°C. Thus the substitution of a sulfonic acid residue for a carboxyl group does not alter the interactions with sulfhydryls. Both PCMB and PCMBS react rapidly with 25 percent of the sulfhydryls of the isolated red cell membrane. The extremely slow penetration of PCMBS has been attributed to its highly ionized state and consequent low lipid solubility. The small uptake of chlormerodrin in the cold is also attributable to a slow penetration and not to differences in chemical reactivity.

The mechanisms of permeation of the mercurials can be evaluated by comparing various mercurials. PCMBS must penetrate as a large polyvalent anion (the Hg^+ is shielded by formation of complexes with anions such as Cl or phosphahte, if present) (29). Penetration in ionic form is very slow and presumably via aqueous pathways involving fixed positive charges of proteins

(13). Two such pathways can be distinguished, the major one being blocked by the amino-reactive agent 4-acetamido-4'-isothiocyanostilbene-2-2'-disulfonic acid (SITS) an inhibitor of anion permeation. For example, in Figure 4–3 the permeation into the membrane is represented by the slow continuing component of uptake, whereas binding to superficial sulfhydryls is represented by the rapid component. The penetration is markedly inhibited by SITS, whereas the binding is not.

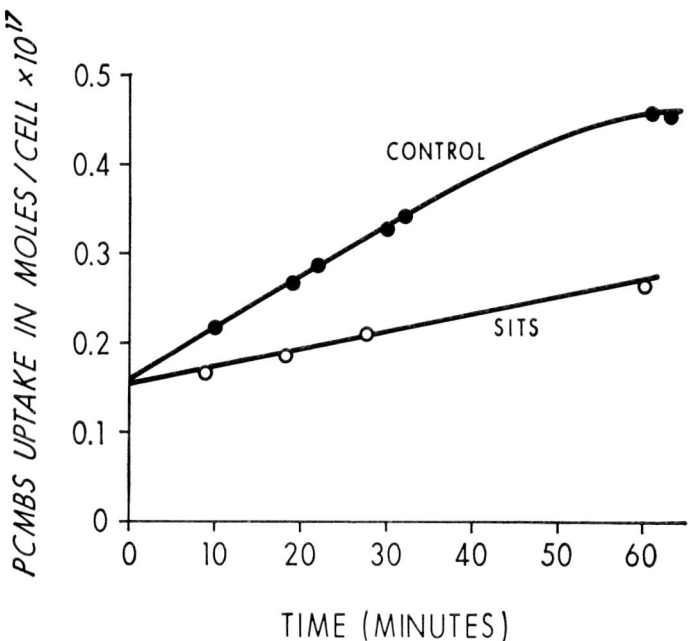

Figure 4–3. PCMBS uptake by normal and SITS-treated human red blood cells (From reference 13).

PCMB penetrates much more rapidly than PCMBS (Fig. 4–2) via a third pathway, lipid solubility. Even at physiological pH a small fraction of PCMB is undissociated and can presumably pass through the lipid. As the pH is reduced, increasing the amount of undissociated PCMB, the rate of penetration via this pathway increases remarkably. The same pH dependence is not evident in the case of PCMBS because its dissociation is not influenced by pH in the area of neutrality (M. Takeshita and

A. Rothstein, unpublished observations). With mercurials possessing no anionic group such as 1-bromomercuri-2-hydroxypropane (BMHP) (4) penetration via the lipid pathway is very rapid, almost entirely via the lipid pathway (20). Methylmercury also passes rapidly into the cells, presumably by the same route (J. White and A. Rothstein, unpublished observations).

On the one hand, those mercurials that penetrate slowly via aqueous channels have a prolonged, potent effect on membrane systems, with only a slow buildup in concentration and of effects within the cell. On the other hand, those mercurials that pass rapidly into the cell via lipid solubility have no effect or a small transient effect on the membrane, but accumulate rapidly in the cytoplasmic systems. For example, the small uptake of the slowly penetrating PCMBS is associated with a large change in cation permeability, the larger uptake of the more rapidly penetrating PCMB is associated with a smaller effect on cations, and the very rapid uptake of BMPH is associated with no effects at all on cations (Fig. 4–4).

The slowly penetrating mercurials can be used to distinguish various populations of sulfhydryls in the membrane, and to determine the functional significance of each population. One population behaves as though it is located on the outer surface of the membrane, directly available for interaction without any intervening diffusion limited step (28). The interaction is rapid and it can be described by a simple 1-to-1 mass law, using the Scatchard Plot. For example, the short-term binding of chlormerodrin to red cells in the cold (to minimize penetration) can be represented by a straight line segment of negative slope representing the mass law component and by a curved line segment representing penetration into the membrane (Fig. 4–5). The former is time-independent and it is rapidly reversed by addition of protein to the medium, whereas the latter is time-dependent and is only slowly reversed by addition of protein. In the case of PCMBS which penetrates more slowly, the second component is not observed in short time periods.

The number of superficial sulfhydryls can be determined by extrapolation of the straight line segment of the Scatchard Plot. It is 1.8×10^{-18} sites per cell or 5 percent of the total membrane

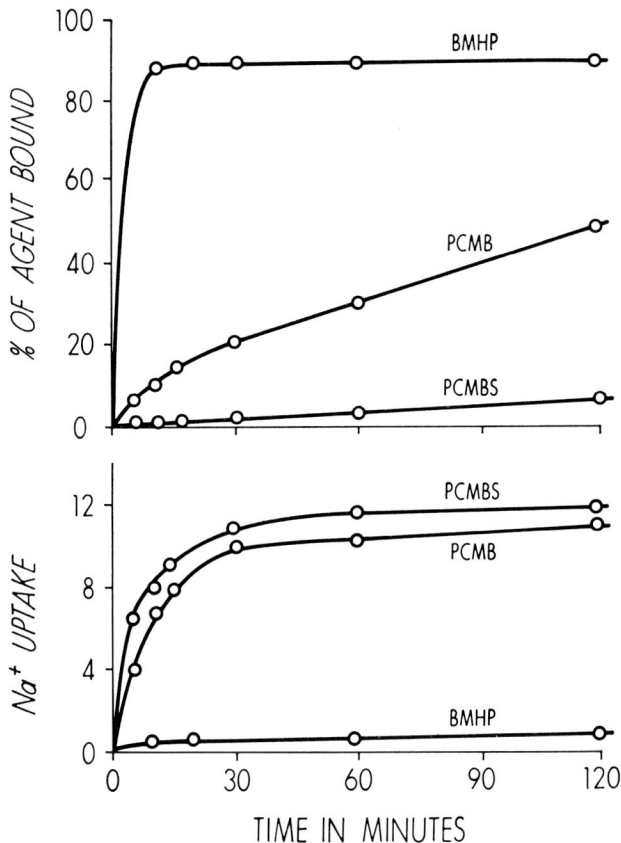

Figure 4–4. The uptake of mercurials and their effects on passive Na uptake by human red blood cells (Redrawn from reference 20).

sulfhydryls titrateable with PCMBS or chlormerodrin, or 1 percent of the total membrane sulfhydryl groups. Among this population are functional sites associated with sugar transfer across the membrane. An experiment to demonstrate this conclusion is given in Figure 4–6. Cells were exposed for 10 minutes to chlormerodrin, taking up 5.1×10^{-18} molecules per cell. The sugar transfer was inhibited over 90 percent. Exposure to added proteins resulted in a complete reversal of the inhibition. The amount of chlormerodrin removed was 1.7×10^{-18}, about equivalent to that associated with the mass law component of Figure 4–5. The chlormerodrin not removed from the cells, 3.4×10^{-18}

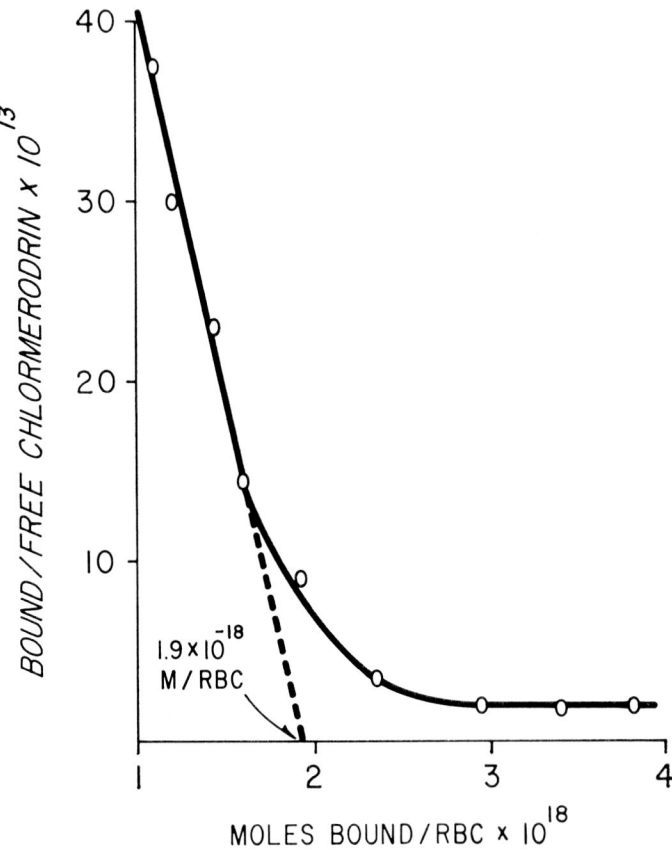

Figure 4–5. A mass law plot of the binding of chlormerodrin by human red blood cells at 4°C (From reference 28).

mole/cell, has penetrated into the interior of the membrane and is associated with sites not related to sugar transfer.

Not all of the superficial sulfhydryl sites are equivalent either in chemical properties or in functional relationships. Of several sulfhydryl reactive agents tested, PCMBS and dinotrofluorobenzene (DNFB) (4) inhibited sugar transfer, whereas SITS 5'-5-dithiobis-(2-nitrobenzoic acid) (DTNB) did not (11). Pretreatment with these agents reduced PCMBS binding to various degrees (Table 4-II). It can be concluded that no two of the agents react with identical populations of sulfhydryls, although some overlap must exist. Furthermore, less than 40 percent of the

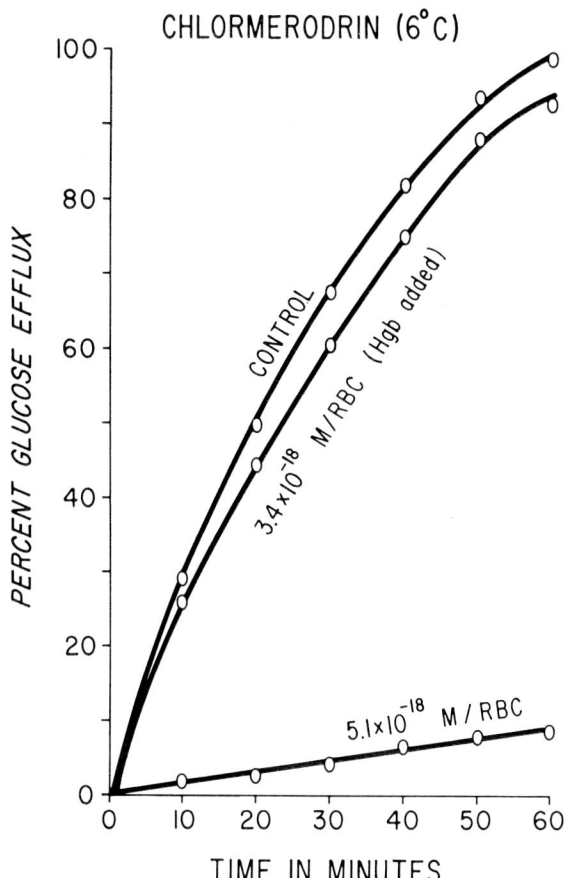

Figure 4-6. The reversal of the inhibitory action of chlormerodrin on glucose efflux from human red blood cells as related to the binding and desorption of the agent (From reference 28).

TABLE 4-II
POPULATION OF SUPERFICIAL SULFHYDRYLS OF RED CELLS REACTIVE WITH DIFFERENT REAGENTS, BASED ON DISPLACEMENT OF PCMBS BINDING

Reagent	% of Sites	Inhibition of Sugar Transfer
PCMBS	100	+
SITS	15	0
FDNB	40	+
DTNB	30	0

sulfhydryls that react with PCMBS can be involved with sugar transfer.

After the initial interaction with the superficial sulfhydryl groups, the slowly penetrating mercurials exemplified by PCMBS, penetrate into the interior of the membrane and ultimately into the interior of the cell (21). The full time course of the uptake is given in Figure 4–7. An initial rapid binding is followed by a slow uptake, and finally by a loss of a fraction of previously absorbed reagent. The sequence is explained as follows:

The initial rapid binding is due to reversible interaction with the superficial sites discussed previously. In the following period of time the agent slowly penetrates into the membrane via

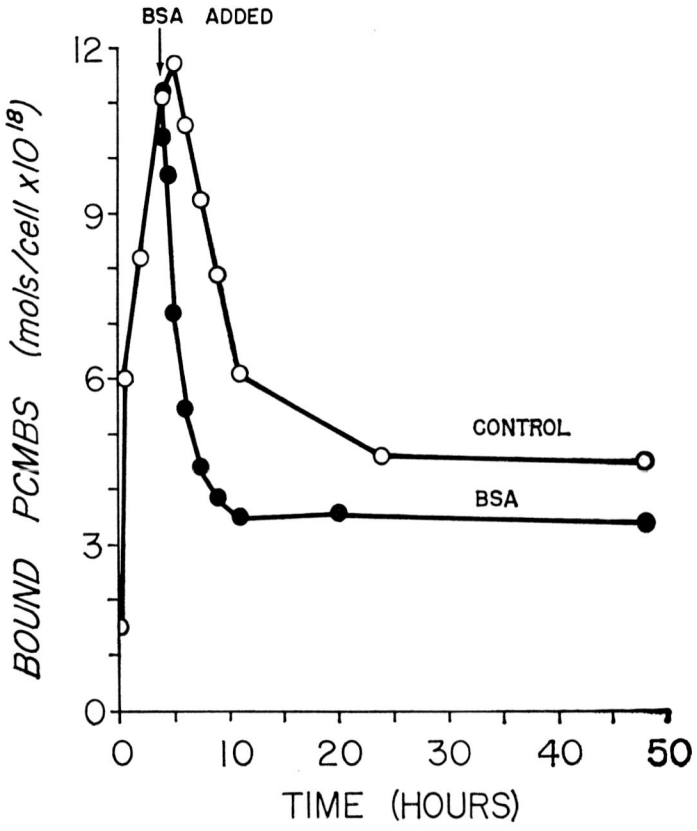

Figure 4–7. The uptake and release of PCMBS by human red blood cells and the effect of added albumin. (From reference 21).

aqueous channels, interacting with sites located in an internal sulfhydryl compartment. Finally after a long period of time, the reagent passes through the membrane to react with sulfhydryl groups of hemoglobin. Superimposed on the processes described above is the complication that the cells slowly release soluble sulfhydryl material that interacts with the remaining PCMBS in the medium. After a few hours the released sulfhydryl groups exceed the concentration of remaining PCMBS so that its concentration is reduced to almost zero. The externally bound PCMBS and a large fraction of that within the membrane (about two-thirds) diffuses slowly into the medium. The remainder diffuses into the interior of the cell. Once reacted with the hemoglobin it is trapped and will not diffuse out at a perceptible rate. The reversal of binding can be initiated earlier by addition of protein to the medium.

Associated with the binding of PCMBS in the internal membrane compartment are two effects, an increased permeability toward Na^+ and K^+ (21) and an inhibition of the active transport system for Na^+ and K^+ (17). The effects on permeability are demonstrated in Figure 4–8 in terms of K^+ leakage. No leak-

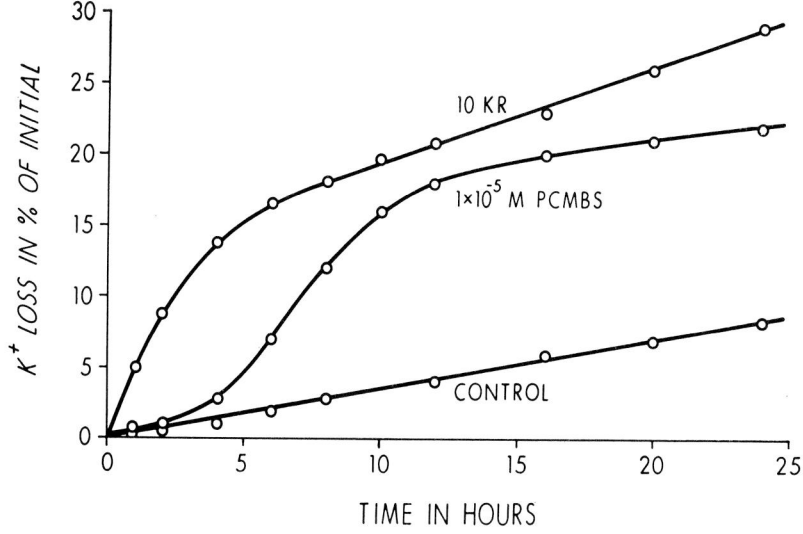

Figure 4–8. The leakage of K^+ from human red blood cells induced by PCMBS and by x-irradiation (Redrawn from data in references 21 and 22).

age is associated with the rapid binding of PCMBS superficial sites, but PCMBS accumulates in the internal membrane component, and the rate of leakage increases until a steady state is reached. The delay due to slow penetration is eliminated if the sulfhydryl sites are modified by x-irradiation (28). When uptake of PCMBS ceases and the agent moves out of the membrane compartment, the K^+ leakage returns to normal. The effect is completely reversible and can be initiated earlier by addition of protein or cysteine. The rate of K-leakage is roughly proportional to the amount of PCMBS in the internal compartment. The maximum number of sulfhydryl sites involved is about 3×10^{-18} M/cell (about 2% of the total sulfhydryl groups) (18).

The effect of PCMBS on active transport involves a different population of internal sulfhydryl sites than does its effect on permeability. The mechanism involves an inhibition of the Na-K-activated membrane ATPase, an essential part of the transport mechanism (17).

The consequences of the increased permeability and decerased transport of Na and K are a dissipation of the normal cation gradients. In the cases of the red blood cell, the colloidal osmotic pressure cannot be compensated. The cells consequently swell, and will ultimately lyse (9). In other cells the dissipation of ion gradients leads to losses of other functions such as conduction in nerve and muscle, failure of fluid absorption in the intestine, and diuresis in the kidney (18).

It is of some interest that PCMBS has no effect on anion permeability, whereas a variety of other chemical agents, especially those reactive with amino groups, influence both the anion and the cation permeabilities (anion permeability is decreased and cation permeability is increased) (4). The difference between PCMBS and the other agents is surprising because PCMBS is itself an anion and might be expected to penetrate the membrane via anion receptive channels. The sensitivity of PCMBS penetration to SITS, an inhibitor of anion permeation (13) as demonstrated in Figure 4–3, indicates that such is the case. The failure of PCMBS to inhibit anion permeability must therefore be attributed to the unresponsiveness of anion permeation to mercaptide formation.

Only the minor component of PCMBS uptake, that which is insensitive to SITS, is associated with the increased permeability cations (13) and with the inhibition of Na^+-K^+-transport (11). Thus the development of the cation leakage occurs at the same time in the presence of SITS as in its absence, despite the fact that only 20 percent as much PCMBS enters the membrane. It can be concluded that only the small fraction of sulfhydryl groups associated with SITS-insensitive pathway is associated with cation permeation and transport.

DISCUSSION AND CONCLUSIONS

The patterns of distribution of mercurials in the tissues, and the patterns of toxicity depend at least in part on their behavior in blood. Although the factors that play an important role can be delineated, quantitative data are scarce. Because the interactions of mercurials with sulfhydryls to form mercaptides are reversible, their distribution tends toward an equilibrium that is determined by concentrations and affinities. Perhaps 90 percent or more of the sulfhydryls of whole blood are within the red blood cells, largely associated with hemoglobin. The relative affinities of the various populations of sulfhydryls are not known but because the amounts of mercurials relative to total sulfhydryl groups are very small, binding will tend to be associated with populations of sulfhydryl groups that have exceptionally high affinities. The available data indicate large differences in distributions between cells and plasma for different mercurials and for different species, indicating that the high affinity populations are not always identical.

Under normal circumstances only the small fraction of mercury that is diffusible can leave the blood stream. The precise amount of diffusible mercury and its form are not known. Low molecular weight sulfhydryl compounds such as cysteine are undoubtedly important but anionic complexes may also be involved. The amount of mercury that deposits in a tissue during a single passage of blood will depend not only on the filtration of diffusible mercury but on the rate at which additional metal can dissociate from the large protein-bound reservoirs in plasma and within the red blood cells. In the case of the cells, the rate of diffusion through the plasma membrane is undoubtedly the limiting factor. Little information is available concerning the rates of redistributions in blood but it is evident that the process is very slow.

The permeation of several mercurials through the membrane of the red blood cell has been studied in some detail. Depending on the charge and lipid solubility of the mercurial, at least three pathways may be involved. The lipid soluble forms penetrate relatively rapidly, presumably by dissolving in the membrane lipids, whereas the ionic forms penetrate slowly. In the latter case, two pathways can be distinguished, one of which is inhibited by the amino-reactive reagent SITS, probably the normal pathway for physiological anions.

The lipid soluble mercurials, because they penetrate rapidly, are largely associated with the sizeable reservoir of sulfhydryl groups of hemoglobin. They have a minimal effect on membrane functions and are expected to have their predominant effect on internal cellular functions. Highly ionized mercurials, in contrast, reach high concentrations in the membrane. For prolonged periods of time they have pronounced effects on sulfhydryl-dependent membrane functions. Three such functions have been studied in red blood cells. Sugar transfer is rapidly blocked by interaction of mercurials with a small population of sulfhydryls that are superficially located on the outside of the membrane. Cation permeability is increased and active transport of cations (Na^+ and K^+) is decreased by interaction of mercurials with small populations of sulfhydryls within the membrane structure. These sites are reached slowly via the SITS-insensitive pathway of permeation. The consequence of these disturbances is a dissipation of the cellular Na^+ and K^+ gradients with ultimate osmotic lysis of the cells. In other types of cells the consequences are disturbances of transport functions as in the kidney or intestine, or of conductive functions as in nerve or muscle.

REFERENCES

1. Berlin, M.: Renal uptake, excretion, and retention of mercury. II. A study in the rabbit during infusion of methyl and phenyl mercuric compounds. *Arch Environ Health*, 6:626, 1963.
2. Berlin, M., Fazackerly, J. and Nordberg, G.: The uptake of mercury

in the brains of mammals exposed to mercury vapor and to mercuric salts. *Arch Environ Health, 18:*719, 1969.
3. Berlin, M. and Gibson, S.: Renal uptake, excretion and retention of mercury. I. A study in the rabbit during infusion of mercuric chlorate, *Arch Environ Health, 6*:617, 1963.
4. Cafruny, E. J.: The site and mechanism of action of mercurial diuretics. *Pharmacol Rev, 20:*89, 1968.
5. Clarkson, T., Gatzy, J. and Dalton, C.: Studies on the equilibration of mercury vapor with blood. URAEP #582, 1961.
6. Diem, K. (Ed.): *Documenta Geigy, Scientific Tables (6th edition).* Ardsley, New York, Geigy Pharmaceuticals, 1962, pp. 553–557.
7. Hayes, A. D. and Rothstein, A.: The metabolism of inhaled mercury vapor in the rat studied by isotope techniques. *J Pharmacol Exp Ther, 138:*1, 1962.
8. Hughes, W. L., Jr.: A physiochemical rationale for the biological activity of mercury and its compounds. *Ann NY Acad Sci, 65:*454, 1957.
9. Jacob, H. S. and Jandl, J. H.: Effects of sulfhydryl inhibition on red blood cells. I. Mechanism of hemolysis. *J Clin Invest, 41:*779, 1962.
10. Kessler, R. H., Lozano, R. and Pitts, R. F.: Studies of the structure and diuretic activity relationships of organic compounds of mercury. *J Clin Invest, 36:*656, 1957.
11. Knauf, P. and Rothstein, A.: Paper in preparation
12. Knauf, P. A. and Rothstein, A.: Chemical modification of membranes. I. Effects of sulfhydryl and amino reactive regents on anion and cation permeability of the human red blood cell. *J Gen Physiol, 58:* 190, 1971.
13. Knauf, P. A. and Rothstein, A.: Chemical modification of membranes. II. Permeation paths for sulfhydryl agents. *J Gen Physiol, 58:*211, 1971.
14. Lundgren, K.D., Swensson, A. and Ulfvarson, U.: Studies in humans on the distribution of mercury in the blood and the excretion in the urine after exposure to different mercury compounds. *Scand J Clin Lab Invest, 20.:*164, 1967.
15. Norseth, T.: Studies on the Biotransformation of Methylmercury Salts in the Rat, Ph.D thesis, University of Rochester, 1969.
16. Passow, H., Rothstein, A. and Clarkson, T. W.: The general pharmacology of heavy metals, *Pharmacol Rev, 13:*185, 1961.
17. Rega, A., Rothstein, A. and Weed, R. I.: Erythrocyte membrane sulfhydryl groups and the active transport of cations, *J Cell Physiol, 70:*45, 1967.
18. Rothstein, A.: Sulfhydryl groups in membrane structure and function. In F. Bonner and A. Kleinzeller (Eds.): *Current Topics in Membrane and Transport.* New York, Academic Press, 1970.
19. Rothstein, A. and Hayes, A. D.: The metabolism of mercury in the rat studied by isotope techniques. *J Pharmacol Exp Ther, 130:*166, 1960.

20. Shapiro, B., Kollman, G. and Martin, D.: The diversity of sulfhydryl groups in the human erythrocyte membrane. *J Cell Physiol*, 75:281, 1970.
21. Sutherland, R. M., Rothstein, A. and Weed, R. I.: Erythrocyte membrane sulfhydryl groups and cation permeability. *J Cell Physiol*, 69:185, 1967.
22. Sutherland, R.M., Stannard, J.N. and Weed, R.I.: Involvement of sulphhydryl groups in radiation damage to the human erythrocyte membrane. *Int J Rad Biol*, 12:551, 1967.
23. Suzuki, T., Miyama T. and Katsumuma, H.; Mercury contents in the red cells, plasma, urine, and hair from workers exposed to mercury vapor. *Ind Health*, 8:39, 1970.
24. Suzuki, T., Miyama, T. and Katsumuma, H.: Comparison of mercury contents in maternal blood, umbilical cord blood, and placental tissues. *Bull Environ Contam Toxicol*, 5:502, 1971.
25. Swensson, A., Lundgren, K. D. and Lindstrom, O.: Distribution and excretion of mercury compounds after single injection. *Arch Ind Health*, 20:432, 1529.
26. Swensson, A. and Ulfvarson, U.: Toxicology of organic mercury compounds used as fungicides. *Occup Health Rev*, 15:5, 1963.
27. Ulfvarson, U.: Distribution and excretion of some mercury compounds after long term exposure in man. *Arch Gewerbpath Gewerbkyg*, 19:412, 1962.
28. Van Steveninck, J., Weed, R. I. and Rothstein, A.: Localization of erythrocyte membrane sulfhydryl groups essential for glucose transport. *J Gen Physiol*, 48:617, 1965.
29. Webb, J.L.: *Enzyme and Metabolic Inhibitors, Vol. II.* New York, Academic Press, 1966, pp. 635–983.
30. Weissman, N., Schoenbach, E. G. and Armistead, E. B.: The determination of sulfhydryl groups in serum. *J Biol Chem*, 187:153, 1950.

DISCUSSION

Hammer: How much work has been done on the toxicity of mercury using nerve cells?

Rothstein: There has been some but not a great deal; mercurials inhibit conduction and dissipate the ionic gradients, depending on the compound used, and the amount. In perfused squid axons mercurials have been applied to the inside or outside surfaces to determine the location of the controlling sulfhydryl groups. Unfortunately the agents used penetrate rather rapidly and little difference was observed.

Barber: Surely if organomercurials like dimethylmercury can penetrate the cell membrane rapidly because of their high lipid solubility, the buildup in concentration of these compounds will occur in the lipid phase of the membrane.

Rothstein: Yes, the concentration of lipid soluble mercurials in the membrane may be high, but it is localized in the lipid phase of the membrane rather than in the aqueous channels. Effects of mercurials on ion permeability or on ion transport seem to be associated only with the aqueous channels reached by the ionic forms of mercurials; the nonionic forms on the other hand, although they penetrate rapidly do not "see" the cation controlling sites. They penetrate rapidly, and are "soaked up" by internal proteins, with little effect or only a transient effect on permeability.

Observer: Did you say that the buildup of PCMBS is actually in aqueous channels?

Rothstein: Yes, PCMBS is a highly ionized form. It penetrates slowly, but a large proportion reaches the cation controlling sites. With PCMB or BMHP, the rapid penetration is via the lipid phase without access to the sulfhydryls that control the cations. For PCMB, only a small fraction that goes through that SITS-insensitive channel as an anion reaches the cation-controlling sites.

Gage: If mercuric ion is added to whole blood and left until equilibrium is reached, is the ratio of mercury in the plasma and red cells commensurate with the sulfhydryl concentrations, leaving out here the kinetic factors?

Rothstein: I don't know that anyone has carefully looked at this problem *in vitro*. In the studies reported here, the cells were suspended in saline with no external protein so that all of the agent added initially is diffusable. After a time the cells leak enough protein to bind much of the residual agent, resulting in some outflow of mercurial, and a reversal of the effect.

In whole blood the amount of diffusible mercury is extremely small and nobody knows exactly how small or its chemical form. The rate of equilibrium would be very much slower, depending on the exact size of the diffusible component. The final equilibrium state would probably be concentration-dependent. The distribution of inorganic mercury and certain other mercurials *in vivo* between cells and plasma is not proportional to the numbers of sulfhydryl groups in the two compartments. It is necessary, therefore, to postulate that the data from animals represent a nonequilibrium condition—that is, a steady-state condition where mercury keeps draining out of the cells into the tissues more rapidly than from plasma proteins, implying that the rate of dissociation from plasma proteins is extremely slow, or it is necessary to postulate that there are some special small populations of sulfhydryls in the plasma that have a much higher affinity for mercury than those of hemaglobin.

Foulkes: Am I correct in interpreting your explanation as postulating different aqueous pores for cations and anions?

Rothstein: Yes. I am certain that the SITS-sensitive channel is the normal anion channel. I believe that the SITS-insensitive channel is the cation channel.

Berlin: We have made some observations on the binding of mercurials by proteins in plasma (unpublished observation); it seems to be dose-dependent. What is known about variations between different kinds of cells in relation to these factors which you studied in the red cells?

Rothstein: Your first comment, the dose dependency, I think would suggest very much that there are different populations of sulfhydryls in blood. The population that will first react with the mercurial is that with the highest affinity. When it becomes saturated the mercury will spill over into other populations with lower affinities, resulting in dose-dependent distributions. Unfortunately there is virtually no data whatsoever about this important point.

In answer to your second question there just isn't enough known about the behavior of mercury in other systems. The only one that has been studied in great detail is the kidney and it is so complex with different populations of cells that it is difficult to understand what happens at the level of a single cell. Thus the only data of consequence is from the red blood cells.

Vostal: Are the PCMBS and the PCMB effects which were studied here *in vitro* and their effect on the erythrocytes typical of an organ which is selective for the transport of the cations? In the kidney PCMBS and PCMB are both completely ineffective as far as the diuresis is concerned, but they can block the diuretic effect.

Rothstein: I suggest that accessibility factors must account for the complex behavior of mercurials in the kidney. The diuretic effect occurs at a particular location in a particular population of cells. Perhaps a two-step reaction is required. The diuretic and nondiuretic agent compete for the first site, but only the diuretic agents are able to reach the second site to give the physiological response. Such an explanation is more of a rationalization of our lack of understanding.

Nechay: There is another difficulty with accessibility factors. If, for instance, nondiuretic mercurials would first saturate the nonspecific sites or the sites which are not responsible for diuresis then you could push it to the point where they would effect the sites that are responsible for diuresis.

Rothstein: True, obviously the existing explanations are inadequate but they are the only ones we have.

Stannard: You mentioned the radiation work briefly. I remember we had anomalies in contrasting the effects of radiation with the organic mercurials in that the mercurials effected glucose transport whereas radiation did not; looking at cation transport they seem to react alike and as you say involve perhaps the same sulfhydryl groups. I wonder now if you could relate the anomaly to these three pathways that you mentioned?

Rothstein: No, I'm afraid not. The failure of radiation to effect the sugar transfer can be explained perhaps on the basis that isolated single sulfhydryl groups are involved that can form mercaptides but cannot form disulfides. In the case of the effects of irradiation on cation transport, the evidence is strong that disulfide formation is involved.

Barber: Is the action of mercury on your net potassium leakage a breakdown in the permeability or is it an effect directly on the sodium pump?

Rothstein: There are two distinct effects on cations. One is on the downhill leak with sodium and potassium permeation about equally affected, but with the permeation of choline, a somewhat larger cation, unchanged. The second effect is on the sodium-potassium active transport system. Each of the two effects involves a different population of internal membrane sulfhydryl groups, that is small in size, perhaps one or two percent of all the sulfhydryl groups in membrane. Both effects lead to dissipation of ion gradients and the ultimate result in the red blood cell is osmotic lysis.

Barber: Was there any stimulation of the Na^+–K^+ pump at low mercury concentrations?

Rothstein: We never saw it.

Barber: It has been reported.*

Rothstein: Yes, I know but we have never really looked for it.

Norseth: I'd like to come back to the species differences regarding methylmercury in red cells and plasma. I have some experiments which indicate that in the rat there are two different distributions of methylmercury. One compound is probably excreted in bile and reabsorbed again; this compound has another red cell to plasma distribution than the originally injected compound, in my case methylmercury chloride. It is very complicated.

Rothstein: What we need now is some relatively simple data about interactions of various mercurials with the various proteins in blood. We need to know something about the nature and amount of diffusible complexes and we need to know something about the kinetics of the movement from one compartment to the other. We can talk about these things and we know that they are important in the distributions, but there's very little hard data.

McDuffie: Would you comment on the role of disulfide in the protein. Does it effect the equilibrium?

Rothstein: The same sulfhydryls are involved in the mercurial effect as in the radiation effect except that the latter could be correlated with the amount of disulfide formed. Normally the membrane is relatively free of disulfides. The slow reversal of enhanced cation permeability after irradiation can be correlated with the spontaneous disappearance of disulfide and the formation of free sulfhydryl.

Magos: When mercury moves from one compartment to another, from plasma to red blood cell and back, would you say it actually moves in a noncharged form as atomic mercury or mercury chloride?

Rothstein: Yes, I would guess so. Inorganic mercury exists as an array of complexes with chloride, some of them neutral. I think that the rapid movement through the membrane may very well be as uncharged mercurichloride, because it is clear with the organic mercurials that ionic forms penetrate slowly, whereas uncharged forms penetrate rapidly.

* Joyce et al.: Br J Pharmacol, 9:463, 1954.

SESSION II
ACTION OF MERCURY ON MEMBRANES AND TRANSPORT PROCESSES

David W. Fassett, *Chairman*

Chapter 5

SITE OF THE FUNCTIONAL LESION RESPONSIBLE FOR AMINO-ACIDURIA AFTER ADMINISTRATION OF ORGANOMERCURIALS AND OTHER METAL COMPOUNDS

E.C. Foulkes

ABSTRACT

Amino-aciduria of heavy metal poisoning results for inhibition of tubular reabsorption of filtered amino acids. In the present study a double indicator dilution technique was adapted to the rabbit in an attempt to localize at the cellular level the site of the functional lesion responsible for the increased amino acid clearance. Effects of heavy metal compounds (organomercurials, uranyl ion) were sought on the two opposite membranes of tubular epithelial cells (the membrane facing the lumen, LM and that exposed to the peritubular interstitial fluid, PTM) as well as on possible intracellular components of the reabsorptive mechanisms. Artery-to-vein transit patterns of amino acids and inulin revealed the presence of amino acid carrier systems at PTM, but these systems cannot be involved in amino acid reabsorption. The urinary volume of distribution of α-aminoisobutyric acid (AIB) remained the same as that for inulin under all conditions tested, even when AIB reabsorption was strongly inhibited. The inhibitory metal compounds must therefore have exerted their effect at LM, and it may be inferred that amino-aciduria of heavy metal poisoning results from a functional lesion at the luminal cell membrane.

REPORT

Introduction

One of the well documented symptoms of intoxication by heavy metals is the inhibition of tubular reabsorption of filtered amino acids. The resultant aminoaciduria has been

---

This work was supported by NIH grants ES-AM 00580 and 5P10 ES00159. It gives me much pleasure to thank Mrs. L. Greenland and Mrs. S. Blanck for their valuable help.

reported after exposure to a variety of metals. The present work is concerned primarily with localization of the action of mercury compounds. Specifically, experiments were performed to determine at the cellular level the site of the functional lesion depressing amino acid reabsorption in renal tubules poisoned with mercury.

Of the many carrier systems studied in the renal tubule some are believed to act at the membrane separating the tubule cell from the peritubular interstitium (PTM). Such is the case with transepithelial movement of sodium, as well as with secretion of para-aminohippurate (PAH) (3). The existence of specific amino acid carrier systems at the peritubular membrane was also recently reported (1). Other carriers may be localized at the cell membrane facing the tubular lumen (LM). Finally, a direct role of intracellular structures in transport processes cannot be excluded a priori. The experiments described here strengthen the conclusion that carrier mechanisms whose activity can be demonstrated at PTM do not contribute to amino acid reabsorption, and that neither their inhibition by mercurials can account for the observed aminoaciduria (1) nor could evidence be obtained for an intrancellular block to amino acid movement. It can thus be inferred that the action of metals responsible for increased amino acid clearances is exerted at the luminal cell membrane.

Methods

Application of the double indicator dilution technique of Silverman et al. (5) to the rabbit has been previously described (2). The experiments consist of an arterial injection of a bolus of ^{3}H-labeled inulin and ^{14}C-labeled amino acid, followed by rapid collection of successive fractions of renal venous blood and urine. Over 90% of the injected inulin which escapes filtration is usually recovered in venous blood within 25 to 30 seconds; several minutes are required for complete urinary recovery of filtered inulin. Fractional recoveries of amino acids were computed on the basis of 100% recovery of non-filtered inulin in blood or of filtered inullin in urine, respectively. Circulating vascular volumes were maintained steady by infusion of 6% dextran in saline. Experiments on artery-to-urine transit times were performed with kidneys perfused at constant flow and normal pressures with blood withdrawal from the aorta. For this purpose a finger pump was inserted into the arterial shunt; perfusion pressure was monitored distally to the pump. Venous effluent was returned from a constant level reservoir into a jugular vein. Recirculation of tracer was prevented by collecting total venous effluent for at least 60 minutes after arterial injection.

The studies were carried out with male albino rabbits weighing on the average 2 kg. Glomerular filtration rate (GFR) and effective renal plasma flow (ERPF) were measured by determination of creatinine and PAH clearances. The animals were infused for 70 to 80 minutes at a rate of 0.5 to 1.0 ml/kg/min with physiological saline containing 10% (W/V) mannitol and suitable amounts of creatinine and PAH. After a first control study, parachlormercuribenzoate (PCMB) or other mercury compounds were injected; twenty to thirty minutes later a second study was carried out. Uranium (1μ mole uranyl acetate/kg) was administered as in earlier work (4) two days prior to the experiment. Materials and methods used in chemical and radiological assays have been previously detailed (1).

Results

Reaction of the peritubular cell membrane especially with dicarboxylic amino acids has already been described (1). The phenomenon is here further illustrated by one of three studies performed with L-glutamic acid (see Fig. 5-1). The lower recovery of glutamate relative to that of inulin in renal venous blood is a measure of uptake of the amino acid across PTM. Just as in the case of aspartate (1) the intravenous administration of PCMB (5 mg Hg/kg) increased the relative venous recovery of glutamate, indicating an inhibition of the glutamate carrier system at PTM. Unlike the transfer of glutamate across PTM, transport of phenylalanine or methionine out of the postglomerular inulin space was not significantly affected by the

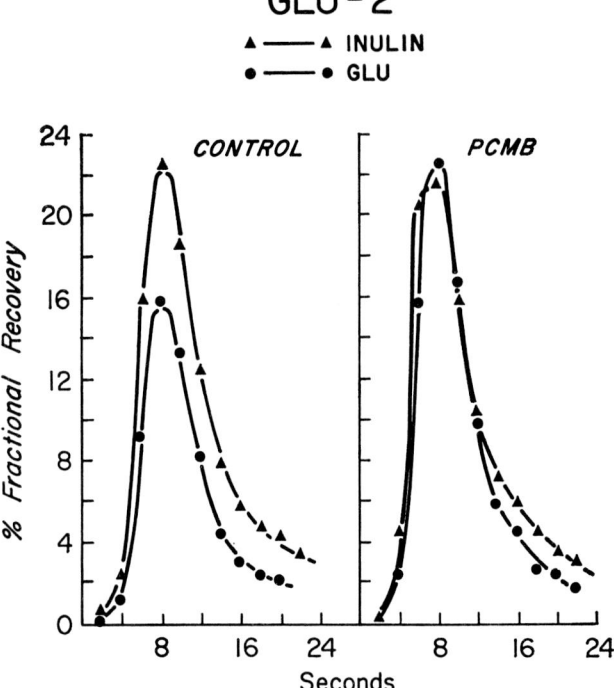

Figure 5–1. Artery-to-vein transit pattern of L-glutamate before and 21 minutes after intravenous injection of 19 mg PCMB/kg body weight. Renal blood flow: control 24.0 ml/min, after PCMB 30.8 ml/min.

intravenous injection of 19 mg PCMB. Figure 5–2 shows the results of one of four experiments carried out with these two neutral amino acids.

In contrast with the dicarboxylic and monocarboxylic amino acids mentioned above, as well as with a variety of other neutral and basic amino acids (unpublished experiments), alpha-amino-isobutyric acid (AIB) in repeated experiments, on the one hand, reacted only slightly with PTM (see Fig 5–3). On the other hand, tubular reabsorption of AIB is strongly depressed by PCMB. This effect is seen in Figure 5–4 which illustrates urinary recovery curves of the amino acid before and after PCMB administration; the results shown are representative of five similar studies, in which arterial administration of 10 mg Hg/kg as PCMB reduced fractional AIB reabsorption by an average of

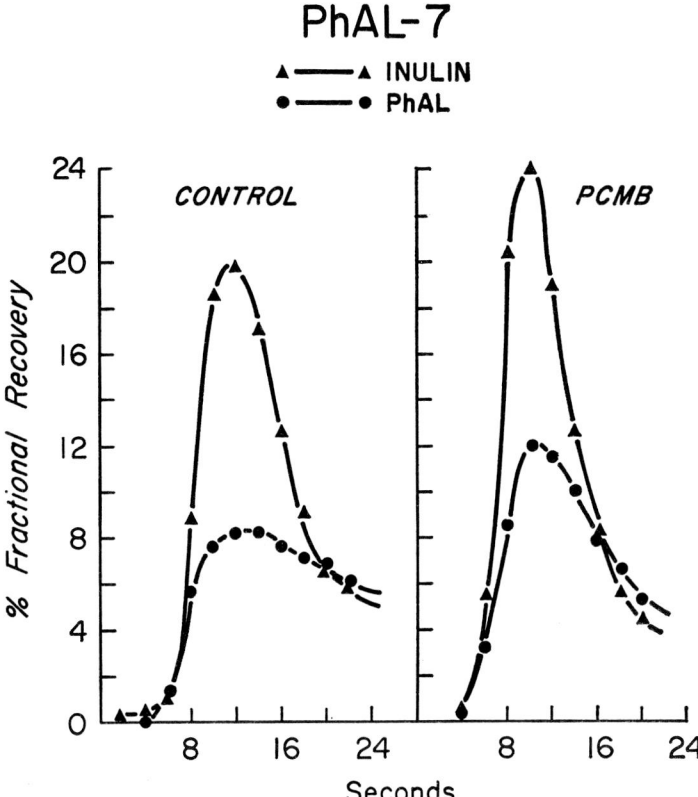

Figure 5–2. Artery-to-vein transit pattern of L-phenylalanine before and 20 minutes after intravenous injection of 19 mg PCMB/kg body weight. Renal blood flow: contol 21.2 ml/min, after PCMB 26.8 ml/min.

75%. In Figure 5–4 the areas between the recovery curves for inulin and AIB provide a measure of the amount of filtered AIB reabsorbed from tubular urine. The obvious inhibition of this reabsorption is not accompanied by a significant separation of the mean transit time ($\bar{t}$) of inulin from that of AIB. Experiments similar to that illustrated in Figure 5–4 were also carried out with three uranium poisoned rabbits. These injected rabbits showed an average fractional AIB reabsorption of only 29%, compared to a value of over 80% obtained repeatedly in control animals. As with PCMB, uranium depression of AIB reabsorption was not accompanied by any increase of $\bar{t}_{AIB}$ over $\bar{t}_{In}$.

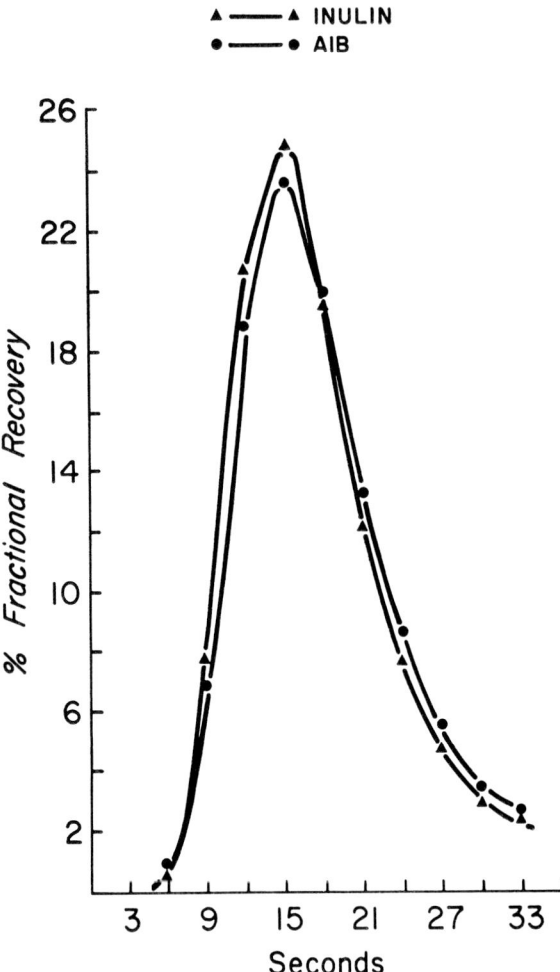

Figure 5–3. Artery-to-vein transit pattern of AIB. Renal blood flow: 14.4 ml/min.

Action of other Compounds of Mercury on PTM

Preliminary experiments have shown that intravenously injected neohydrin (9 mg/kg) exerts an effect on transfer of dicarboxylic amino acids across PTM similar to that of PCMB. On the one hand, two further studies showed a significant effect

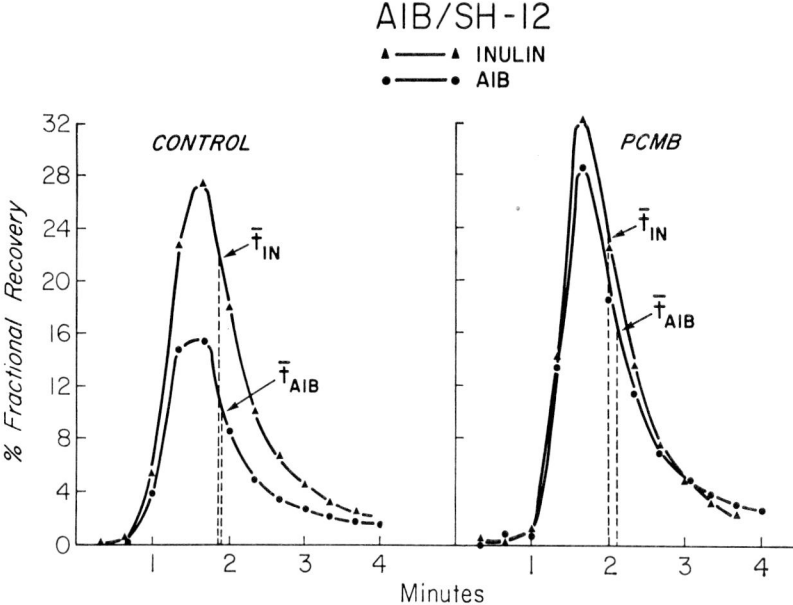

Figure 5–4. Artery-to-ureteral urine transit pattern of AIB, before and after intraarterial injection of 19 mg PCMB/kg body weight. Mean transit times ($\bar{t} = \dfrac{\Sigma\, C_t \cdot T \cdot \Delta T}{\Sigma\, C_t \cdot \Delta T}$, where C_t = concentration after elapsed time T, ΔT = 20 seconds) are indicated for inulin and AIB. Urine flow: control 1.0 ml/min, after PCMB 0.9 ml/min.

with intravenous injection of 5 mg Hg/kg as $HgCl_2$. On the other hand, monomethyl mercury (13 mg/kg) caused no inhibition of aspartate transport at PTM after intravenous injection; some inhibition was observed after intraarterial administration.

DISCUSSION

Results described in this paper confirm the previous observation (1) that specific mechanisms catalyze amino acid transport from the peritubular interstitium across the peritubular cell membrane (PTM) into tubular epithlial cells of rabbit kidneys. As also reported in the earlier publication these transport mechanisms differ in their affinities for various amino acids. To this distinction between transport mechanisms on the basis of substrate specificity may now be added the observation that they differ also in their sensitivity to PCMB (see Figs. 5–1 and 5–2). The reaction of glutamate with PTM is depressed by the mercurial, as was previously illustrated for aspartate (1); no effect is seen under these conditions on the reaction of PTM with phenlyalanine. The carrier system responsible for the transfer of dicarboxylic amino acids across PTM is clearly distinct from that responsible for movement of neutral amino acids.

Different compounds of Hg appear to affect the peritubular membrane in a similar manner to that of PCMB. However, a quantitative comparison of the sensitivity of the dicarboxylic amino acid carrier system at PTM to different mercurials is difficult under present conditions. Undoubtedly major differences exist between the respective volumes of distribution of injected drugs and between their intrarenal concentrations at the time of measurement. This view is underscored by the result of intraarterial as opposed to intravenous administration of monomethyl mercury. Preliminary studies suggest that methylmercury resembles PCMB also in its action on the luminal cell membrane.

In contrast to the rapid penetration of PTM by dicarboxylic amino acids, phenylalanine, methionine, and several other neutral and basic amino acids (unpublished results) the reaction of alpha-amino isobutyric acid (AIB) with PTM is much less significant. However, in spite of this absence of rapid transfer of AIB across PTM the compound is readily reabsorbed from glomerular filtrate. An additional observation is that of the PCMB sensitivity of dicarboxylic amino acid transport across PTM under conditions where their fractional reabsorption is not affected by the mercurial. We are again led to the earlier conclusion (1) that the mechanisms at PTM cannot be involved in the process of amino acid reabsorption; the question of their physiological significance must remain open at this time. Finally it follows that the aminoaciduria of heavy metal poisoning cannot be attributed to the kind of lesion here described at the PTM after administration of PCMB.

Increased urinary recovery of AIB in poisoned animals (see Fig. 5–4) reflects inhibition of AIB reabsorption from tubular fluid. Figure 5–4 further illustrates the fact that in spite of this inhibition no significant differences arose between the mean artery-to-urine transit times of AIB and inulin. In other words the volumes of distribution of the two excreted solutes remain identical within experimental precision. Now even if we conservatively assume that only 15% by volume of the renal cortex consists of proximal intraceullar spaces, addition of this 15% in the average experiment would have almost doubled the tubular fluid and urine volume calculated as urine flow $\times \bar{t}_{In}$. The present results therefore imply that AIB which has crossed into tubular cells from the luminal side cannot have diffused back into tubular fluid. Such backflow would, of course, result from a block of amino acid reabsorption beyond LM, either at some hypothetical intracellular site or at PTM. It follows therefore that the amino acid movement must be bocked at or near the level of the luminal cell membrane. This conclusion does not necessarily implicate a direct reaction of the metal with LM. It is conceivable that the membrane lesion could, indeed, result indirectly from some other reaction of Hg with the tubule cell; the present results cannot help evaluate this possibility. It can be stated with confidence, however, that the aminoaciduria which follows administration of PCMB results from a lesion at the luminal membrane.

The same conclusion can also be drawn from the kinetics of AIB excretion in uranium poisoned animals. Injection of uranyl acetate did depress tubular reabsorption of filtered AIB, but caused no increase in $\bar{t}_{AIB}$ compared to $\bar{t}_{In}$. Again, inhibition of transport at LM best explains these findings. Other heavy metals known to produce aminoaciduria in all probability also exert their inhibitory action on the luminal side of the tubule cells.

REFERENCES

1. Foulkes, E.C.: Effects of heavy metals on renal aspartate transport and the nature of solute movement in kidney cortex slices. *Biochim Biophys Acta*, 241:815, 1971.
2. Foulkes, E.C.: Glomerular filtration and renal plasma flow in uranium poisoned rabbits. *Toxicol Appl Pharmacol*, 20:380, 1971.
3. Foulkes, E.C., and Miller, B.F.: Steps in PAH transport by kidney slices. *Am J Physiol;* 196:86, 1959.

4. Nomiyama, K., and Foulkes, E.C.: Some effects of uranyl acetate on proximal tubular functions in rabbit kidney, *Toxicol Appl Pharmcol, 13*:89, 1968.
5. Silverman, M., Aganon, M.A., and Chinard, F.M.: Specificity of monosaccharide transport in dog kidney. *Am J Physiol, 218*:743, 1970.

DISCUSSION

Goldwater: Are the mechanisms for aminoaciduria different for different metals?

Foulkes: No. The localization of action of these compounds is the same judging on the experimental basis I just presented, and the mechanism of action presumably is also the same. It is very difficult however to answer that definitively because it is difficult to make quantative comparisons between the different mercury compounds.

Klein: Is this a distal tubular effect?

Foulkes: No, amino acid reabsorption occurs proximally.

Fang: Are the losses in the recovery of aspartic acid due to metabolic conversion?

Foulkes: This is probably not metabolism for several reasons; firstly, we can load the animal with aspartic acid and up to certain plasma concentrations the recovery curve remains unchanged.

Further, we can show the same effect with a whole variety of amino acids. If one wished to argue that they all are metabolized to the same extent this would be a possibility I could not deny.

Fang: We do not find mercury affecting metabolic breakdown of γ-aminobutryic acid.

Foulkes: This would fit in then with my interpretation.

Fang: It is also possible these effects are due to aspartic acid breakdown.

Foulkes: But let me add one point. I didn't show you the complete curves. What I showed you was the inulin curve; and the aspartic acid curve will tail out because it will include reabsorbed amino acid. This of course reappears in plasma after a very definite tubular delay. It may take about ten seconds for the material to pass through the cells from lumen to blood side.

Gage: Are the functional changes you describe reversible and is the subsequent tubular necrosis explicable by a sustained change in membrane permeability?

Foulkes: The only answer that I can give you, Dr. Gage, is an indirect one. We have taken uranium poisoned animals and permitted them to recover from acute renal failure. Then we did this kind of study and found that the kidneys behaved essentially normally. I cannot answer the question with regard to mercury because as I pointed out these are acute preparations.

Site of the Functional Lesion

Nechay: Is it mechanically significant that PCMB interferes with the reabsorption of amino acids which are carboxylic acid? This leads of course to what every pharmacologist would suggest immediately, namely control compounds. How about other carboxylic acids which are not mercurial? Would they do the same?

Foulkes: I don't think I can answer that. We have tested arginine and lycine; they are transported by a system specific for basic amino acids.

Nechay: Do you know how they are reabsorbed? I'm raising the point as to whether or not you are dealing with carboxylic acid or whether you are dealing with mercury.

Foulkes: Methylmercury exerts the same effect as inorganic mercury.

Nechay: But this does not exclude that carboxylic acid of another type does not do the same thing.

Foulkes: I think it's a valid suggestion but would you agree with me when I answer you by saying that if we can at least qualitatively obtain the same effect with different non-ionic mercury compounds then the chances of what we are seeing being primarily an anion effect are small?

Nechay: It helps.

Kazantzis: Another alternative explanation to a membrane holdup is the possibility of attachment of the aspartic acid within the cell for a period during which it is retained and then released.

Foulkes: That the aspartic acid would enter and remain . . . ?

Kazantzis: Yes, it's mopped up by a sponge as it were and then released more slowly.

Foulkes: The only answer I can give is that we know from the artery-to-vein transit patterns of amino acids that it takes something of the order of ten seconds for them to move from tubular lumen back to blood.

Let us assume at a point between tubular fluid and blood a hypothetical sink for amino acids. The compounds would have to stay in this sink for at least 15 seconds, that is until the point is reached where the inulin recovery curve drops below that of the amino acid.

Frankly, I don't think it's terribly likely but I can't deny the possibility. Do you think its likely?

Kazantzis: It's an act of timing really.

Foulkes: I would agree.

Rothstein: Did you make some measurements on the transit of PCMB?

Foulkes: No. Transit from where to where?

Rothstein: Well, if you had the same kind of curves for PCMB in blood and in urine you'd have some idea how much is being left in the kidney and where.

Foulkes: This might tell us something about where the PCMB has gone.

Rothstein: Yes. Was the PCMB injected as a pulse?

Foulkes: No, the PCMB was given intravenously about ten minutes before the analysis.

Rothstein: So it was already mixed and partially equilibrated. This might be a very good system then for looking at the factors in blood that may determine where in the kidney the PCMB goes.

Clarkson: Is there any effect or are there any reasons to expect an effect of mercury on the urinary amino acid pattern?

Foulkes: There are changes in the amino acid pattern during the aminoaciduria. We are in the process of setting up some of these assays because we would like to know which of the amino acids are primarily affected. We're also doing this kind of work with cadmium at the moment but for technical reasons this is a little bit more difficult.

Magos: If you accept that PCMB or any mercury compound forms a bond with a carboxyl group, this doesn't contradict your theory on the role of carboxyls in the process.

Fassett: Have you studied a diuretic such as mercaptomerin in which the mercury is attached to a sulphur and not to a carbon?

Foulkes: Yes, but the only diuretic we studied was chlormerodrin.

Gage: Would you care to make any observations on the proteinuria observed in mercury poisoining?

Foulkes: I don't think I could add any useful comments.

Rothstein: Did you ever try putting a second pulse of materials through to see whether the lesion had reversed? In other words, see if it was a transient change?

Foulkes: No, but we are limited because this is an acute preparation.

Clarkson: It might be worthwhile to look at different complexes of mercury in your system to determine the effect of mercury cysteine rather than mercury chloride.

Chapter 6

ACTION OF MERCURY ON RENAL SODIUM TRANSPORT AND ADENOSINETRIPHOSPHATASE ACTIVITY

BOHDAN R. NECHAY

ABSTRACT

The molecular basis for the interference with the renal transport of Na^+ by certain compounds of mercury is not known. So far unsuccessful attempts were continued to differentiate between the inhibition of $Na^+ + K^+$ ATPase by diuretic and nondiuretic mercurials. *In vitro* studies with various ATPase preparations from the dog kidney have shown the following: (1) It was confirmed that both the diuretic and nondiuretic mercurials are inhibitors of the enzyme; (2) The inhibition by a diuretic mercurial mercaptomerin was augmented by acid pH. However, this was not true for other diuretics chlormerodrin or mersalyl or nondiuretics PCMB or PCMPS. Thus, there was no positive correlation with the known potentiation of mercurial diuresis by acidosis; (3) Changes in concentrations of cysteine, bovine albumin, Na^+, Mg^{++}, ATP or $SO_4^=$ did not selectively influence the inhibition by chlormerodrin or PCMPS; (4) High concentrations of four out of five diuretic mercurials tested prevented a rise in the enzyme activity upon preincubation. In similar experiments four out of five nondiuretic mercurials tested did not do so. The kidneys from dogs pretreated with two diuretics, two nondiuretics or two combinations of diuretic and nondiuretic mercurials had about equally reduced enzyme activities. This was in contrast to the known antagonism of PCMB or PCMPS to mercurial diuresis. An ATPase preparation from human kidney, which was highly dependent upon Na^+ and K^+ for activity, was sensitive to inhibition by mersalyl chlormerodrin.

I am indebted to James Arly Nelson for some of the data incorporated in this presentation. His dissertation includes relevant experiments with mercury (19).

This work was supported by U.S. Public Health Service, National Institutes of Health Grant AM 13019.

From the point of view of renal electrolyte transport there are two kinds of mercurials: those that interfere with the tubular reabsorption of Na^+ and those that do not. Mercuric chloride, chlormerodrin, mersalyl, mercaptomerin and meralluride are examples of commonly referred to diuretic agents. p-Chloromercuribenzoic acid (PCMB), p-chloromercuriphenyl-sulfonic acid (PCMPS), ethylmercury and methylmercury compounds are examples of agents which do not cause diuresis when given intravenously to dogs (28). In analogy to the kidney, meralluride reduces the secretion of a salty fluid from the salt glands of sea gulls while PCMB is without effect (14). In contrast to the differential effect on transepithelial Na^+ transport in the kidney and the salt gland, both chlormerodrin and PCMB inhibit the translocation of Na^+ across the bladder of the toad (22). The mercurial structure-natriuretic activity relationships were subject to intensive studies (8, 28) and reviews (2, 4, 13, 23, 25, 27).

There is a growing evidence that the Na^+ and K^+ activated adenosine triphosphate hydrolyzing system (24)–$Na^+ + K^+$ ATPase; ATP phosphohydrolase, EC 3. 6. 1. 3–participates in the renal reabsorption of Na^+ (5, 7, 11, 16, 20, 21). The fact that mercurials are inhibitors of the $Na^+ + K^+$ ATPase suggests that this enzyme system may be the receptor for the natriuretic action of certain compounds of mercury (6, 15, 26). However, since the nondiuretic mercurials are equally active inhibitors of the $Na^+ + K^+$ ATPase it appears imperative to distinguish between the diuretic and the nondiuretic mercurials on biochemical and pharmacological level before the natriuretic action of mercury can be ascribed to the inhibition of this enzyme system. The present data fail, as did our previous work (15), to implicate the $Na^+ + K^+$ ATPase as the natriuretic receptor for several of the common mercurial diuretic agents.

We have previously reported that the inhibition of $Na^+ + K^+$ ATPase by mercaptomerin increases with decreasing pH and decreases with increasing pH of the medium (15), a situation which correlates with the potentiation of mercurial diuretics in acidosis and the reduction of their effect in alkalosis (1, 3, 9, 10). In current experiments, shown in Table 6-I, the results with mer-

TABLE 6-I
EFFECT OF pH ON % INHIBITION OF TOTAL ATPase ACTIVITY IN VARIOUS DOG RENAL PREPARATIONS BY SOME COMPOUNDS OF MERCURY

Drug	Preparation	Buffer	pH	Hg, M				Ouabain $10^{-4} M$	n
				10^{-3}	10^{-4}	10^{-5}	10^{-6}		
Mercaptomerin	Cortex homogenate	Histidine	6.8	54 ± 2	38 ± 1	12 ± 2		38 ± 2	8
		Histidine	8.0	49 ± 2	26 ± 2	7 ± 2		38 ± 2	8
	Whole kidney homogenate	Histidine	6.8	43 ± 1	34 ± 4	13 ± 3	1 ± 2	42 ± 1	4
		Histidine	8.0	41 ± 2	29 ± 6	10 ± 2	1 ± 1	39 ± 1	4
Mersalyl	Cortex microsomes	Histidine	6.8	74	76	65	25	36	2
		Histidine	8.0	75	74	70	34	32	2
	Cortex homogenate	Imidazole	6.8	48 ± 3	40 ± 1	33 ± 1	5 ± 1	19 ± 2	3
		Imidazole	8.0	56 ± 2	43 ± 2	33 ± 2	3 ± 4	15 ± 3	3
Chlormerodrin	Whole kidney homogenate	Histidine	6.8	74 ± 1	72 ± 1	48 ± 3	7 ± 4	42 ± 1	4
		Histidine	8.0	76 ± 1	68 ± 2	52 ± 1	8 ± 4	39 ± 1	4
	Cortex microsomes	Imidazole	6.8		87 ± 1	68 ± 2	14 ± 4	86 ± 1	3
		Imidazole	7.4		91 ± 1	70 ± 2	18 ± 3	88 ± 1	3
		Imidazole	8.0		87 ± 1	73 ± 1	16 ± 1	85 ± 1	3
PCMPS	Cortex microsomes	Histidine	6.8	89	84	62	15	67	2
		Histidine	8.0	91	80	62	19	67	2
PCMB	Cortex microsomes	Histidine	6.8		77	67	15	43	2
		Histidine	8.0		79	67	18	42	2

captomerin were confirmed. However, unfortunately for the positive correlation, we were unable to extend this observation to two other mercurial diuretic agents, mersalyl and chlormerodrin. In agreement with our previous observation the enzyme inhibition by PCMB was not sensitive to changes in pH which was extended to another nondiuretic mercurial PCMPS. In the experiments of Jones *et al.* there was a failure of the acid pH to enhance the absolute inhibition of the ATPase system by meralluride (6).

If mercurials produce their diuretic effect by inhibition of the $Na^+ + K^+$ ATPase, it is conceivable that certain cellular materials may protect the enzyme *in vivo* against inhibition by the nondiuretic but not by the diuretic mercurial agents. Since mercuric cystiene is a much more potent diuretic than is mercuric chloride (27), it was of interest to determine whether or not cysteine might have a selective influence on the enzyme inhibition by different types of mercurials. Figure 6-1 shows that cysteine prevented the inhibition of a purified $Na^+ + K^+$ ATPase preparation produced by chlormerodrin or PCMBS in a similar manner. At lower mercury concentrations there was some enzyme stimulation by PCMBS but not by chlormerodrin. The significance of this detail is not clear.

In experiments similar to those shown in Figure 6-1 there was no differential effect on ATPase inhibition by chlormerodrin or PCMBS in presence of the following substances in following concentration ranges: Mg^{++} 1.5 to 20 mM; ATP 1.5 to 10 mM; Na^+ 10 to 100 mM; $SO_4^=$ 15 to 100 mM; bovine albumin.

Since there is a delay in diuretic response to diuretic mercurial agents the effects of preincubation of various inhibitors with the ATPase preparation were tested. The results with five diuretic (upper row) and five nondiuretic (middle row) mercurials are shown in Figure 6-2. For structures of these agents see the paper by Weiner *et al.* (28). Chlormerodrin, mersalyl, Win 8518 or mercuric chloride, when present at 10^{-3} and 10^{-4} M concentrations, prevented the increase in ATPase activity observed with increasing time of preincubation. On the other hand, PCMBS, PHMB (except for one point), PHMBS or PCMB did not prevent the increase in the enzyme activity even when present

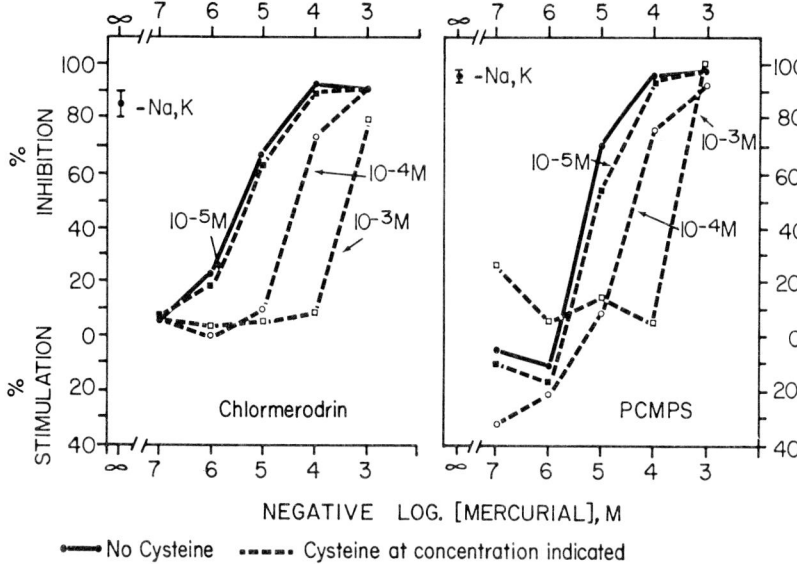

Figure 6–1. Effect of cysteine on inhibition of ATPase activity by chlormerodrin or PCMPS. Percent inhibition or stimulation of total ATPase activity is indicated on the ordinate. Omission of Na$^+$ and K$^+$ from the incubation medium is indicated by −Na, K. The source of enzyme was the NaI-treated microsomal fraction from the dog kidney. The enzyme suspensions were assayed at pH 6.8 with tris buffer.

at these high concentrations. The diuretic, mercaptomerin, and the nondiuretic, PMA, were exceptions to this generalization.

In similar experiments which are not shown, two hour preincubation of homogenates in the presence of chlormerodrin and PCMPS combined at 10^{-4} M concentration, each produced less ATPase inhibition than 10^{-4} M chlormerodrin alone. This observation is of interest since PCMPS antagonizes mercurial diuresis (12). On the other hand, we have previously reported an additive inhibitory effect of mercaptomerin and PCMB without preincubation *in vitro* (15).

In further experiments not shown, the increase in ATPase activity observed with preincubation of the dog cortex homogenates appeared to be located in the lower g fractions (i.e. nuclear and mitochondrial). PCMPS and chlormerodrin produced

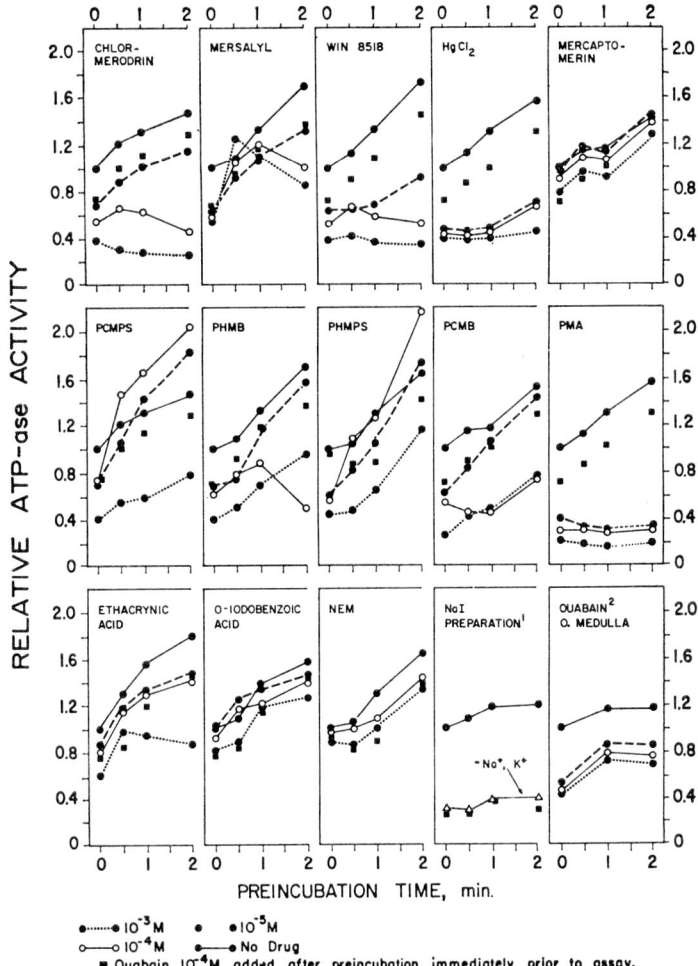

Figure 6-2. Effect of preincubation at 37°C on ATPase activity in the presence or absence of various inhibitors. The source of the enzyme was an homogenate of dog kidney cortex prepared in sucrose—Tris solution. The preincubation medium was composed of the following: 100 μmol NaCl; 20 μmol KCl; 3 μmol MgCl$_2$; 3 μmol histidine; and 40 μmol Tris buffer at pH 7.4 in a total volume of 0.9 ml. At the times shown, 3 μmol ATP was added to make the final volume of 1.0 ml. The Pi released during the ten minute period following addition of ATP represents the ATPase activity. ATPase at all times is compared to the drug activity without preincubation (0 preincubation). PCMPS and PHMPS = p-chloro and p-hydroxymercuriphenyl sulfonic acids. PCMB and PHMB = p-chloro and p-hydroxymercuribenzoic acids. PMA = phenylmercuric acetate. NEM = N-ethyl maleimide. [1]Cortex 80,000 x g fraction treated with NaI, n = 6. [2]Outer medulla whole tissue homogenate. Ouabain was present during the entire preincubation period.

the same degree of inhibition in preincubated and not preincubated homogenates. Thus, the reduced effectiveness of PCMPS as an inhibitor when the homogenates were preincubated with the mercurial may be due to factors such as drug metabolism, redistribution or alteration of the enzyme protein by the drug rather than a change of enzyme induced by the preincubation itself.

Jones et al. (6) reported that in rats pretreated with mercaptomerin or meralluride the renal ATPase was reduced but remained unchanged after pretreatment with PCMB. In our previous experiments we have observed no inhibition of the renal enzyme following injection of chlormerodrin, mercaptomerin, PCMB or PCMPS to dogs (15). To compare these experiments it is important to realize that on the one hand the enzyme preparation used by Jones et al. (6) was not very active and consequently required relatively little dilution before assay. On the other hand, we have used highly active microsomal fractions so that 1 to 4:000 dilutions were required for the enzyme assay. The high dilution might have contributed to the dissociation of the enzyme-inhibitor complex which would account for the lack of observable enzyme inhibition in our previous experiments.

In the following experiments the enzyme was prepared similar to the method of Jones et al. (6) except that we have used a Waring Blendor for initial homogenization of the tissue followed by the use of all glass tissue grinders. Figure 6-3 shows the scheme of enzyme preparation. Jones et al. used plastic tissue grinders only. Again, the dog was the experimental animal in our work since most renal work on the mechanism of action of mercurials was done in this species.

Table 6-II shows that infusion of ouabain into the left renal artery, reduced the ATPase activity in the left kidney. The dose of ouabain used was the largest dose compatible with consistent survival of the preparation, and caused an ipsilateral reduction of Na^+ reabsorption by 30 to 40 percent which was probably the maximal effect for the drug (20). This experiment indicated that the enzyme preparation contained ouabain sensitive ($Na^+ + K^+$ dependent) ATPase. Since no desoxycholate or ethylenedi-

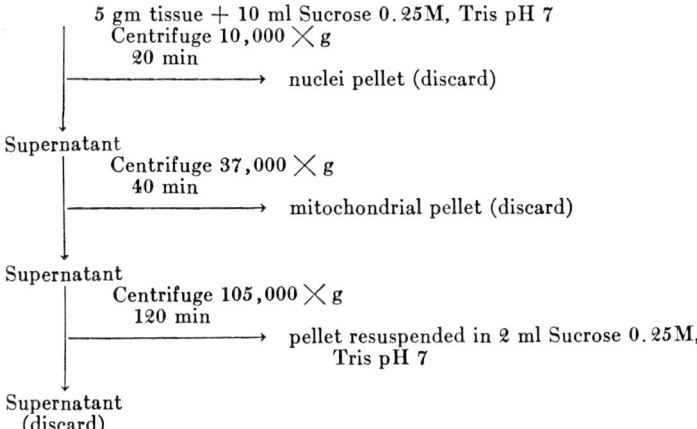

Figure 6–3. Schematic representation of renal tissue fractionation in experiments shown in Tables 6-II and 6-III.

TABLE 6–II
EFFECTS OF PRETREATMENT WITH OUABAIN ON TOTAL ATPase ACTIVITY IN 105,000 × g FRACTIONS FROM CANINE KIDNEYS

Tissue	Total ATPase
	μmol Pi/mg prot/hr
Untreated cortex	27.6 ± 0.5*
Right cortex	26.1 ± 1.4
Left cortex	18.5 ± 2.2
Untreated outer medulla	28.2 ± 0.8†
Right outer medulla	33.8 ± 0.8
Left outer medulla	17.5 ± 0.6

* 22 kidneys.
† 12 kidneys.
Ouabain was infused into the left renal artery at a dose of 1.3μg/kg body weight/min for 1 hour. The kidneys were removed 1 hour after the end of infusion. Values are for 5 dogs ± standard error.

amine tetracetic acid was used in preparation of the enzyme it was not inhibitable by ouabain when added *in vitro*.

According to the results in Table 6-III pretreatment with either diuretic or nondiuretic mercurials reduced the renal ATPase to approximately the same degree. The combined administration of diuretic and nondiuretic mercurials produced an inhibition similar to that caused by single agents. These results are in agreement with our previous experiments *in vitro* (15) but do not correlate with the fact that PCMB or PCMPS prevent or abolish mercurial diuresis (12). It is clear that the present experiments

Action of Mercury on Renal Sodium

TABLE 6-III
EFFECTS OF PRETREATMENT WITH MERCURIALS ON TOTAL ATPase ACTIVITY IN 105,000 × g FRACTIONS FROM CANINE RENAL CORTEX

Drug		Total ATPase	n*
mg/kg of Hg I.V.		μmol Pi/mg prot/hr	
None		27.6 ± 0.8	20
Mercaptomerin	4	15.4 ± 0.5	6
Mersalyl	4	15.6 ± 1.2	6
Chlormerodrin	4	16.9 ± 0.2	6
PCMB	4	15.8 ± 1.2	18
PCMPS	4	19.9 ± 0.5	6
Mersalyl 2 + PCMB 2		15.0 ± 0.6	6
Mercaptomerin + PCMPS 2	2	10.4 ± 0.3	6

* number of kidneys.
The kidneys were removed 1 hour after injection of drugs.

do not necessarily correspond to the situation *in vivo* because the redistribution of mercury might have occurred during enzyme isolation procedures.

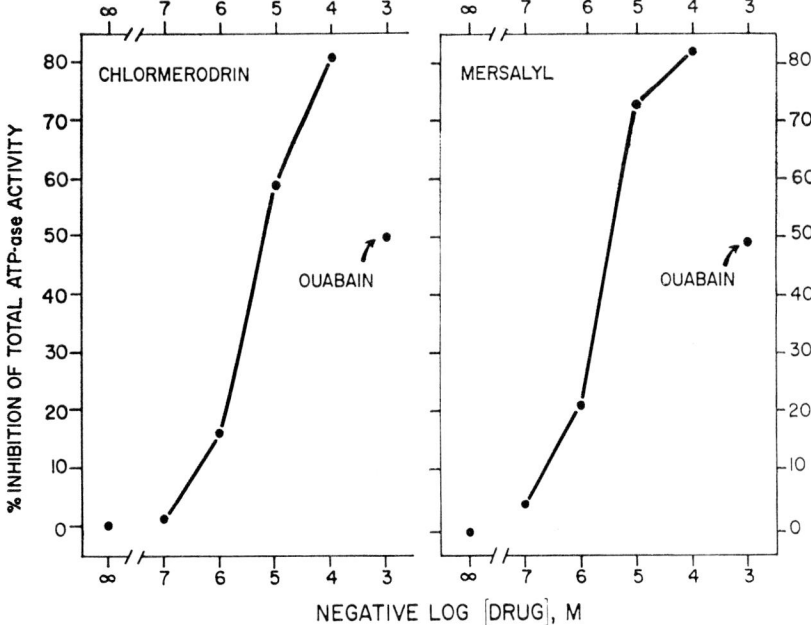

Figure 6-4. Inhibition of microsomal ATPase from human renal cortex by chlormerodrin and mersalyl. The enzyme was prepared as described previously (15). The specific activity was 48μmol Pi/mg protein/hr. The pH of the medium was 7.4.

Characteristics, distribution and sensitivity to ouabain of $Na^+ + K^+$ ATPase isolated from human kidneys are very similar to those for the dog kidney enzyme (17, 18). As could be predicted the human enzyme is highly sensitive to representative mercurial agents (Fig. 6-4), and the inhibition pattern is similar to that observed with canine enzyme preparations (15).

REFERENCES

1. Axelrod, D. R., and Pitts, R. F.: The relationship of plasma pH and anion pattern to mercurial diuresis. *J Clin Invest, 31:*171, 1952.
2. Cafruny, E. J.: The site and mechanism of action of mercurial diuretics. *Pharmacol Rev, 20:*89, 1968.
3. Ethridge, C. B., Myers, D. W., and Fulton, M. N.: Modifying effect of various inorganic salts on the diuretic action of salyrgan. *Arch Intern Med, 57:*714, 1936.
4. Heidenreich, O.: Quecksilberhaltige Diuretica. *Handbook of Exp Pharmacol, 24:*62, 1969.
5. Hendler, E. D., Toretti, J., and Epstein, F. H.: The distribution of sodium-potassium-activated adenosine triphosphatase in medulla and cortex of the kidney. *J Clin Invest, 50:*1329, 1971.
6. Jones, V. D., Lockett, G., and Landon, E. J.: A cellular action of mercurial diuretics. *J Pharmacol Exp Ther, 147:*23, 1965.
7. Katz, A. I., and Epstein, F. H.: Physiologic role of sodium-potassium-activated adenosine triphosphatase in the transport of cations across biological membranes. *N Engl J Med, 278:*253, 1968.
8. Kessler, R. H., Lozano, R., and Pitts, R. F.: Studies on structure diuretic activity relationships of organic compounds of mercury. *J Clin Invest, 36:*656, 1957.
9. Levy, R. I., Weiner, I. M., and Mudge, G. H.: The effects of acid-base balance on the diuresis produced by organic and inorganic mercurials. *J Clin Invest, 37:*1016, 1958.
10. Maren, T.H.: Renal carbonic anhydrase and the pharmacology of sulfonamide inhibitors. *Handbook of Exp Pharmacol, 24:*195, 1969.
11. Martinez-Maldonado, M., Allen, J.C., Eknoyan, G., Suki, W. and Schwartz, A.: Renal concentrating mechanism: Possible role for sodium-potassium activated adenosine triphophatase. *Science, 165:*807, 1969.
12. Miller, T.B. and Farah, A.E.: Inhibition of mercurial diuresis by nondiuretic mercurials. *J Pharmacol Exp Ther, 135:*102, 1962.
13. Miller, T.B. and Farah, A.E.: The mechanism of action of mercurial and thiazide diuretics. *Biochem Clinics, 2:*443, 1963.

14. Nechay, B.R., Larimer, J.L. and Maren, T.H.: Effects of drugs and physiologic alterations on nasal salt excretion in sea gulls. *J Pharmacol Exp Ther, 130:*401, 1960.
15. Nechay, B.R., Palmer, R.F., Chinoy, D.A., and Posey, V.A.: The problem of $Na^+ + K^+$ adenosinetriphosphatase as the receptor for diuretic action of mercurials and ethacrynic acid. *J Pharmacol Exp Ther, 157:*599, 1967.
16. Nechay, B.R. and Nelson, J.A.: Renal ouabain-sensitive ATP-ase activity and Na^+ reabsorption. *J Pharmacol Exp Ther, 175:*717, 1970.
17. Nechay, B.R., Sarles, H. E., Remmers, A.R. Jr., Fish, J.C., Beathard, G.A., Nelson, J.A., Lindley, J.D., Lerman, M.J., and Contreras, R.R.: Human renal ATP-ase: A preliminary report on activity and characteristics. *Am Soc Nephrology,* Abstracts, 1970, p. 58.
18. Nechay, B.R., Sarles, H.E., Remmers, A. R. Jr., Fish, J.C., Beathard, G.A., Nelson, J.A., Lindley, J.D., Lerman, M.J., and Contreras, R.R.: Characteristics and distribution of $Na^+ + K^+$ ATP-ase in human kidney. *Fed Proc, 30:*332, 1971.
19. Nelson, J.A.: The role of $Na^+ + K^+$ ATP-ase in the renal reabsorption Na^+. Ph.D dissertation, The University of Texas Medical Branch, Galveston, Texas, 1970.
20. Nelson, J.A. and Nechay, B.R.: Effect of cardiac glycosides on renal adenosine triphosphatase activity and Na^+ reabsorption in dogs. *J Pharmacol Exp Ther, 175:*727, 1970.
21. Nelson, J.A. and Nechay, B.R.: Interaction of ouabain and K^+ *in vivo* with respect to renal adenosine triphosphatase activity and Na^+ reabsorption. *J Pharmacol Exp Ther, 176:*558, 1971.
22. Pendleton, R.G., Sullivan, L.P. and Tucker, J.M.: The effect of chlomerodrin on the isolated toad bladder. *J Pharmacol. Exp Ther, 164:*362, 1968.
23. Pitts, R.F.: *The Physiological Basis of Diuretic Therapy.* Springfield, Thomas, 1959.
24. Skou, J.C.: Enzymatic basis for active transport of Na^+ and K^+ across cell membrane. *Physiol Rev, 45:*596, 1965.
25. deStevens, G.: *Diuretics, Chemistry and Pharmacology.* New York, Academic Press, 1963.
26. Taylor, C.B.: The effect of mercurial diuretics on adenosinetriphosphatase of rabbit kidney *in vitro. Biochem Pharmacol, 12:*539, 1963.
27. Weiner, I. and Farah, A.: Pharmacology of mercurial diuretics, Proceedings 3rd International Pharmacological Meeting, New York, Pergamon Press, 8:15, 1968.
28. Weiner, I.M., Levy, R.I. and Mudge, G.H.: Studies on mercurial diuresis: Renal excretion, acid stability and structure-activity relationship of organic mercurials. *J Pharmacol Exp Ther, 138:*96, 1962.

DISCUSSION

Rothstein: Your system can't distinguish different mercurials and the same is true of the red cell ATPase or any other ATPase. With no distinction between diuretic and nondiuretic materials at the enzyme level, don't you almost have to fall back on the idea that the nondiuretic mercurials simply don't reach the enzyme in the *in vivo* situation?

Nechay: This could very well be. The nondiuretic mercurials like PCMB are known to be secreted by the renal tubule; they probably get very well to at least proximal tubules, the site of secretion of organic acids. It is possible they do not get this well into the distal tubule as the diuretic mercurials do.

Clarkson: Weren't some of the organomercurials given *in vivo*?

Nechay: Yes, in some experiments.

Clarkson: Was the same inhibition observed with a diuretic mercurial given *in vivo*?

Nechay: Yes, but it is still possible that PCMB is sitting somewhere close to the enzyme and when you start processing the kidney it will redistribute or attach to the enzyme.

Rothstein: I'll give a different answer to that; every cell has ATPase and in grinding up the whole kidney you are looking at the mixed average of all the ATPases of all the cells that you grind up. It's a little difficult to get around the fact that it's only a small population of those ATPases in particular cells that are involved in the reabsorption process. There's no way of avoiding this problem because you cannot isolate the particular cells that are doing the reabsorbing and look at the particular ATPases.

Nechay: I thought of this too but the point is that the bulk of the work of the kidney is sodium reabsorption.

Rothstein: I didn't say that. The gospel says . . .

Nechay: The gospel says that it is but I'm not convinced that it is. The bulk of the sodium and potassium ATPase should be functional in the reabsorption of sodium because there doesn't seem to be all that much excess of it in the kidney provided the enzyme is functional in the reabsorption of sodium.

Foulkes: The work especially of Dr. Barger's[*] group at Harvard Medical School shows that one of the important determinants of sodium excretion may in many cases be the intrarenal blood flow distribution. One can alter this distribution quite considerably.

Nechay: I am quite aware of the fact that there are many ways that you can influence the excretion of sodium by the kidney and obviously blood flow is one of them. Metabolic manipulation of the renal cell is one of the major ways that can influence the reabsorption of sodium. Mercury, of course, can have vasconstricting effects in the kidney. However, mercury

---

[*] Barger: *N Engl J Med*, 284:482, 1971.

compounds reduce the reabsorption of filtrate in single tubules, an effect independent of blood flow.

There is another possibility that I haven't mentioned. Mercurials may affect the passsive permeability of the peritibular membrane to make it more diffcult for the sodium to be reabsorbed.

Gage: Has a histochemical method been used to study the inhibition of ATPase in the tubular membrane?

Nechay: Yes, but there is a lot to be desired in histochemical techniques.

Chapter 7

EFFECT OF ORGANIC MERCURIAL COMPOUNDS ON RENAL ORGANIC ION TRANSPORT

JERRY B. HOOK AND GERALD H. HIRSCH

ABSTRACT

A great deal of controversy has arisen concerning the effect of the organic mercurial diuretics on renal organic anion transport. In certain species these compounds depress active secretion of p-aminohippurate (PAH). Yet only the diuretics that are organic anions are capable of being actively transported and thereby competitively inhibiting PAH transport. It has been suggested that suppression of PAH transport by mercurial diuretics might be related to a general depression of cellular metabolism by these compounds rather than a specific effect on anion transport. This hypothesis was tested by determining the effect of several mercurial compounds on the simultaneous accumulation of PAH and the organic cation, N-methylnicotinamide (NMN) by an *in vitro* slice technique. Meralluride, mercaptomerin, chlormerodrin and p-chlormercuribenzoate were added to beakers containing renal cortical slices from rats or dogs. After incubating under oxygen for 90 minutes at 25°C both tissue and medium were analyzed for PAH and NMN. The results demonstrate that diuretic organic mercurials inhibit PAH and NMN accumulation by renal cortical slices to the same degree. Similar data were obtained in tissue from dogs and rats. The inhibitory effect of chlormerodrin on PAH accumultaion was not altered by changes in medium pH. During short periods of incubation the uptake of PAH into rat renal cortical slices was inhibited by chlormerodrin and mercaptomerin in a noncompetitive fashion. It is concluded that the inhibitory effect of organic mercurial compounds on PAH transport is nonspecific, reflecting general depression of metabolic activity by these compounds.

Supported by U.S. Public Health Service Grant AM–10913.

REPORT

Introduction

In addition to their well-documented effect on renal transport inorganic ions the mercurial diuretics also have pronounced effects on other renal functions. One of the systems affected by these drugs is responsible for the active secretion of organic anions by the cells of the proximal tubule. The prototype and most widely studied compound handled by this active secretory mechanism is para-aminohippurate (PAH). A considerable literature has developed concerning the effect of organic mercurial compounds on this system and yet the data are not definitive. Certain organic mercurials, such as mersalyl, inhibit renal transport of PAH in man (1, 3) and comparable results have been seen in rats (10). However, this inhibition is not so readily seen in the dog (1, 11). It has been suggested that mercurial diuretics are capable of being transported and are, therefore, acting as competitive inhibitors of PAH transport (3). Certain of these compounds are organic acids and data are available suggesting that they are capable of being secreted (7, 8, 19). However, chlormerodrin inhibits PAH transport, yet this compound is not an organic acid (20). Cafruny and Gussin (6) and Cafruny et al. (5) demonstrated that though chlormerodrin accumulates in renal tissue, the accumulation does not appear to be via an active transport mechanism. Thus, it is not clear whether the inhibitory effect of mercurial diuretics on PAH transport is truly a competitive phenomenon or merely a reflection of a more generalized metabolic alteration in proximal tubular function. Certain mercurials inhibit glucose transport as well as the secretion of organic anions (18, 21). This led Cafruny (4) to suggest that depression of PAH transport might reflect a general metabolic depression rather than a specific effect of mercurial compounds on anion transport. The purpose of our investigation was to test this hypothesis.

In the cells of the proximal tubule there are two active secretory mechanisms for organic ions: one for the transport of anions, such as PAH, and a separate, but parallel system, responsible for the transport of organic cations (15). In our laboratory we have taken advantage of the specificity of the two systems to study the effects of certain manipulations on the transport of organic anions. For instance, the addition of uremic serum from nephrectomized animals to an *in vitro* slice preparation will inhibit the transport of PAH, but not that of the organic base N-methylnicotinamide (13). Others have shown that this inhibition is due to the presence of organic acids in the serum (2). Secondly, substrate stimulation of PAH transport has no effect on simultaneously measured NMN transport (12). Thus, selectivity of effect can be demonstrated with this preparation. It was of interest to use this slice technique to determine the specificity of the inhibitory effect of mercurial diuretics on anion transport. The working hypothesis was that a mercurial compound that specifically inhibited PAH transport would have no effect on the transport of NMN.

The effect of four different compounds on the *in vitro* accumulation of PAH and NMN was determined. We utilized the anionic mercurials meralluride and mercaptomerin, the nonionic mercurial diuretic chlormerodrin, and the nondiuretic mercurial *p*-chlormercuribenzoate. Inasmuch as there appears to be a marked species' difference in the effect of mercurials on PAH transport the effect of these compounds *in vitro* was determined in tissue from two different species, dogs and rats.

Methods

Dogs were anesthetized with pentobarbital sodium and rats were stunned by a blow on the head. Kidneys were quickly removed and placed in ice-cold saline (0.9 percent NaCl). Renal cortical slices were prepared freehand and briefly kept in cold saline until incubated. Slices were incubated in 2.7 ml of the phosphate buffer as devised by Cross and Taggart (9), which contained 7.4×10^{-5}M PAH and 6.0×10^{-6}M (2.5×10^{-2} μC/ml) NMN-^{14}C. Duplicated incubations were carried out in a Dubnoff apparatus at 25°C under a gas phase of 100 percent oxygen at pH 7.4. The duration of incubation was 90 minutes except in those instances where the rate of uptake was being determined when incubation times were two and 12 minutes. Following incubation the slices were quickly removed from the beakers, blotted and weighed. Both the tissue and a 2 ml aliquot of medium were treated as outlined by Cross and Taggart (9) and Hook and Munro (13). PAH was estimated by the method of Smith *et al.* (17), while 1.0 ml of slice and media homogenate was added to 10 ml of modified Bray's solution (2.5 gm 2,5-diphenyloxazole and 100 gm napthalene per 1 of dioxane) and the amount of ^{14}C NMN was counted in a Beckman LS-100 liquid scintillation counter. Results were expressed as slice/medium (S/M) ratio where S equals mg/gm or DPM/gm tissue and M equals mg/ml or DPM/ml medium.

The effect of mercurial compounds on PAH and NMN accumulation by renal cortical slices when added directly to the incubation beakers was determined at final concentrations ranging from 10^{-5} to 10^{-3}M. In a single experiment slices from both kidneys of a dog or the kidneys of four rats were pooled. Slices were then distributed (100 to 200 mg) to beakers containing various concentrations of the mercurial compound and to drug-free controls. All incubations were conducted in duplicate and each experiment was replicated five times.

The effect of chlormerodrin and mercaptomerin was determined on the rate of uptake of PAH into rat kidney slices. Kidneys from five rats were pooled, slices were prepared and equally divided into two groups of 16 beakers. Duplicate incubations were conducted at four PAH concentrations (1, 2, 4 and 8 × 10^{-4}M). The uptake of PAH at each concentration was determined after two and 12 min of incubation. The difference in uptake between two and 12 min divided by ten was used as the minute rate of uptake of PAH at that concentration. The rate of uptake at each concentration was determined and the data were plotted on a double reciprocal plot. Similar experiments were then conducted in the presence of 5×10^{-4}M chlormerodrin and mercaptomerin.

All data are reported as the mean ± standard error. Statistical analyses employed are described by Lewis (14).

Results

When slices of rat renal cortex were incubated for 90 minutes under oxygen in a dilute solution of PAH and NMN, the tissue actively accumulated both ions. After incubation the S/M ratio for PAH in rat kidney slices was 11.8±1.1 and in the same slices the ratio for NMN was 6.2±0.4. The ratios for dog cortical slices incubated similarly were 5.9±0.2 and 3.8±0.2, respectively (Fig. 7–1). The organic mercurial diuretic meralluride produced a dose-dependent inhibition of ion accumulation in tissue from both species. As can be seen in Figure 7–1, meralluride significantly reduced PAH and NMN S/M ratios in tissue from both species at 10^{-4}M. A concentration of 10^{-3}M effectively inhibited

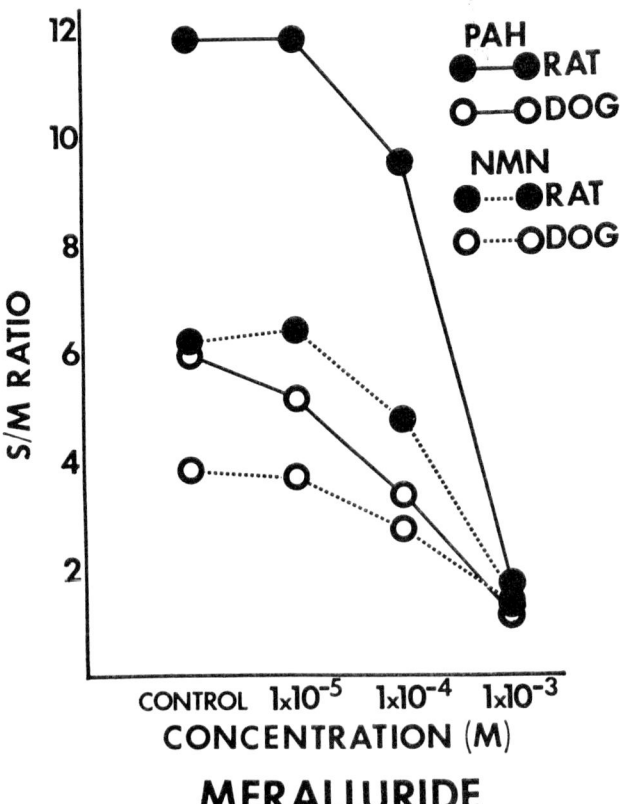

Figure 7–1. Effect of meralluride on accumulation (S/M ratio) of PAH and NMN by renal cortical slices from dog and rat. Points represent means ± S.E. from five separate experiments. When no vertical lines are shown the variation is within the radius of the circle. All incubations were 90 minutes in duration.

Effect of Organic Mercurial Compounds 129

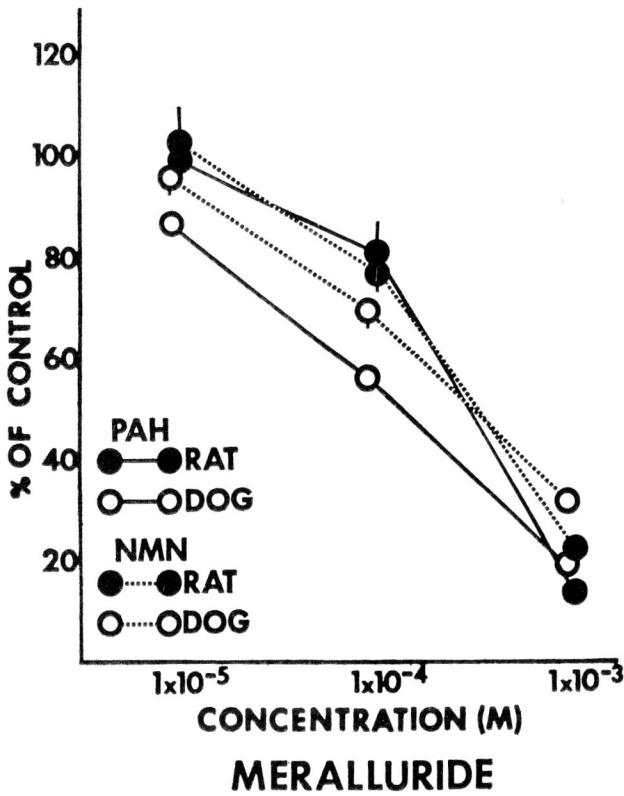

Figure 7–2. Effect of meralluride on accumulation (S/M ratio) of PAH and NMN by renal cortical slices from dog and rat. All data are expressed as percent of control (S/M ratio with no drug present). Points represent means ± S.E. of five separate experiments. When no vertical lines are shown the variation is within the radius of the circle. These data are the same as those shown in Figure 7–1.

accumulation of both compounds. The dose-dependent nature of these responses is better illustrated in Figure 7–2 where it appears that the system most sensitive to meralluride was that for PAH in dog tissue. Similarly, tissue accumulation of both PAH and NMN was inhibited by the addition of mercaptomerin to the incubating medium (Fig. 7–3). The data suggest that rat tissue was more susceptible to the drug in that 50 percent inhibition of both PAH and NMN S/M was observed at approximately

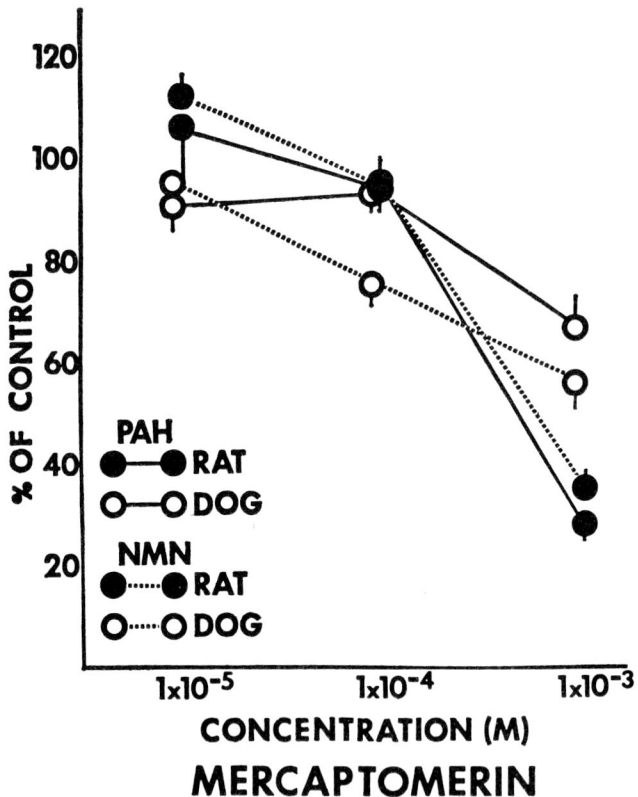

Figure 7-3. Effect of mercaptomerin on accumulation of PAH and NMN by renal cortical slices from dog and rat. Data are expressed as percent of control in five separate experiments. Points represent means ± S.E.

5×10^{-4}M in rat tissue whereas in the dog there was still considerable accumulation of both ions at 10^{-3}M. Mercaptomerin appeared to have a greater effect on NMN than on PAH in dog tissue (Fig. 7-3). The non-ionic mercurial diuretic, chlormerodrin, significantly inhibited NMN accumulation at all concentrations employed (Fig. 7-4). Approximately 50 percent inhibition of NMN S/M was seen at 8×10^{-5}M and 50 percent inhibition of PAH was seen at approximately 5×10^{-4}M. The effect of chlormerodrin does not appear to be species' specific in that tissues from the dog and rat were affected in a similar manner. A nondiuretic compound, p-chlormercuribenzoate

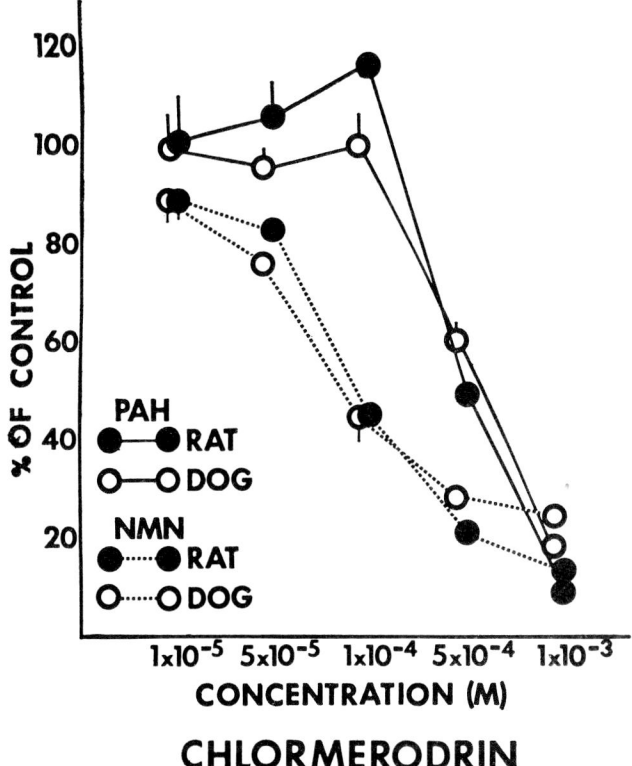

Figure 7–4. Effect of chlormerodrin on accumulation of PAH and NMN by renal cortical slices from dog and rat. Data are expressed as percent of control in five separate experiments. Points represent means ± S.E.

(PCMB) inhibited accumulation of organic ions only at the highest concentration employed (10^{-3}M). NMN S/M ratio in rat tissue appeared to be enhanced at the highest concentration of PCMB (Fig. 7–5).

Tissues from control rats were incubated at varying concentrations of PAH for two and 12 minutes. The difference in uptake was then used as an index of the rate of transport into the tissue slice. When these data were plotted on a double-reciprocal plot, a straight-line function was produced (Fig. 7–6). Comparable uptake studies were done in the presence of 5×10^{-4}M chlormerodrin and mercaptomerin. The uptake of PAH was inhibited by both mercurial compounds. Chlormerodrin and

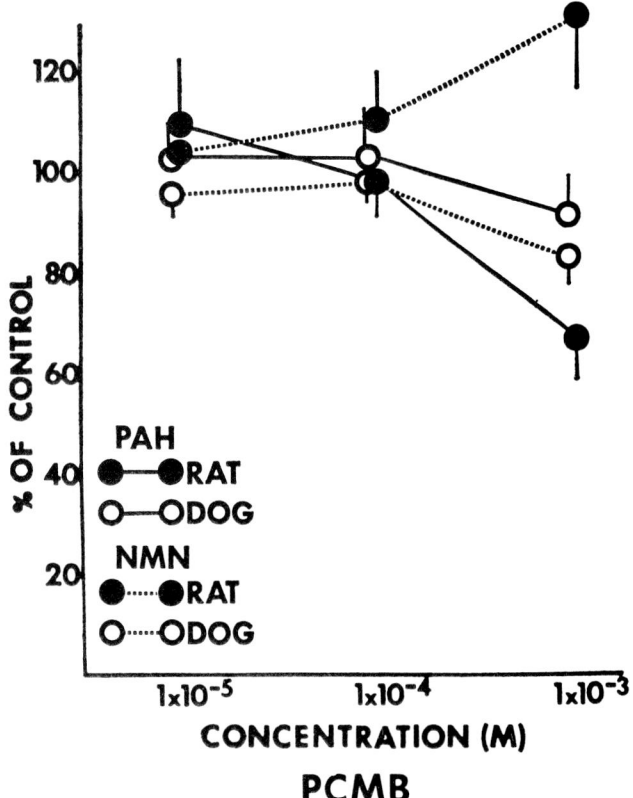

Figure 7-5. Effect of para-chlormercuribenzoate (PCMB) on accumulation of PAH and NMN by renal cortical slices from dog and rat. Data are expressed as percent of control in five separate experiments. Points represent means ± S.E.

mercaptomerin reduced the maximal uptake rate of PAH as evidenced by the elevation in the y-intercept on the graph. Furthermore, mercaptomerin also appeared to alter the x-intercept.

In a last series of experiments the effect of reducing medium pH on PAH S/M ratio in rat renal cortical slices was determined. Although PAH S/M in control beakers was slightly decreased (from 13.8 to 11.8) the inhibiting action of 5×10^{-4}M chlormerodrin remained unaltered at pH 7.0 (Table 7-I).

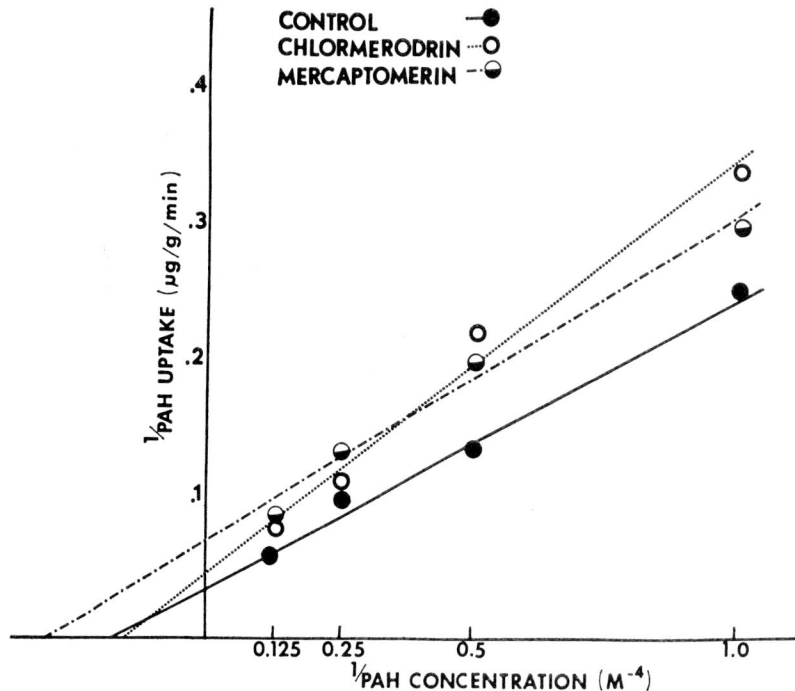

Figure 7-6. Effect of chlormerodrin and mercaptomerin on PAH uptake between two and 12 minutes of incubation. Points represent means of five separate experiments. Both drugs were in a concentration of 5×10^{-4}M.

TABLE 7-I
EFFECT OF MEDIUM pH ON CHLORMERODRIN INHIBITION OF PAH ACCUMULATION (S/M) IN RAT RENAL CORTICAL SLICES

	Control	Chlormerodrin	% Change
pH 7.4	13.8	6.9*	51.2
	(0.8)	(0.8)	(7.9)
pH 7.0	11.8	5.9*	50.4
	(0.9)	(0.5)	(2.7)

Renal cortical slices from eight pooled rat kidneys were equally distributed into four groups; two at pH 7.4 and two at pH 7.0, two with no drug, and two containing chlormerodrin in a final concentration of 5×10^{-4}M. Incubations (90 min) were run in duplicate and the values averaged. Values in the table are means ($\pm$ S.E.) of five replications.
* Significantly different (p < .05) from nondrug control.

DISCUSSION

The data reported in this paper demonstrate the indiscriminate action of organic mercurial diuretics on active transport of PAH.

The observation that two acidic mercurials inhibit accumulation of the organic base NMN to the same degree as that of PAH in tissue from two different species support this position. Furthermore, the observation that the nonacidic mercurial, chlormerodrin, reduced PAH accumulation and had an even more striking effect on NMN also substantiates the argument. The inhibitory potency of all three diuretics was approximately the same (50 percent inhibition near $5 \times 10^{-4}M$). The nondiuretic compound PCMB had very little effect at $10^{-3}M$ (which approached the limit of solubility of the compound). However, inhibition of PAH and NMN accumulation are not restricted to diuretic compounds. Mercuric chloride and methylmercury produce effects similar to those seen with the diuretics. Both produced 50 percent inhibition of PAH and NMN accumulation in rat renal cortical slices at approximately $10^{-4}M$. (G. Hirsch, unpublished observations).

The natriuretic effect of organic mercurial diuretics is greatly potentiated in acidosis (4, 20). This has been attributed to enhanced rupture of the carbon-mercury bond, releasing free mercuric ion (4, 20). Altering the pH of the medium (and presumably increasing the propensity to ionize) did not alter the effect of chlormerodrin on PAH accumulation (Table 7-I). This could indicate that mercuric ion is not responsible for inhibition of organic ion transport. Alternatively, it is possible that mercury enters the cells as part of the parent compound and, if necessary, carbon-mercury cleavage occurs intracellularly. Effective buffers could maintain intracellular pH resulting in no change in availability of mercuric ion when pH of the medium was reduced. The carbon-mercury bond of PCMB is resistant to rupture *in vitro* and this stability has been used to support the concept of the mercuric ion as the active form of the diuretics (4, 20). If one accepts this reasoning, the minimal effect of PCMB on organic ion accumulation favors the view that this effect also is due to mercuric ion.

Inasmuch as the S/M ratio for PAH and NMN is measured in a steady-state, it was considered possible that the effect of the mercurials was on retention of ions rather than directly on the transport process. Ross and Farah (16) demonstrated that the rate of accumulation of PAH during very early times of incubation can adequately reflect the rate of transport of the material. This is possible, for in these early periods of time the intracellularly accumulated PAH is not of sufficient concentration to alter the net flux. Therefore, the effect of mercurials was determined during short periods of incubation to determine the effect of these compounds on the uptake process. Both mercurials, the organic acid mercaptomerin and the nonacidic chlormerodrin, reduced the rate of PAH entry into rat renal cortical slices (Fig. 7-6). Granted, the extrapolation of PAH uptake into a renal slice is a long distance from pure enzyme kinetics. Mercurials have a high affinity for protein and tenaciously bind to tissue (20). It is possible that the mercurials (or mercuric ion) indiscriminately bind to intracellular proteins resulting in the inactivation of many enzymatic reactions. These changes would be nonreversible in terms of classical enzyme kinetics, thus emphasizing the fruitlessness of a rigorous kinetic analysis of these data. Both diuretics produced an inhibition of PAH uptake that was not in any way similar to classical competitive antagonism in that the y-intercept was elevated (Fig. 7-6). Thus, these data do not support the idea that organic mercurial compounds competitively interfere with PAH transport. Changes such as those seen would be consistent with a general depression of metabolism or interference with some energy process required for active transport rather than competition for transport sites.

REFERENCES

1. Berliner, R.W., Kennedy, T.J., Jr. and Hilton, J.G.: Salyrgan and renal tubular secretion of para-aminohippurate in the dog and man. *Amer J Physiol,* 154:537, 1948.
2. Bourke, E., Preuss, H.G., Rose, E., Weksler, M.E. and Schreiner, G.E.: Effects of neomycin in impaired PAH uptake by renal tubules incubated with uremic serum. *Fed Proc,* 26:265, 1967.
3. Brun, C., Hilden, T. and Raaschow, F.: On the effects of mersalyl on the renal function. *Acta Pharmacol Toxicol,* 3:1, 1947.
4. Cafruny, E.J.: The site and mechanism of action of mercurial diuretics. *Pharmacol Rev,* 20:89, 1968.
5. Cafruny, E.J., Cho, K.C. and Gussin, R.Z.: The pharmacology of mercurial diuretics. *Ann NY Acad Sci,* 139:362, 1966.

6. Catruny, E.J. and Gussin, R.Z.: Renal tubular excretion of mercurials in the aglomerular fish, *Lophius americanus*. *J Pharmcol Exp Ther*, *155*:181, 1967.
7. Campbell, D.E.S.: The excretion of mercaptomerin and its diuretic effect modified by bromcresol green and probenecid. *Acta Pharmacol Toxicol*, *16*:151, 1959.
8. Campbell, D.E.S.: Modification by bromcresol green or probenecid of the excretion and diuretic effect of three mercurial diuretics, diurgin®, ch'ormerodrin and mercumatilin. *Acta Pharmacol Toxicol*, *17*:213, 1960.
9. Cross, R.J. and Taggart, J.V.: Renal tubular trasnport: accumulation of p-aminohippurate by rabbit kidney slices. *Am J Physiol*, *161*: 181, 1950.
10. Dicker, S.E.E The action of mersalyl, calomel and theophylline sodium acetate on the kidney of the rat, *Br J Pharmacol Chemotherap*, *1*:194, 1946.
11. Handley, C.A., Telford, J. and LaForge, M.: Xanthine and mercurial diuretics and renal tubular transport of glucose and p-aminohippurate in the dog. *Proc Soc Exp Biol Med*, *71*:187, 1949.
12. Hirsch, G.H. and Hook, J.B.: Maturation of renal organic acid transport: substrate stimulation by penicillin and p-aminohippurate (PAH). *J Pharmacol Exp Ther*, *171*:103, 1970.
13. Hook, J.B. and Munro, J.R. Specificity of the inhibitory effect of "uremic" serum on p-aminohippurate transport. *Proc Soc Exp Biol Med*, *127*:289, 1968.
14. Lewis, A.E.: *Biostatistics*. New York, Reinhold Publishing Corporation, 1966.
15. Pitts, R.F.: *Physiology of the Kidney and Body Fluids*. Chicago, Year Book Medical Publishers, Inc., 1963.
16. Ross, C.R. and Farah, A.: p-Aminohippurate and N-methylnicotinamide transport in dog renal slices—an evaluation of the counter-transport hypothesis. *J Pharmacol Exp Ther*, *151*:159, 1966.
17. Smith, H.W., Finkelstein, N. Aliminosa, L., Crawford, B. and Graber, M.: The renal clearance of substituted hippuric acid derivatives and other amino acids in dog and man. *J Clin Invest*, *24*:388, 1945.
18. Vander, A.J.: Effects of zinc, cadmium, and mercury on renal transport systems. *Am J Physiol*, *204*:781, 1963.
19. Weiner, I.M., Burnet, A.E. and Rennick, B.R.: The renal tubular secretion of mersalyl (salyrgan) in the chicken. *J Pharmacol Exp Ther*, *118*:1470, 1956.
20. Weiner, I. and Farah, A.: Pharmacology of mercurial diuretics. Proc. Third Intern. Pharmocol. Congress, July, 1966.
21. Weston, R.E., Grossman, J, Edelman, I.S., Escher, D., Leiter, L, and Hellman, L.: Renal tubular action of diuretics, II, effects of mercurial diuresis on glucose reabsorption. *Fed Proc*, *8*:164, 1949.

DISCUSSION

Vostal: What is the ratio between the amount of mercury in the tissue and in the solution? High concentrations of mercury in the incubation media *in vitro* must mean that the tissue concentration of mercury is very high.

Hook: Yes, it probably is.

Vostal: Do you think that we can correlate your results quantitatively with the effects observed *in vivo* with much lower concentrations?

Hook: Yes. Frequently the concentration of a drug required to produce an effect *in vitro* is greater than that *in vivo* following systemic administration. This is related to a host of factors such as nutrition, oxygen delivery to the tissues, etc. Since depression of PAH transport also occurs *in vivo* and the effects we observed *in vitro* are related to concentration of drug we feel our data can justifiably be correlated with effects seen *in vivo*.

Vostal: True, but we don't see such a deep depression of overall metabolism or general functions of the tubule *in vivo* when we inject organomercurial diuretics in therapeutic doses.

Hook: Actually the data from intact animals do suggest that general depression of metabolism occurs *in vivo*. As I mentioned, mercurials have been shown to interfere with transport of glucose and amino acids as well as PAH and inorganic ions.

Magos: Was there any protein loss from the slice during incubation and if so, did you measure it?

Hook: There is usually a small amount of protein loss during incubation. In this particular study we didn't measure it. We have in others; it didn't interfere with our assay.

Foulkes: Are you certain that you are dealing only with inhibition? On a wet weight basis, swelling of the tissue in the presence of a mercurial could mislead one into believing he is dealing with inhibition.

Hook: Marked swelling of the tissue could reflect a general depression of cellular metabolism as we have suggested to be occurring. However, no gross changes in slice size were noted during these experiments.

Rothstein: Was there some potassium leakage in your supernates? You may be at a level where you get a very leaky membrane and potassium leaks out; that could explain a whole array of effects. You wouldn't have to invoke some special metabolic effects.

Hook: If a lot of potassium is lost a metabolic effect results.

Fassett: How good is the evidence that the mercury-carbon bond ruptures? This is a very fundamental question I would think with regard to the toxicity of methylmercury.

Hook: The question of rupture of the carbon mercury bond is fundamental as to how the diuretics block sodium transport. In his review, Cafruny* summarized the development of the "intact molecule" and "mercuric ion" hypotheses.

* Cafruny: *Pharmacol Rev.* 20:89, 1968.

Fassett: What is the current knowledge about the excretion of the organic bases, which are important in so many drugs?

Hook: Many organic bases are actively transported by a system similar to, but distinct from, that which transports organic acids. NMN is used as an example of the actively transported bases. Many drugs, such as the ganglionic blocking agents, curare, phentolamine, and others are organic bases and are transported by this system. In the acid urine these exist as cations and thus are retained in solution in the urine. The combination of active secretion and high water solubility lead to rapid removal of these compounds from the plasma.

Chapter 8

MERCURIC ION, ORGANOMERCURIALS AND RENAL DIURESIS

Jaroslav J. Vostal and Thomas W. Clarkson

ABSTRACT

Organomercurial diuretics, although generally displaced by newer, nonmercurial, oral diuretics, constitute still a useful model for the study of general mechanisms for the biotransformation and turnover of all organomercurial compounds. In their specific action on sodium transport in the kidney they offer, besides the possibility of elucidating the mechanisms of mercurial diuresis, a scientific tool for studying the conditions responsible for the release of mercury from its organometallic bonds and for explaining different toxicities of various organomercurials with respect to the type of intramolecular mercury linkage.

Stability *in vitro* and *in vivo*, biotransformation in the experimental animal, and renal effects of three typical representatives of diuretic and nondiuretic mercurials, mersalyl, chlormerodrin and p-mercuribenzoate, labeled with radioactive mercury (^{203}Hg) were studied in acute experiments on dogs. The results clearly demonstrate that all studied representatives of organomercurials deposit inorganic mercury in the renal tissue when injected *in vivo* and release mercuric ion *in vitro* under physiological conditions, without close relationship to their diuretic effects. Monothiol compounds, penicillamine and N-acetylpenicillamine, although structurally similar to cysteine and glutathione which are well known for their potentiating effects on mercurial diuresis, were shown to counteract the diuretic effects of mercuric ions or organomercurials.

The results are discussed in the terms of existing hypotheses on the mechanism of action of organomercurials.

This paper is based on work performed under contract with the U.S. Atomic Energy Commission at The University of Rochester Atomic Energy Project, was partially supported by U.S.P.H.S. grant GM15190 and has been assigned Report No. UR–49–1413.

REPORT

Organomercurial diuretics have been nearly displaced in the modern therapy of fluid and solute retention by newer, nonmercurial, oral diuretics. However, organomercurial diuretics still constitute an important and challenging chapter in today's medicine as a typical illustration of how an effective drug can stimulate a rapid development of other types of clinically successful compounds. Furthermore, organomercurial compounds with their organometallic bond and clearly defined target (sodium transport in the body) are a useful tool for studying the conditions generally responsible for the release of mercury from its organometallic bond and for studying general biological actions of organometallic compounds.

The first use of mercury in the treatment of edema is usually ascribed to Paracelsus and beneficial effects of mercury compounds on urine flow have been since that time rediscovered by several European medical schools (9). The last rediscovery occurred in 1920 when organomercurial compounds replaced calomel and other inorganic mercury salts in diuretic therapy and started a new era of scientific screening for new and more effective mercurial diuretics (16). Despite the fact that organomercurials dominated the treatment of sodium and fluid retention for more than 20 years and were exposed to intensive clinical and experimental research efforts for more than 40 years their mechanisms of action remained unsolved.

The obvious similarity between the effects of new organomercurial diuretics and previously well-known actions of ionizable inorganic mercury was early emphasized by clinical observations that mercurials have to produce much higher levels of mercury in the urine before diuretic effects comparable to those of inorganic mercury can be observed. This led one group of investigators to believe that organomercurial compounds do not act directly but through dissociation of small amounts of ionized inorganic mercury (15). This theory was seriously challenged by Kessler's *et al.* studies (11) on the structure-activity relationship of mercurial compounds. Their conclusions suggested that there are at least three *structural* requirements for a mercurial compound to produce diuretic properties: first, a terminal mercury-carbon linkage; second, a free mercury valence; and third, a hydrophyllic substituent three carbons removed from mercury. The investigators proposed a critical lock-and-key relationship between this type of compound and two of the theoretic receptor sites for diuresis in the tubular cell (Fig. 8–1A).

Practically all of the clinically useful mercurials possess this steric configuration and the entire group of compounds can be characterized by general formula of a substituted mercuri-propyl compound having the following basic structure:

$$R-CH_2-CH(OY)-CH_2-Hg-X$$

where X denotes a dissociable substituent (usually halogen or theophylline) having no effect on diuretic potency but influencing the solubility or toxicity of the compound. The nature of the OY group is determined by the solvent in which mercuration of the organic moiety is performed (usually methoxy- or ethoxy-group when the solvent is an alcohol or hydroxyl group if the solvent is water) and generally considered to be without any effect on diuretic potency or toxicity. However, the presence and character of this substituent might have obvious influence on the ease of splitting the adjacent mercury-carbon linkage and consequently on the rate of biotransformation of the compound in the living organism. Finally, R usually denotes a group of a

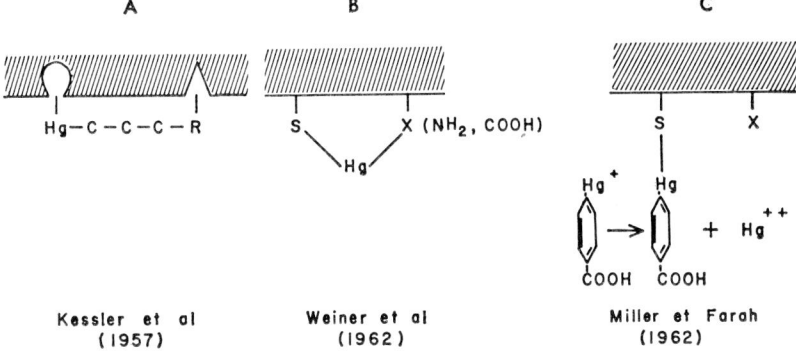

Figure 8–1. Presumed renal receptor for mercurial diuretics and theoretical mechanism of blocking effects of p-mercuribenzoate.

complex nature varying from such a simple structure as urea to substituted aromatic rings and aromatic acids.

Changes in the acid-base balance have been recognized for many years as one of the most important factors influencing the magnitude of the diuretic response to organic mercurials. This fact was explained only in part by the structural theory. Consequently, later on the old-fashioned mercuric ion hypothesis received new support from the studies published by Weiner et al. (20). These authors were able to distinguish the diuretic from the nondiuretic material in more than 30 compounds on the basis of two *functional* prerequisites: first, the compound *in vitro* should release inorganic mercury in the presence of a thiol group at pH 4.0; and second, after intravenous injection into dogs, it should be rapidly excreted in the urine. Weiner et al. proposed furthermore that the *in vivo* splitting of the carbon-mercury linkage occurs in close proximity to the specific receptor consisting of two sites and that both valences of the mercuric ion are bound to these groups; at least one of them is assumed to be a sulfhydryl group. The nature of the other ligand is unknown; it could be a free carboxyl or amino-group in the near vicinity of the first site (Fig. 8–1B). The theory was further extended by Miller's et al. finding (12) that mercurial diuresis could be prevented or stopped by the administration of an acid-stable nondiuretic mercurial. Also, these authors concluded that the single point attachment of the only one available valence of mercury in the stable mercurial would not produce diuresis, and that relatively high concentrations of stable mercurial in the vicinity of the receptor could successfully compete for the receptor site with mercuric ion released from the labile molecule (Fig. 8–1C). These two groups (Weiner et al. and Miller et al.) based their interpretations of the diuretic mechanisms on the acid-lability of organomercurials *in vitro* and assumed that these compounds are able to release inorganic mercury under mild conditions of physiological pH. If this scheme is correct, nondiuretic mercurials should give *in vivo* levels of inorganic mercury significantly lower than those seen after diuretic doses of mercuric ion; and similarly, lower levels should be found in diuretic mercurials during metabolic alkalosis when the resulting diuretic effects are low. A direct answer to this question has been for many years limited by methodological problems; small concentrations of ionized mercury in the presence of intact organomercurial compounds were not available for routine analytical methods with low resolution power. The recent development of selective analytical methods using either radioactive labeling or cold vapor absorption for final detection of trace amounts of mercury (3, 5, 8) offered an experimental possibility for reinvestigating this problem (17).

Acute experiments were performed on anaesthetized mongrel dogs, using standard experimental protocol and techniques for the production of alkalosis and acidosis similar to those used by Weiner et al. (20). Radioactive mercury ^{203}-labeled chlormerodrin (1–3–chloro–2–methoxypropyl-urea, Amersham-Searle, spec. act. 500–600 mCi/g Hg) and mersalyl (sodium O-(3–hydroxy mercuri–2–methoxypropyl) carbamyl–phenoxyacetate, Amersham/Searle, spec. act. 50–60 mCi/g Hg) were selected as representatives of diuretic materials (Fig. 8–2); both were shown to

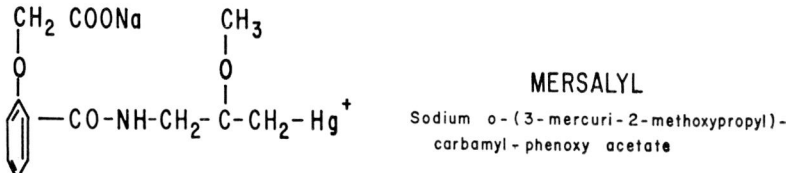

Figure 8–2. Chemical structure of studied representatives of diuretic and nondiuretic mercurials.

be acid-labile *in vitro* and to have high diuretic activity. However, their concentrations in the renal cortex at the time of diuresis differ significantly, mersalyl concentrations being low and chlormerodrin concentrations being high (Table 8-I). Sodium p-chlormercuribenzoate, used as a typical nondiuretic compound, was labeled with radioactive mercury^{203} by isotopic exchange from cold p-chloromercuribenzoic acid (K&K Laboratories) and radioactive mercuric nitrate (Tracerlab, spec. act. 2.3 mCi/mg Hg) according to a modification of Cross and Pinajian's method (6) and purified by repeated precipitation. Final levels of inorganic impurities were less than 0.7 to 0.8 percent in all compounds and were tested for prior to each experiment. All three compounds were compared with minimal diuretic doses of mercuric cysteine (Table 8-I).

TABLE 8–I

IN VIVO AND IN VITRO CHARACTERISTICS AND RENAL UPTAKE OF STUDIED MERCURY COMPOUNDS

	Diuretic Activity in vivo	Acid Lability in vitro	Renal Concentrations
Mercuric cysteine	high	—	high
Chlormerodrin	high	high	high
Mersalyl	high	high	low
p-Mercuribenzoate	none	low	medium

Sodium and chloride excretion were used as an index of mercurial diuresis, and a steady urine flow was maintained by reinfusion of net urinary losses of sodium, chloride and water during the entire experiment. A dose of 2 mg Hg/kg body weight, injected slowly into the jugular vein during the first five minutes of the experimental period in the form of a mixture with a tenfold excess of cysteine, was selected as an arbitrary dose to be compared with minimal diuretic levels of inorganic mercuric cysteine. One, three and five hours after administration of mercurials or inorganic mercury, aliquots of the renal cortex were homogenized, counted for total mercury and analyzed for the presence of inorganic mercury by the method described by Clarkson *et al.* (5). This method selectively measures radioactive inorganic mercury in the presence of intact organomercurial compound. The principle of the method is based on selective reduction of inorganic mercury by stannous chloride to elemental vapor; the latter is swept from the sample by an air stream, collected on a specific absorbent and its activity recorded by a gamma scintillation rate meter.

Figures 8–3 and 8–4 illustrate the results obtained in our studies (17); the highest concentrations of total mercury were found after chlormerodrin, the lowest after mersalyl and the concentrations after p-mercuribenzoate were approximately in the middle. The time course of tissue concentrations showed an early accumulation in the first hour and later on, the concentrations in the renal cortex were steady or slightly reduced in all studied mercurials. In comparison, tissue levels of inorganic mercury were low and in all experiments showed a continual increase during the entire observation period, indicating a continuous

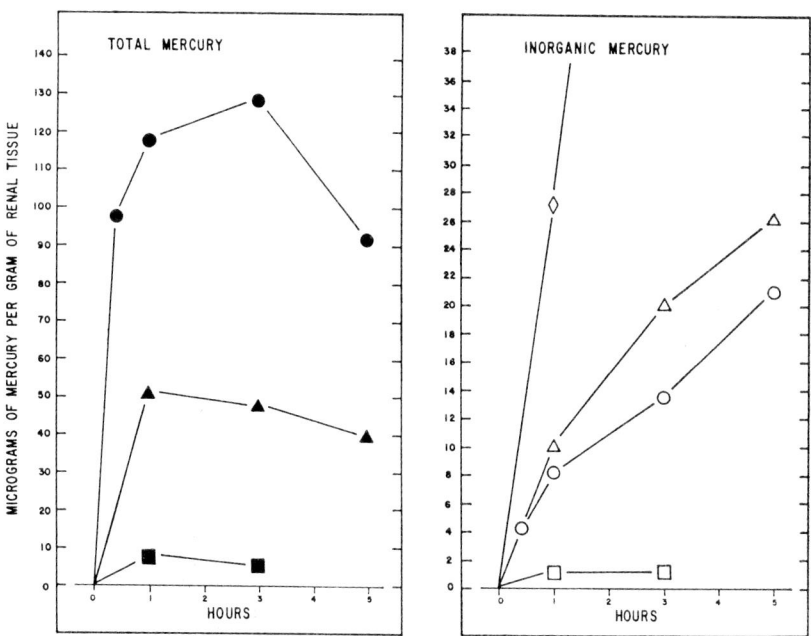

Figure 8–3 (left) and 8–4 (right). Tissue levels of total and inorganic mercury in the renal cortex of acidotic dogs following minimal diuretic dose of cysteine ($\Diamond$) and single injections of chlormerodrin ($\bullet$, $\bigcirc$), p-mercuribenzoate ($\blacktriangle$, $\triangle$) and mersalyl ($\blacksquare$, $\square$) in the dose of 2 mg Hg/kg body weight with ten-fold excess of cysteine.

breakdown of organomercurial compounds. Mersalyl again gave the lowest level of inorganic mercury, the chlormerodrin levels were about five times higher and p-mercuribenzoate produced the highest level.

Table 8-II compares the tissue levels of total and inorganic mercury in acidosis and alkalosis. Total mercury concentrations in alkalotic dogs did not significantly differ from the levels in acidotic dogs. This is in agreement with previous reports in literature (2, 19) indicating small effects of acid-base changes on total mercury levels in the renal cortex. Similarly, inconsistent changes were found in organic mercury between acidosis and

TABLE 8–II
EFFECTS OF CHANGES IN ACID-BASE CONDITIONS ON TOTAL AND INORGANIC MERCURY IN THE RENAL CORTEX OBSERVED IN INDIVIDUALLY PAIRED DOGS WITH IDENTICAL ACID OR ALKALOTIC LOADS AND ANALYZED 60 MINUTES AFTER ADMINISTRATION OF MERCURIALS

	Total Mercury $\mu g/g$ Tissue		Inorganic Mercury $\mu g/g$ Tissue	
	Acidosis	Alkalosis	Acidosis	Alkalosis
Chlormerodrin	127.0	119.0	8.5	8.6
p-Mercuribenzoate	38.0	68.0	6.2	16.4
Mersalyl	9.3	6.5	1.0	1.2

alkalosis in dogs injected with mersalyl or chlormerodrin; tissue levels of mercuric ion in alkalotic dogs injected with p-mercuribenzoate were higher than in acidosis. P-mercuribenzoate, however, produced no indication of diuresis in any of the dogs over the five hour period of the studies. In contrast, both chlormerodrin and mersalyl elicited the typical mercurial diuresis in acidotic dogs.

The actual point of initiation of diuresis was estimated by extrapolation of the excretion data to the time when the first increase in the excretion of solutes and water was higher than corresponding control values and levels of inorganic mercury were read at the time of the onset of diuresis and compared with those obtained after injection of minimal diuretic doses of mercuric cysteine (Table 8-III). Onset of diuresis following injection of minimal diuretic dose of mercuric cysteine resulted in cortical levels of $6\mu g$ of inorganic mercury per gram of cortical

TABLE 8-III
ESTIMATED LEVELS OF INORGANIC MERCURY IN THE RENAL CORTEX AT THE TIME OF ONSET OF DIURESIS PRODUCED BY SINGLE INJECTION OF DIURETIC MERCURIALS IN ACIDOTIC DOGS AND COMPARED WITH SIMILAR CONDITIONS IN ALKALOSIS AND AFTER ADMINISTRATION OF NONDIURETIC ORGANOMERCURIALS

Compound	Dose mg Hg/kg BW	$\mu g\ Hg^{++}/Tissue$	
		Acidosis	Alkalosis
Mersalyl	2.0	0.5	0.7
Chlormerodrin	2.0	2.0	1.8
p-Mercuribenzoate	2.0	2.1	4.8
Mercuric cysteine	0.3	6.3	—

tissue. Chlormerodrin and mersalyl produced diuretic effects only in acidotic dogs and at tissue levels of approximately 2.0μg and 0.5μg, respectively, per gram wet weight. Values for p-mercuribenzoate and acidosis, read at corresponding time intervals after the administration, were nearly identical with the levels after chlormerodrin injection but no diuretic effects were observed.

In conclusion, the results appear to contradict the mercuric ion hypothesis at least in two main points: firstly, the minimal diuretic dose of mercuric cysteine gives levels of inorganic mercury in the renal cortex substantially higher than the levels of inorganic mercury released to the kidney tissue by equally diuretic doses of chlormerodrin and mersalyl; secondly, chlormerodrin and mersalyl failed to produce diuresis in alkalotic dogs. The levels of inorganic mercury at the time when diuresis usually starts as well as during the entire experimental period were, however, identical or similar to those obtained in acidotic dogs.

Although many theoretical considerations can still be made in defense of the mercuric ion hypothesis, namely that [1] the amount of inorganic mercury attached to the diuretic receptor is only a small fraction of the observed levels of inorganic mercury and therefore the bulk of inactive mercuric ions can not be correlated with diuretic effects (19, 20) or that [2] the diuretic mercurials are able to release inorganic mercury so close to the receptor site that there is no need for high levels of inorganic mercury similar to those observed after mercuric cysteine (10, 20), at least one incontroversial fact can be deduced from the present experimental evidence, i.e. the reason for the lack of diuretic activity is obviously not related to the stability of organomercurial compound in the living organism as shown with the example of p-mercuribenzoate. This is not surprising in view of the fact that practically all mercurials studied *in vivo* release inorganic mercury and differ only in the rate of breakdown of the carbon mercury linkage (3).

In consideration of the underlying mechanisms for the splitting of the carbon-mercury bond, it must be emphasized that organic compounds of mercury differ radically from all other compounds having a metal attached to the carbon. Contrary to other organometallic bonds and also to nitrogen-mercury bonding the linkage of mercury to carbon is capable of much greater stability. The carbon-mercury bond can be attacked by inorganic acids but this action generally takes place only with hot and concentrated mineral acids or their mixtures. However, even this rule is not generally valid and the stability of the bond can vary extensively with different types of organomercurial compounds (21). Thus, for example, unsubstituted short-chain mercurials of the type of methyl- and ethyl-mercury are well-known representatives of acid stable compounds and although proved to be capable of depositing inorganic mercury in the living organisms (14) release inorganic mercury *in vitro* slowly even under exposure to mixtures of concentrated mineral acids. The presence of oxygen or nitrogen atoms on the beta carbon of the alkylmercury chain seems to decrease the acid resistance considerably and methoxy-derivatives of methyl- and ethyl-mercury are in comparison with the behavior of their unsubstituted parent molecules easily split *in vitro* and *in vivo* (7). A key position of these substituents for the lability of the carbon-mercury bond can be demonstrated by Weiner's *et al.* study (20) where 21 of 24 studied alkylmercury compounds having an oxygen atom on the second carbon (OY) were shown to be acid-labile *in vitro* although unsubstituted methyl- and ethyl-mercury as well as 3-hydroxypropylmercury resisted rupture even with conditions of pH 3.0 and an excess of thiol compounds.

Similar effects of the adjacent amino or hydroxyl group can be found even in arylmercurials; p-hydroxyphenyl and p-aminophenyl are *in vitro* acid-labile although their parent molecule, phenylmercury, is rather stable under the acid conditions (20). The interpretation of *in vitro* resistance is, however, difficult as some of the apparently stable mercurials, especially those containing free carboxyl groups, can be dissolved only under extreme conditions or precipitate under acid conditions (21). p-Mercuribenzoate can be mentioned as a representative of this group. Clarkson *et al.* (4) proved that mercuric ion is released *in vivo* after the injection of p-mercuribenzoate in the rat and our experiments fully confirm the evidence in the dog. This obviously contradicts its *in vitro* acid-resistance found by Weiner *et al.* (20). The stability of p-mercuribenzoate *in vitro* was therefore reinvestigated under conditions similar to those of the *in vivo* study and compared again with the behavior of representatives of diuretic mercurials, chlormerodrin and mersalyl.

Experiments were performed using the techniques similar to those used by Benesch *et al.* (1) and Weiner *et al.* (20) except that instead of an indirect measuring of the decrease of the concentration of the original compound, the release of mercuric ion was analyzed using the method described by Clarkson *et al.* (5) Parent compounds, chlormerodrin, mersalyl and p-mercuribenzoate, were the same as in the previous experiments and labeled with radioactive mercury (^{203}Hg). The labeled compounds were dissolved in a small volume of water with minimal additions of 1 N NaOH to increase their solubility, mixed with a freshly prepared ten-fold excess of cysteine (L-cysteine base, Sigma) and adjusted to final volume and pH with 0.1 N NaOH, HCl or universal buffer. Solutions were left at room temperature with occasional stirring. At selected time intervals aliquots were taken and immediately analyzed. Concentrations of inorganic mercury were expressed in percent of total mercury in the sample and plotted against time.

The results are presented in Figure 8–5 and Tables 8-IV and 8-V. The exposure of chlormerodrin to pH 4.0 in the presence of excess cysteine results in nearly 17 percent of the carbon-mercury bonds being split in four hours; this is in agreement with previous reports (20). Much smaller rates of breakdown occur at pH 6.0 and pH 8.0 (Table 8-IV). In contrast, the release of inorganic mercury by p-mercuribenzoate was practically negligi-

150 Mercury, Mercurials And Mercaptans

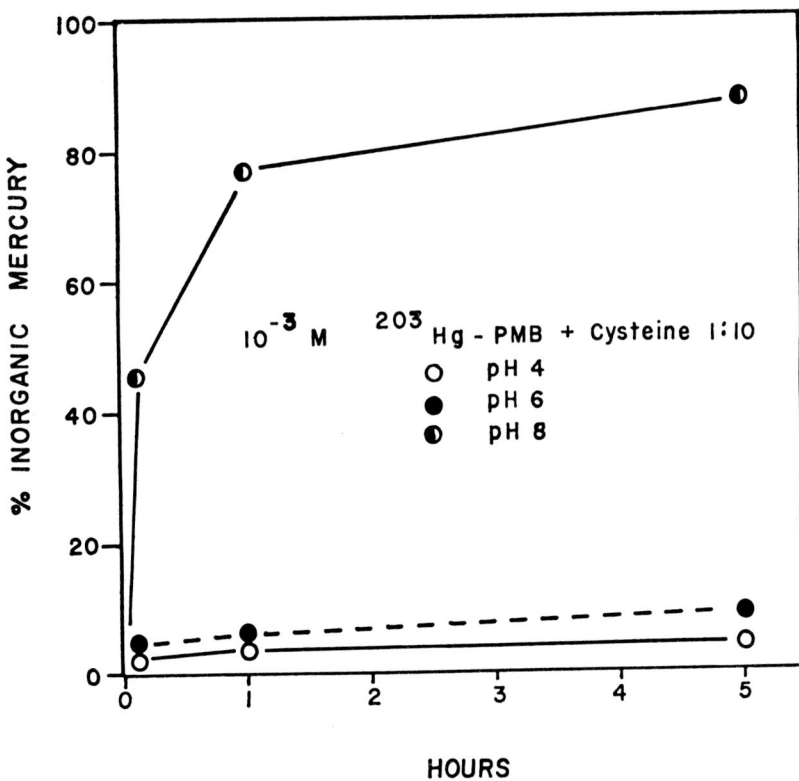

Figure 8–5. The effect of pH and excess of cysteine on the rate of breakdown of p-mercuribenzoate *in vitro*. Inorganic mercury is expressed as fraction of total mercury concentration.

TABLE 8-IV
EFFECT OF pH ON RELEASE OF INORGANIC MERCURY FROM 10^{-6}M SOLUTION OF ^{203}Hg-CHLORMERODRIN IN THE PRESENCE OF TENFOLD EXCESS OF CYSTEINE

Time		% Inorganic Mercury	
	pH 4.0	pH 6.0	pH 8.0
0	0.78	0.78	0.78
2 min	5.8	4.6	4.6
1 hr	8.2	6.2	6.2
4 hrs	16.9	6.8	7.0
25 hrs	49.6	9.5	7.3

TABLE 8-V
RELEASE OF INORGANIC MERCURY FROM 0.05 M SOLUTION OF
^{203}Hg-MERSALYL ADJUSTED TO pH 7.8 IN THE PRESENCE OF
TENFOLD EXCESS OF CYSTEINE

Time	% Inorganic Mercury
0	0.58
15 min	0.92
1 hr	1.7
4 hrs	4.3
7 hrs	6.6
8 hrs	7.3

ble at acid pH 4.0 and 6.0; however, at pH 8.0 nearly 80 percent of the mercury was split from the parent molecule in the first hour (Fig. 8–5). Solutions of mersalyl released only 5 to 7 percent of the carbon-bound mercury after four to eight hours of incubation at pH 7.8 under similar conditions (Table 8-V).

These results, showing the lability of p-mercuribenzoate under conditions not too different from physiological ones, explain the apparent contradiction between the acid-resistance of p-mercuribenzoate in the test *in vitro* and the levels of inorganic mercury found after administration of this organomercurial in the rat and in the dog, and warn against a mechanistic interpretation regarding the capability of mercurials to release inorganic mercury on the sole basis of acid-lability test. Obviously, the lability of any organomercurial *in vivo* is a function not only of pH and of its diuretic properties but also of the specific structure of the individual molecule; this again is in variance with the mercuric ion hypothesis.

Neither can the structural lock-and-key theory (11) completely explain the mechanism of mercurial diuresis; one of the main difficulties of this theory is why the mercuric ion has a high diuretic effect. The discrepancy is further supported by the evidence that [1] the mercuric ion is the most potent diuretic substance when compared with organomercurials on the basis of total mercury levels in the renal cortex, i.e. shows the maximum effect with the lowest tissue levels; and [2] contrary to acid-labile mercurials the diuretic effects of inorganic mercury are only slightly modified with changes in acid-base balance (20). Although the latter fact was recently challenged by the demonstration that even mercuric ion diuresis undergoes considerable change between metabolic acidosis and alkalosis (13), the effects of inorganic mercury and mercuric cysteine or glutathionate constitute still serious objections for the structural theory. Kessler *et al.* (11), in formulating steric requirements for the action of organomercurials, explained the diuretic activity of inorganic mercury with a postulate that diuresis is not produced by the mercuric ion itself but by its complex with cysteine or acetylcysteine formed in the body and described as a chemical form in which mercury is excreted into the urine. According to Kessler's *et al.* claim the distance between the mercury atom and the second carbon atom in the structurally active form of the three-carbon chain may be replaced with a sulfur atom; then the final structure of one sulfur and two carbons between the mercury and the hydrophylic amino-group would constitute the usual three units and fit into the lock-and-key mechanism (Fig. 8–6).

If this is correct, other sulfhydryl compounds structurally similar to cysteine should be at least equally active in producing diuresis. Penicillamine and N-acetylpenicillamine, both mono-

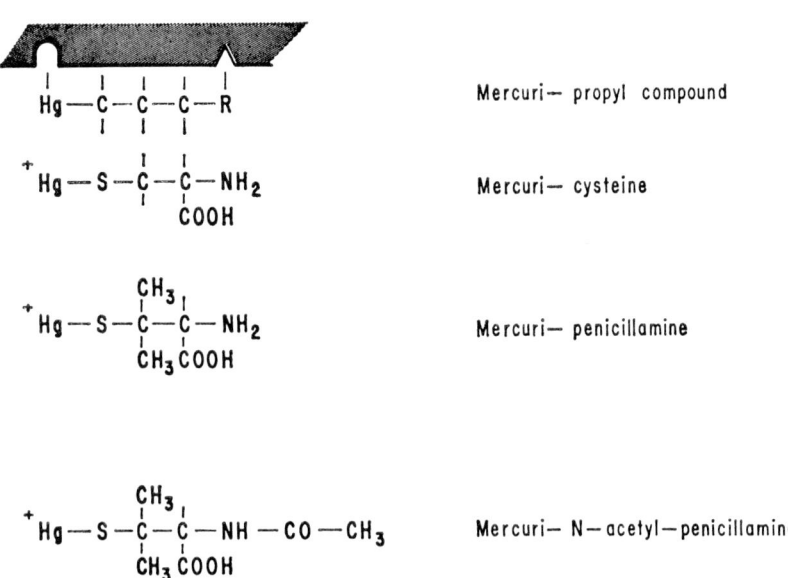

Figure 8–6. Presumed lock-and-key receptor for organomercurials and mercuric cysteine (11), and molecular structure of studied congeners of cysteine.

thiol compounds and congeners of cysteine, were originally isolated as metabolities of penicilline and later on, successfully used in the treatment of Wilson's disease and heavy metal intoxications. They differ from the structure of cysteine only by two methyl groups situated on the sulhydryl-bearing carbon atom and theoretically fulfill the requirement of a three member chain distance between the mercury and the amino-group (Fig. 8–6). However, when tested on the effect on mercuric ion diuresis in dogs, both compounds instead of potentiating the diuretic response were shown to abolish or completely prevent mercurial diuresis (18).

Experiments were performed on anaesthetized female dogs kept in a steady hydration state by continuous replacement of urinary losses of water, sodium and chloride. Changes in urine flow, urinary excretion of sodium and chloride were used as indicators of mercury-induced diuresis. Mercuric chloride labeled with ^{203}mercury and administered with a ten-fold excess of cysteine was injected intravenously to produce mercurial diuresis. DL-penicillamine (Sigma) and N-acetylpenicillamine (Sigma) were administered into another vein at the peak of diuresis or simultaneously with mercury in doses not smaller than a molar ratio of at least 2:1 excess mercury:cysteine.

Figures 8–7 and 8–8 show the effects of both compounds on developed mercurial diuresis elicited by mercuric cysteine with a dose of 0.3 mg or 1.0 mg Hg/kg body weight. A rapid decrease in urine flow acompanied by the return of chloride excretion to values prior to administration of mercury was observed after administration of both forms of penicillamine along with a transient increase in the excretion of urinary excretion of mercury; this illustrates redistribution within the renal tissue and the animal body. Similar effects have never been observed after administration of cysteine. It was concluded, therefore, that contrary to cysteine or glutathione, penicillamine derivatives can counteract by an unknown mechanism the diuretic effect of mercuric cysteine and consequently do not comply with the postulates of the lock-and-key theory on the renal effects of mercury.

Thus, unfortunately neither one of the mentioned hypotheses

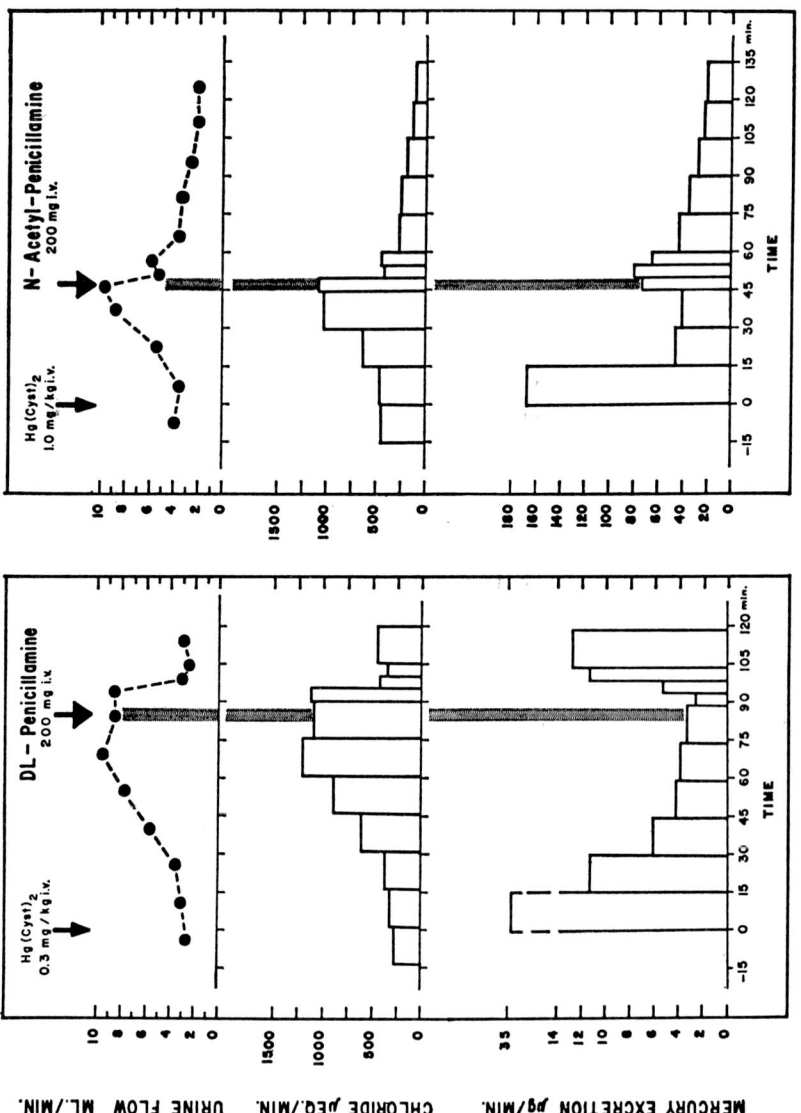

Figure 8-7 (left) and 8-8 (right). Urine flow and the excretion of chloride and total mercury following injections of mercuric cysteine in the dose of 0.3 or 1.0 mg Hg/kg BW resp., prior and after intravenous administration of 200 mg of D,L-penicillamine (left) or N-acetyl-D,L-penicillamine (right).

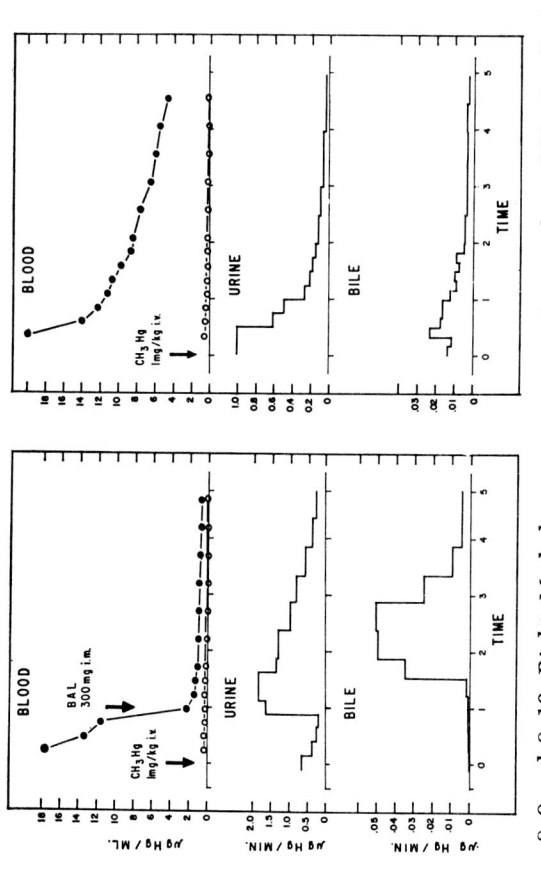

Figure 8–9 and 8–10. *Right*, Methylmercury concentrations in the red blood cells (●) and plasma (○) and the urinary and biliary excretion in a dog after intravenous injection of 1 mg Hg/kg in the form of ^{203}Hg-labeled methylmercury. The time is given in hours after administration. *Left*, Identical parameters in another experiment where 300 mg of dimercaptopropanol was injected intramuscularly 55 minutes after the injection of methylmercury. The time is given in hours after administration.

Vostal: I really do not know.

Berlin: Our studies by autoradiography have shown that the methylmercury-complex with dimercaptopropanol seems to travel rather freely through the membranes especially in the brain. The most likely explanation seems to be that dimercaptopropanol enters the cell and forms a more stable complex with methylmercury than with hemoglobin. The complex distributes more evenly between plasma and cells so that more mercury is present in the plasma.

Vostal: Surprisingly, there was not as much methylmercury transferred from erythrocytes into the plasma as we observed with diuretic mercurials, and part of this methylmercury was recovered in the simultaneous increase of biliary excretion.

Berlin: There might be an increase in the excretion of mercury but there is also more mercury in the tissue, in the brain, muscles and liver after dimercaptopropanol administration.

SESSION III
PHARMACOKINETICS OF MERCURY METABOLISM
J. Henry Wills, *Chairman*

Chapter 9

INTRODUCTION TO SESSION III

J. Henry Wills

I should like to open this session on the Pharmacokinetics of Mercury Metabolism by showing you a diagram (Fig. 9-1) that shows many of the loci at which rates of transport or reaction are important. The first process is entrance of a compound into the body by passage through an epithelial barrier into interstitial fluid (chyle or lymph), and thence into the blood. Once it enters the blood, the chemical may bind to large molecules of the plasma (proteins—particularly globulins, lipoproteins, etc.) or to the formed components of the blood (erythrocytes, leucocytes, platelets). That portion of the chemical that is not bound to formed elements or to large molecules within the blood is capable of being forced out of the blood within a capillary bed by the hydrostatic pressure differential across the walls of the capillaries, and into the interstitial fluid in immediate proximity to the capillary network. Once within the interstitial fluid, the chemical comes into contact with parenchymal cells of various

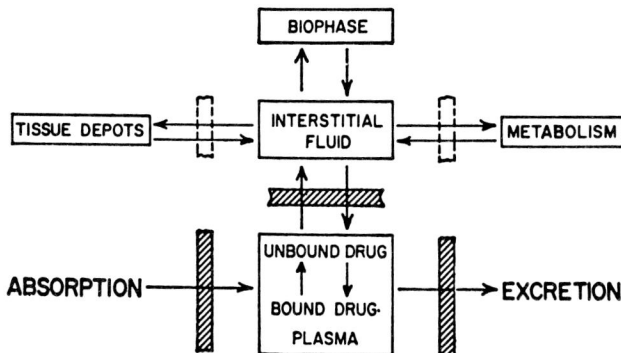

Figure 9-1. Compartments, boundaries, and general processes encountered by a chemical entering the animal body (From Csaky: *Introduction to General Pharmacology,* 1969. Courtesy of the Meredith Corporation.)

sorts. Depending in part on the precise type of cell with which the chemical makes contact, it may be stored, metabolized, or excreted.

The second figure (Fig. 9–2) shows curves for the cumulative excretion through the two principal routes for the mercury in three different organic molecules: A = mercuric acetate, P = phenylmercuric acetate, and E = ethylmercuric chloride. All compounds were given orally. The excretion of all three forms

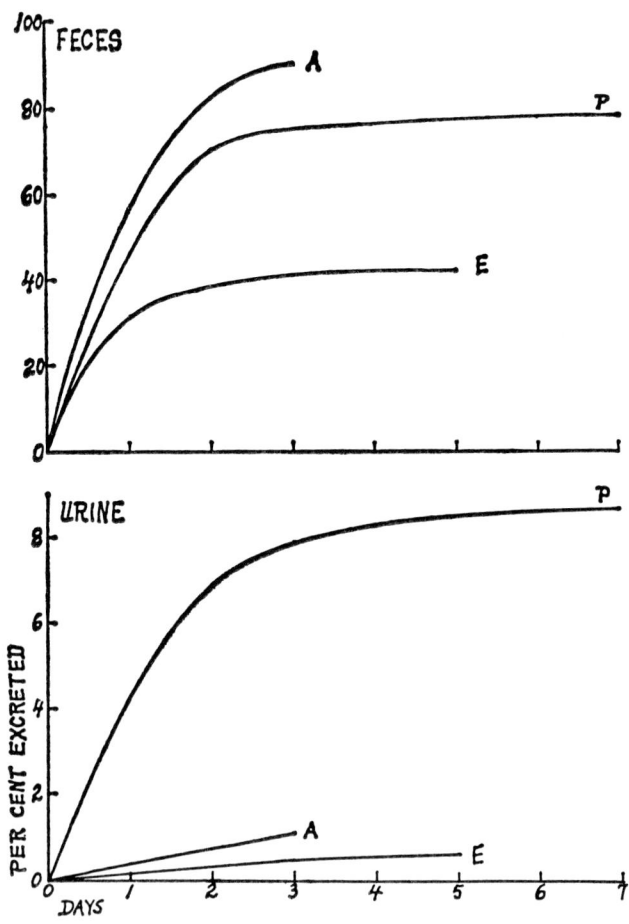

Figure 9–2. Cumulative excretions in urine and feces of single oral doses of mercuric acetate (A), ethylmercuric chloride (E), and phenylmercuric acetate (P) (Constructed with data of S.C. Fang.)

of mercury in the feces was much more important quantitatively than that in the urine. One sees from such curves that the combined excretions of mercuric acetate and of phenylmercuric acetate account for most of the mercury in the oral doses of these compounds, but for only about 41 percent of the intake of mercury as ethylmercuric chloride. This latter finding means that much more of the mercury from ethylmercuric chloride remains within the animal, so that the rates of fixation on, or of solvation from, different cells may be important factors for consideration in regard to the toxicology of this compound.

My third, and final figure (Fig. 9–3) shows the rates of removal from human brain, kidney, and liver of mercury deposited in these organs during Minamata disease. This figure shows that kidney loses initially its load of mercury more rapidly than the

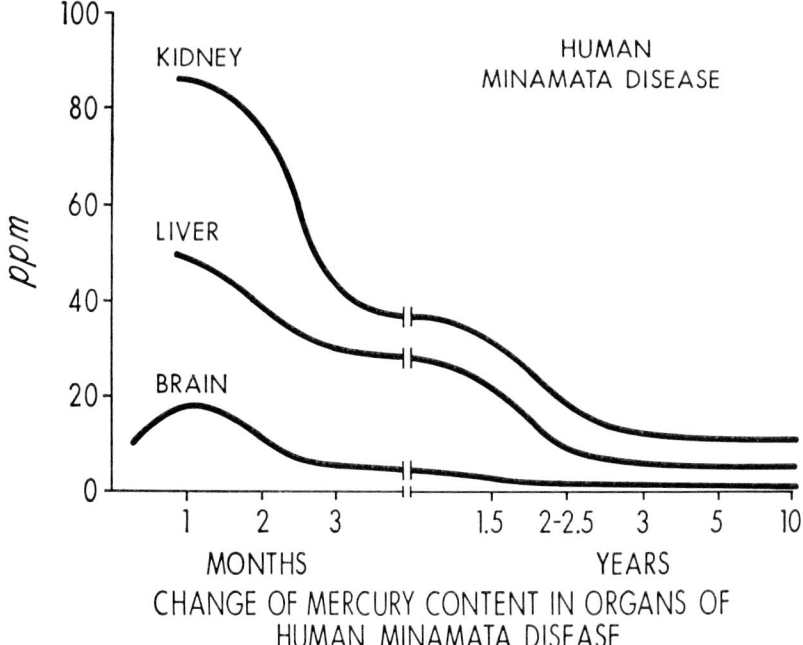

Figure 9–3. Time courses of removal of mercury from kidney, liver, and brain of humans with Minamata disease (From Takeuchi, T.: *Biological Reactions and Pathological Changes of Human Beings and Animals under the Condition of Organic Mercury Contamination*. Internat. Conf. Env. Mercury Contam., Ann Arbor, 1970.)

other two organs, and that brain tends to accumulate some of the mercury lost from kidneys and liver during the early part of the solvation process. Swensson and Ulfvarson, in comparing mercuric nitrate, phenylmercuric hydroxide, and methylmercuric hydroxide, found that brain has not only the greatest avidity for methylmercury but also a greater ability to release mercury in this form than in either of the other forms (Swensson, Å. and U. Ulfvarson: *Acta Pharmacol Toxicol, 26*:273–283, 1968).

The point of all this is that in considering the kinetics of various processes undergone by a substance during and after entry into the animal body, one must look not only at the discrete processes but also at the resultant of all these processes. This may be measured by the concentrations within key target organs, tissues, or cells of the substance administered or of some more active metabolic product of that substance.

The possession by some compound of a high rate of uptake does not guarantee that the compound will be hazardous. If the compound has also a high rate of removal (by storage, metabolic alteration, or excretion), it may be relatively safe despite its rapid initial uptake. One must consider the relative rates of uptake by crucial and not-so-crucial tissues in addition to the general processes that modify the concentration of substance impinging upon the crucial tissues.

Now I shall stop and let the scheduled speakers talk. I hope that you and they will not think I have usurped too much of their time.

Chapter 10

FACTORS AFFECTING THE UPTAKE AND RETENTION OF MERCURY BY KIDNEYS IN RATS

Laszlo Magos

ABSTRACT

Experimental studies indicate that the kidneys play a dual role in the toxicology of mercury. Their remarkable ability to concentrate mercury implies that the organ is always a potential target for damage, at the same time protecting other tissues from mercury damage. This double role of victim and protector emphasizes the importance of studies on the uptake, storage and release of mercury by kidneys. The present paper gives experimental illustrations of the toxicological factors which influence the kidney level of mercury. These are as follows:

1. *Absorption from the injection site.* Ascorbic acid pretreatment delays, cysteine given with mercury facilitates removal from the injection site resulting in differences in the level of mercury in blood available for kidney uptake.

2. *Chemical form of mercury.* The renal uptake after the administration of metallic mercury is slowed down as the metal rapidly diffuses from blood to every tissue. After doses of organomercurials, the blood level remains high, but as mercury is mainly bound in the blood cells, it is less available for filtration and active transport.

3. *Complexing agents.* Depending on experimental circumstances and the type of compound, complexing and chelating agents either increase temporarily or decrease mercury levels in the kidneys.

4. *Kidney metabolism.* Dinitrophenol inhibits the transport of mercury from the peritubular capillaries to tubular cells. Sodium maleate aids diffusable thiol compounds to remove mercury from renal binding sites.

5. *Desquamation of tubular cells.* Tubular necrosis caused either by the mercury itself or by other tubulotoxic agents like sodium fluoride accelerate the loss of cells loaded with mercury.

REPORT

Introduction

Depending on the chemical form at entry into the body, the distribution of mercury shows significant differences. After exposure to metallic mercury or alkylmercurials, kidneys accumulate less and brain more mercury than after the administration of inorganic mercury salts. However, even in the former cases, the concentration of mercury in the kidneys is higher than in any other organs including the brain. Therefore it seems reasonable to say that kidneys are always a potential target for mercury even when they are not the site of the most important changes. The explanation for this discrepancy must be that kidney tissue is relatively insensitive to the toxic effects of mercury.

Adopting a renocentric view, the kidneys might be considered as protectors of more sensitive organs against mercury sometimes at risk to themselves. The kidneys play this protective role best when confronted with inorganic mercury and worst when confronted with metallic mercury or alkylmercurials. Whether the purpose of a therapeutic intervention is to protect the kidneys or other organs, its efficiency must be judged in terms of the behavior of mercury in relation to kidneys.

The paper presented here attempts to review and interpret experimental results on a wide variety of factors which influence the kidney level of mercury.

Absorption from Injection Site

Acute renal failure associated with tubular injury caused by inorganic mercury salts continues to provide a practical therapeutic problem. A therapeutic success against poisoning by inorganic mercury salts in animal experiments should be judged on the following criteria:

1. Increased survival or decreased level of tubular damage after the administration of a dose less than the LD_{100}. (When LD_{100} is used, only the success of different treatments can be

compared but complete lack of any therapeutic effect cannot be established.)

2. After the administration of toxic doses the kidneys might be protected by one or more of the following three mechanisms: (a) a shift in the distribution of mercury from kidneys to other organs; (b) storage of mercury in the kidneys in a less toxic form; and (c) increase in the excretion of mercury resulting in a smaller body burden.

The lack of any effect on the survival of the test animals or on the distribution of mercury in ascorbic acid pretreated rats described by Swensson and Ulfvarson (35) was probably due to the fact that the dose used by them for survival studies killed all the control animals and the dose used for distribution studies was well below the toxic dose. A definite protective effect was observed by Carroll *et al.* (9) who administered toxic doses of mercuric chloride subcutaneously to ascorbic acid pretreated rats. However, they did not establish whether any of the three criteria listed under the second point was responsible for the observed protection. In experiments described here I have found that ascorbic acid was able to protect rats against lethal doses of mercuric chloride, as judged from the increase of LD_{50} from 3.1 mg/kg (95 percent confidence limits 2.3 to 4.2) to 5.0 (confidence limits 3.8 and 6.6). However this protection was not due to any therapeutic effect. The data in Table 10-I show that after the subcutaneous administration of 2, 4 or 8 mg/kg mercury such as $HgCl_2$ not only the kidney level of mercury but also the liver and spleen levels of mercury were lower in rats treated one hour before mercury with 900 mg/kg neutralized ascorbic

TABLE 10-I
EFFECT OF 900 mg/kg ASCORBIC ACID GIVEN ONE HOUR BEFORE THE SUBCUTANEOUS INJECTION OF $HgCl_2$ ON MERCURY CONCENTRATION IN ORGANS OF ANIMALS KILLED 24 HOURS AFTER MERCURY

Dose of Hg in mg/kg	Pretreatment	No. of Rats	Mercury Conc. in µg/g Tissue ± SEM		
			Liver	Spleen	Kidneys
2	−	9	2.5 (0.3)	2.1 (0.2)	28 (1)
2	+	9	1.2 (0.1)	1.3 (0.1)	30 (1)
4	−	6	7.1 (0.9)	4.8 (0.5)	37 (1)
4	+	6	3.7 (0.3)	2.8 (0.2)	31 (1)
8	−	9	12.3 (0.8)	6.7 (0.5)	56 (2)
8	+	9	8.2 (0.7)	5.8 (0.7)	45 (3)

acid compared with control rats. This indicates that there was no shift in the distribution of mercury from kidneys to other organs. The curves in Figure 10–1 show that when mercury concentration in the kidneys was plotted against dose, kidney uptake was not significantly different in ascorbic acid and control rats indicating that mercury in the kidneys of ascorbic acid treated rats was not less toxic than in the kidneys of control rats. Figure 10–2 shows that the body burden was higher in ascorbic acid treated rats than in controls, so that ascorbic acid actually decreased

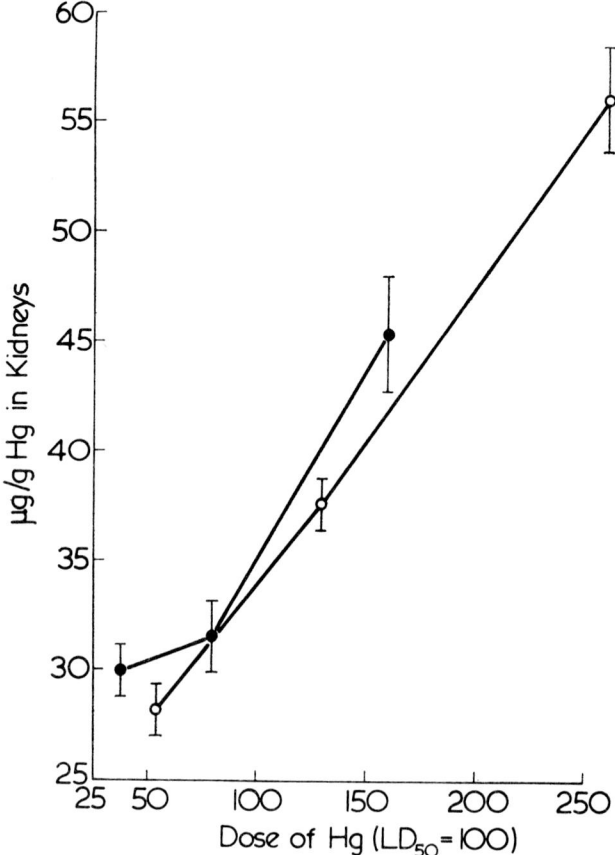

Figure 10–1. Effect of the dose of mercury on the concentration of the metal in kidneys in ascorbic acid pretreated and control rats. Mercury doses are expressed as a percentage of the LD_{50}. Ascorbic acid treated rats, solid circles; control rats, empty circles.

Factors Affecting the Uptake and Retention 171

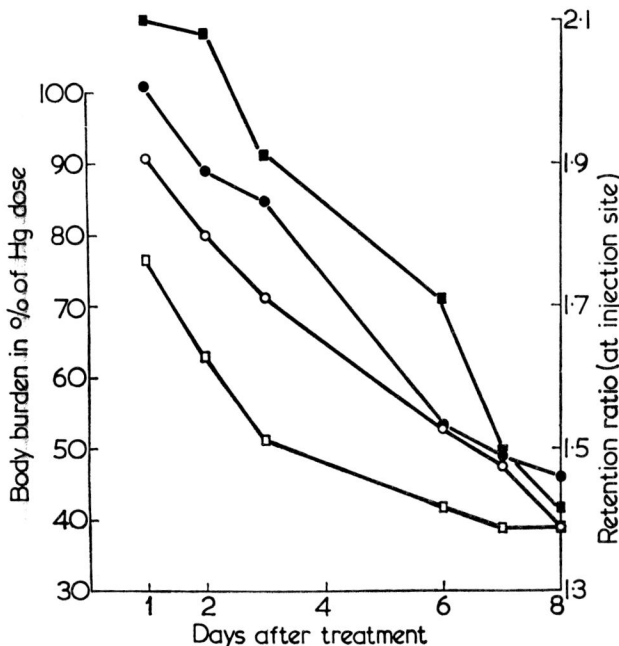

Figure 10–2. Effect of ascorbic acid pretreatment on the body burden and on the retention of mercury at the subcutaneous injection site. Retention ratio is approx. 1, if all the injected mercury is absorbed. Ascorbic acid pretreated rats, solid symbols; control rats, empty symbols; body burden, circles; retention ratio, squares.

excretion. Mercury levels in kidneys and other organs were decreased only because absorption of mercury from the injection site was slowed down by treatment with ascorbic acid as the data in Figure 10–2 show. Retention ratios in Figure 10–2 were calculated by counting whole body activity in the annular well of a gamma scintillation counter in one of two positions. In the "n" position, the animal was placed around the annular well of a gamma scintillation crystal (NS 69 Lead assembly, EKCO, fitted with a N609/End On Type crystal, EKCO and connected through N691A counter with a scaler) with its injected leg near the crystal to give a count rate of C_n. In the "f" position, the injected leg was held away for the crystal to give a lower count rate (C_f). The ratio of $C_n : C_f$, called retention ratio, was approximately 2.5 after injection, and progressively decreased to approach unity. Though it is possible that absorption of mercury from the injection site was decreased by the formation of an insoluble mercury compound, it was striking (Table 10-II), that the edema of the injected leg, judged from the weight difference between injected and untreated leg, was more pronounced in untreated than in ascorbic acid treated rats probably due to decreased capillary permeability. Consequently any changes in the amount of mercury absorbed from the injection sites of rats pretreated with ascorbic acid might be related to changes in capillary permeability or to local chemical changes or to both.

TABLE 10-II
EFFECT OF ASCORBIC ACID PRETREATMENT ON OEDEMA FORMATION AT THE SUBCUTANEOUS INJECTION SITE 24 HOURS AFTER INJECTION OF MERCURIC CHLORIDE

Mercury Dose (mg/kg)	No. of Rats per Group	Fluid Retention in the Injected Leg in g			
		Ascorbic Acid Pretreated Rats		Control Rats	
		Mean	Extreme Values	Mean	Extreme Values
2.0	2	1.6	1.0–2.3	2.2	1.7–2.6
2.8	4	3.3	3.2–3.7	5.1	4.4–5.9
4.0	6	3.7	1.0–4.4	4.5	2.1–7.9
8.0	2	3.1	2.1–4.0	4.8	3.6–5.9

The Chemical Form of Mercury

From the pathological point of view the most important single factor which determines whether the central nervous system (CNS) or kidneys will be damaged is the chemical form of mercury. The reason why metallic mercury vapor attacks the CNS and inorganic mercury salts the kidneys was obscure until Berlin *et al.* (6) reported that the brains of mice of 20 g body weight after exposure to mercury vapor contained ten times more mercury than the brain of mice given intravenous mercuric chloride in a dose equal to the amount of mercury vapor absorbed. However, even in mice exposed to vapor, the kidneys contained 12.5 times more mercury than the brain, and in 10 g mice with a higher brain weight to body weight ratio the proportion of kidney uptake to brain uptake after vapor exposure remained as high as 4.4 (23).

Following the mercury distribution after the intravenous injection of $0.1\mu g$ Hg either as metallic mercury or as inorganic mercury salt, Magos (24,25) found that 30 seconds after injection of metallic mercury that the blood mercury level was only 1/10 of the blood mercury level of rats injected with the salt suggesting that the diffusion of mercury in the metallic form is an extremely rapid process. Kidneys took up less mercury as less was available in the blood. However, despite the fact that the kidney level of mercury started from a lower point in rats injected with metallic mercury, their mercury uptake curve ran parallel to that of the controls treated with mercuric chloride as the curves in Figure 10–3 show. After mercury becomes oxi-

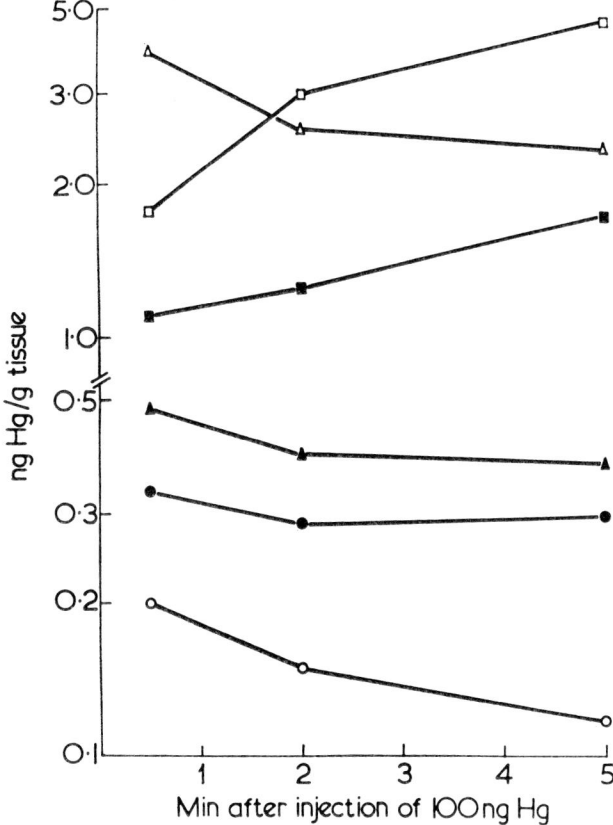

Figure 10–3. Mercury concentration in the kidneys, brain and blood after the intravenous injection of metallic mercury or mercuric chloride. Metallic mercury, solid symbols; mercuric chloride, empty symbols; kidneys, squares; brain, circles; blood, triangles.

dized, kidneys handled mercury in the usual way as was predictable from the *in vitro* studies of Clarkson *et al.* (11).

Metallic mercury disappears within seconds from the blood and this is why kidneys are able to extract less mercury from the blood than after the injection of inorganic mercury salts. The situation is quite different with alkylmercury compounds. Friberg (21) reported that after ten daily doses of methyl-mercury-dicyandiamide or mercuric chloride the blood level of mercury was 100 times higher in the organomercurial group than in the inorganic group and the kidney level of mercury was only half in the former group as in the kidneys of rats treated with mercuric chloride. One of the reasons why kidneys are not as efficient in extracting mercury from the blood after doses of alkylmercurials is that these organomercurials are carried mainly by the red blood cells (22,39) whereas inorganic mercury is distributed nearly equally between plasma and red blood cells (4,34). In spite of the shift in distribution and decrease in the renal concentration of mercury after doses of organomercurials, the kidneys still accumulate more mercury per gram wet weight than every other organ. Kidney mercury consists of organic and inorganic components. The data in Table 10-III show that when rats were given by stomach tube a daily dose of 0.85 mg/kg mercury such as methylmercury dicyandiamide, five times a week, the mercury concentration in kidneys increased more rapidly than in the brain or the liver, even if we neglect the mercury which was present in the inorganic form. After the cessation of nine daily doses of 3.4 mg/kg mercury such as methylmercury dicyandiamide the brain level of mercury decreased slightly from four to six days as compared to the 18 to 21 day period. The mercury concentration in the liver dropped more than 50 percent, but in the kidneys mercury concentration showed a 15 percent increase due to the accumulation of inorganic mercury. Kidney uptake of mercury after the administration of ethylmercurials follows the same pattern (36) as after methylmercury, though some difference in the brain uptake was reported (33).

As methoxyethylmercury is rapidly broken down in the tissues with a half-life of about one day to yield ethylene and inorganic mercury (17), it is not surprising that the methoxyethylmercury content of kidneys never exceeds 3 percent of the dose (17) and the whole distribution follows the same pattern as after the administration of inorganic mercury salts (39). Phenylmercury behaves like methoxyethylmercury (19,22,36) indicating a rapid cleavage of the metal to carbon bond. Whether the mercury-carbon bond is broken down in every organ or mainly in the liver is not known. Daniel et al. (17) suggested that the breakdown of methoxyethylmercury is not catalysed by an enzyme system. In rats 24 hours after the administration of 5 mg/kg methyl-

TABLE 10-III
TOTAL MERCURY CONCENTRATION AND THE PROPORTION OF ORGANOMERCURIAL TO TOTAL MERCURY (IN BRACKETS) IN THREE DIFFERENT ORGANS IN RATS TREATED FIVE TIMES A WEEK BY GAVAGE WITH 0.85μg/kg Hg AS METHYLMERCURIC DICYANDIAMIDE AND IN RATS KILLED AT DIFFERENT INTERVALS AFTER NINE DAILY ADMINISTRATIONS OF 3.4 mg/kg Hg AS METHYLMERCURIC DICYANDIAMIDE

Treatment No. of Doses (0.85 mg/kg Hg per day)	No. of Rats	Mercury Conc. in μg/g Tissue		
		Brain	Liver	Kidneys
9	2	2.5 (95%)	8.4 (92%)	32 (71%)
18	2	3.5 (98%)	12.0 (91%)	66 (59%)
29	2	4.9 (97%)	15.0 (87%)	97 (70%)
No. of days after 9 daily doses of 3.4 mg/kg Hg				
4-6	3	13.0 (96%)	44.0 (96%)	84 (77%)
18-21	6	8.6 (99%)	20.0 (94%)	97 (62%)

Estimations were carried out by a selective atomic absorption method (Magos, L.: Selective atomic absorption determination of inorganic mercury and methylmercury in ingested biological samples. *Analyst*, 1971, in press).

mercury dicyandiamide, the liver of rats with induced drug metabolizing enzymes contained 40 percent more mercury than the liver of control rats, but the difference disappeared in the following 48 hours. Neither the inorganic : organic mercury ratio nor the kidney uptake of mercury was affected. That organomercurials can break down in the kidneys resulting in a diuretically effective mercuric ion concentration was substantiated experimentally by Clarkson et al. (16) and Clarkson and Greenwood (12). Berlin and Ullberg (8) suggested that the breakdown of methylmercury may occur in the liver or kidneys as part of a process of excretion. However as in the plasma of methylmercury treated rats, 22 to 27 percent of the mercury is present in the inorganic form (29,30) some of the inorganic mercury in the kidneys must be formed elsewhere.

Complexing Agents

Mercurials are often stated to be the most specific thiol reagents and conversely thiol compounds are the most specific complexing agents for mercurials. Their effects on mercury *in vivo* are manifold. They increase absorption from the injection site, and depending on the experimental circumstances they either increase kidney uptake of mercury or liberate mercury from the kidneys.

The data in Table 10-IV show that if mercury is injected subcutaneously as a cysteine complex, it is absorbed from the injected site within one hour and kidneys take up more than half of the dose. Three hours after injection the mercury level in kidneys reaches 60 to 70 percent of the injected dose. However, 24 hours after injection of mercuric chloride, little more than 50 percent of the injected dose was left at the injection site and even after three days approximately 20 percent remained in the injected left hind thigh. A quite different thiol effect is shown in Figure 10-4. When BAL was injected six days after the administration of 100µg mercury as HgC_{-2} urinary excretion of mercury showed a dose dependent increase and kidney levels of mercury decreased. There was also a slight increase in the

TABLE 10-IV
EFFECT OF CYTEINE ON ABSORPTION OF MERCURY FROM THE SUBCUTANEOUS INJECTION SITE AND ON THE MERCURY CONTENTS OF KIDNEYS

Treatment	Time after Treatment	No. of Rats	Mercury		
			Retention* Ratio	Retained at the Injection Site	Taken up by Kidney
				µg Hg	
100µg Hg as HgCl$_2$	24 hr	4	1.5 (1.4–1.6)	44.0 (37.0–43.0)	31 (29–33)
	72 hr	4	1.4 (1.2–1.6)	19.0 (14.0–28.0)	41 (41–43)
100µg Hg as Hg(cyst)$_2$	1 hr	3	1.0 (0.9–1.1)	1.9 (1.7–2.2)	52 (49–54)
	3 hr	3	1.1 (1.0–1.1)	1.4 (0.7–1.9)	64 (60–65)

* The retention ratio was measured as described in the text. The numbers in parenthesis are the extreme values of the retention ratio.

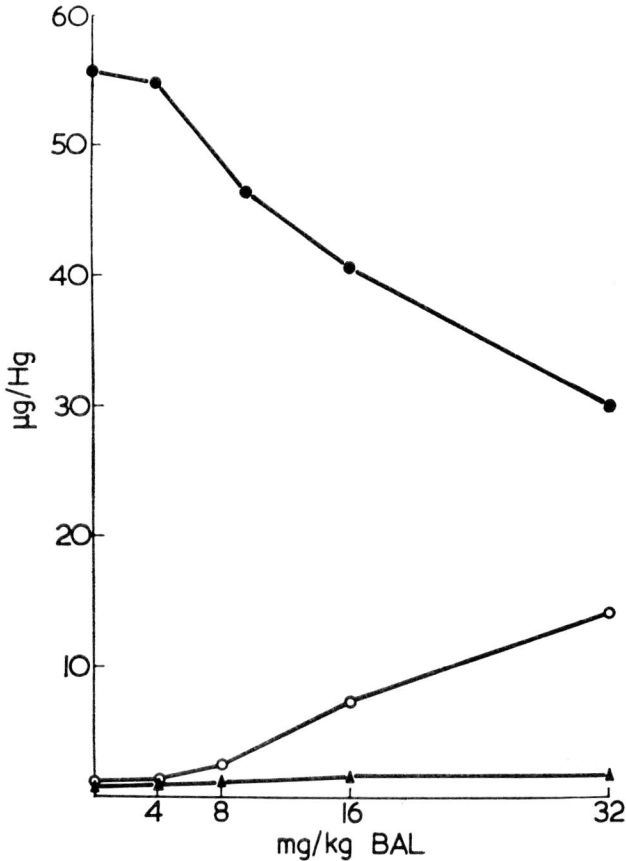

Figure 10–4. The effect of BAL on the mobilization of mercury from kidneys. Kidneys, solid circles; liver, solid triangles; urine, empty circles.

blood and liver levels of mercury (25). The shift in the distribution of mercury was greater if BAL was given shortly after the injection of mercury. If the treatment is repeated (25) or if repeated daily injections of $HgCl_2$ are each accompanied by treatment with BAL (7) the brain level of mercury increases by 25 percent. BAL treatment also accelerates the brain uptake of mercury after injection of methylmercury (5). Actually it seems there is no way to liberate mercury from kidneys without increasing at least temporarily the extrarenal concentration of mercury (14,15,20,27,29).

The other thiol compounds used effectively against the lethal effect of mercuric chloride (1,2) are not as effective as BAL in accelerating the excretion of mercury (27).

Metabolic Effects

The effect of thiol compounds on the storage of mercury in kidneys depends on the concentration of protein thiol groups in the kidneys, on the concentration of diffusible thiol compounds which cross the tubular cells and on their relative affinity to mercury. It has been shown that kidneys have two different binding sites for mercury, one class having a chemical affinity for mercury one hundred-fold greater than the other class (13). The concentration of the strong binding sites is 1.0×10^{-7} and the concentration of the weak binding sites is 30×10^{-7} mole of Hg/g wet weight. When $HgCl_2$ in equal concentration with the strong binding sites was added to kidney homogenate, 7500 times more D-penicillamine was needed to liberate 50 percent of the mercury from protein binding (13). It might be that metallothionein suggested recently as a compound responsible for the binding and retention of mercury in kidneys is identical with the strong binding sites (40).

Any change which prevents mercury from becoming attached to the strong binding sites in the kidney or helps diffusible thiol compounds to compete with strong binding sites from mercury should be of great theoretic and perhaps therapeutic value. That is why it is so important to know more about the mechanism involved in the kidney uptake, storage and release of mercury. If we approach the problem from a purely anatomical point of view, there are two possibilities for mercury uptake by kidneys. Either mercury is first filtered through the glomerulus prior to uptake by renal tissues (41) or kidney uptake might proceed by transport from the peritubular capillaries (3,28) or the combination of these two mechanisms is involved in the process (18). Our experiments carried out in the past few years with Dr. Clarkson indicated that mercury is taken up by kidneys both through the brush border and the basal membrane (15) but also that a disturbance in kidney metabolism can lead to the mobilization of kidney mercury without adding exogenous thiol compounds (14).

Rats pretreated with dinitrophenol before the administration of 100μg Hg either as $HgCl_2$ or $Hg(cyst)_2$ showed a dose dependent decrease in the kidney levels of mercury sacrificed 24 hours after injection. The data in Table 10-V show that DNP did not increase the urinary excretion of mercury which would be the case if DNP inhibited uptake from the tubular urine (15). As urinary excretion was not increased, DNP must inhibit the transport process from the peritubular capillaries to the tubular cells. If DNP was administered 13 days after mercury it had no effect on the kidney level of mercury indicating that DNP was

TABLE 10-V

EFFECT OF DINITROPHENOL GIVEN 90 MINUTES BEFORE THE SUBCUTANEOUS ADMINISTRATION OF 100μg Hg ON THE 24 HOUR URINARY EXCRETION AND OF THE KIDNEY CONTENT OF MERCURY

Treatment	No. of Rats	Kidney Content of Hg in μg (a)	Urinary Hg in μg (b)	a + b
$HgCl_2$	6	34	1.3	35
DNP + $HgCl_2$	6	25	1.4	26
$Hg(cyst)_2$	6	59	8.8	67
DNP + $Hg(cyst)_2$	6	38	9.2	47

not able to interfere with mercury binding sites occupied by mercury (15). Another metabolic inhibitor, sodium maleate, had quite a different effect. Maleate was able to release mercury from the kidneys days after the administration of mercury and to increase the mercury excretion either given before mercury or days after the administration of mercury (14). The curves in Figure 10-5 show that maleate exerted this effect by aiding diffusible thiol compounds to compete for mercury with tissue binding sites (14,21). It is important to point out that the concentration of protein thiol groups was not affected by maleate (32). Without an exogenous thiol compound, higher doses of maleate were necessary to decrease the mercury concentration in kidneys and to increase urinary excretion. However, when an exogenous thiol compound was given to the animals in the form of penicillamine which in itself had only a slight effect on mer-

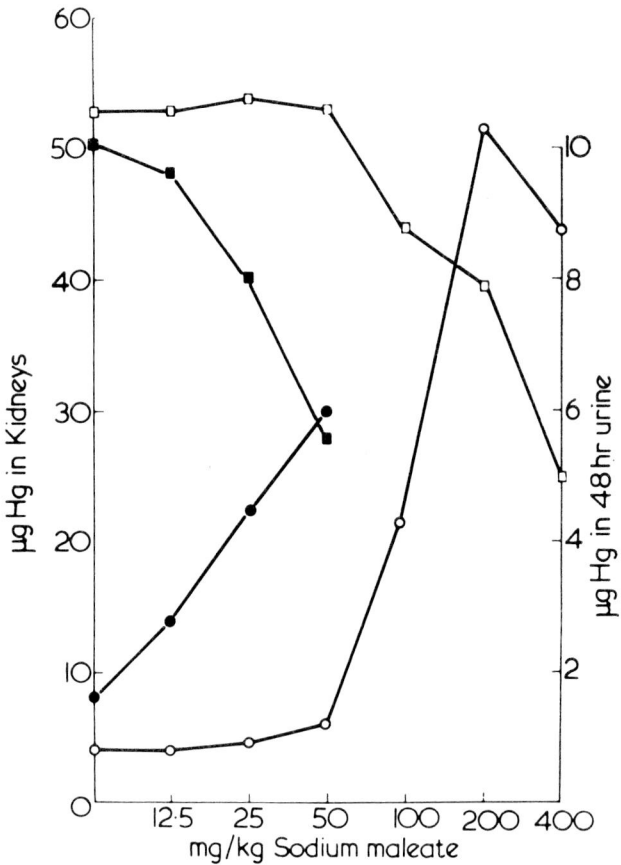

Figure 10-5. The effect of sodium maleate on the mobilization of mercury from kidneys in penicillamine treated and control rats. 35.8 mg/kg penicillamine was injected twice a day intramuscularly for two days. Penicillamine treated rats, solid symbols; controls, empty symbols; kidneys, squares; urine, circles.

cury, the maleate effect became apparent at doses otherwise ineffective (27). Maleate was able to increase the effect of penicillamine but not the effect of BAL which reacts with maleate (32). Increased urinary excretion due to maleate was absent even when BAL was administered three hours after maleate, through the overwhelming proportion of maleate had already passed through the kidneys (32). Moreover BAL given

Factors Affecting the Uptake and Retention

before penicillamine was able to abolish the effect of maleate on the latter compound (32).

Desquamation of Tubular Cells

High doses of sodium maleate caused considerable histological changes in the tubular cells, though proteinuria and mercury excretion had a different time curve indicating that mercury was not lost with kidney proteins (14). Doses used in combination with penicillamine, which were effective in mobilizing mercury, caused no histologically detectable alterations in the kidneys. The correct interpretation of experimental results makes it paramount in every experiment to exclude the possibility that a change in the kidney level of mercury, or increase in the urinary excretion of mercury is caused by the desquamation of tubular cells. Nevertheless some of the mercury is released from the kidneys with the normal cellular turnover of the tubular cells, but if mercury is given in cytocidal quantities a significant part of the mercury is excreted with the desquamated tubular cells (10). Besides this mercury inflicted desquamation, tubular necrosis might be induced sometimes in the completely healthy kidneys of rats given nontoxic doses of mercury by different treatments.

The data in Table 10-VI show that sodium fluoride given in daily doses of 40 mg/kg on the sixth and seventh day after the injection of 100µg Hg such as $HgCl_2$ increased more than fourfold the excretion of mercury and caused more than 30 percent decrease in the mercury content of kidneys. Histological investigation of the kidneys carried out with Dr. W.H. Butler in our

TABLE 10-VI
EFFECT OF TWO DAILY DOSES OF 40 mg/kg SODIUM FLUORIDE GIVEN SIX AND SEVEN DAYS AFTER THE ADMINISTRATION OF 100µg Hg AS $HgCl_2$ ON THE KIDNEY CONTENTS AND URINARY EXCRETION OF MERCURY

NaF Treatment	No. of Rats	48 hrs Urinary Excretion of Hg in µg	Kidney Contents of Hg in µg	Tubular Damage
−	4	2.0	56	None
+	4	8.8	36	In proximal tubule and loop of Henle

laboratory revealed necrosis involving the proximal and Henle tubules indicating that the release of mercury was not the direct consequence of an interference on kidney metabolism by fluoride but it was due to the desquamation of tubular epithelium. A coincidence between tubular necrosis and mercury excretion was reported lately by Trojanowska et al. (38) with thioacetamide.

REFERENCES

1. Aposhian, H.V.: Protection by D-penicillamine against the lethal effects of mercuric chloride. *Science, 128:*93, 1958.
2. Aposhian, H.V. and Aposhian, M.M.: N-acetyl-D-penicillamine, a new oral protective agent against the lethal effects of mercuric chloride. *J Pharmacol Exp Ther, 126:*131, 1959.
3. Bergstrand, A., Friberg, L., Mendel, L. and Odelblad, E.: The localization of subcutaneously administered radio-active mercury in rat kidney. *J Ultrastruct Res, 3:*328, 1959.
4. Berlin, M. and Gibson, S.: Renal uptake, excretion and retention of mercury. I. A study in the rabbit during infusion of mercuric chloride. *Arch Environ Health 6:*617, 1963.
5. Berlin, M., Jerskell, L.G. and Nordberg, G.: Accelerated uptake of mercury by brain caused by 2,3-dimercaptopropanol (BAL) after injection into mouse of methylmercuric compound. *Acta Pharmacol Toxicol, 23:*312, 1965.
6. Berlin, M., Jerskell, J.G. and Ubisch, H.: Uptake and retention of mercury in the mouse brain. *Arch Environ Health, 12:*33, 1966.
7. Berlin M. and Lewander, T.: Increased brain uptake of mercury caused by 2,3-dimercaptopropanol (BAL) in mice given mercuric chloride. *Acta Pharmacol Toxicol, 22:*1, 1965.
8. Berlin, M. and Ullberg, S.: Accumulation and retention of mercury in the mouse. III. An autoradiographic comparison of methyl mercuric acetate with inorganic mercury. *Arch Environ Health, 6:*610, 1963.
9. Carrol, R., Kovacs, K. and Tapp, E.: Protection against mercuric chloride of the rat kidney. *Arzneimittel-Forsch, 15:*1361, 1965.
10. Cember, H.: The influence of the size of the dose on the distribution and elimination of inorganic mercury, $Hg(NO_3)_2$, in the rat. *Am. Ind Hyg Assoc J, 23:*304, 1962.
11. Clarkson, T.W., Gatzy, J. and Dalton, C.: AEC Research and development. Rep. UR–582, *1:* 1961.
12. Clarkson, T.W. and Greenwood, M.: The mechanism of action of mercurial diuretics in rats; the renal metabolism of p-chloromercuribenzoate and its effects on urinary excretion. *Br J Pharmacol, 26:*50, 1966.

13. Clarkson, T.W. and Magos, L.: Studies on the binding of mercury in tissue homogenates. *Biochem J, 99:*62, 1966.
14. Clarkson, T.W. and Magos, L.: The effect of sodium maleate on the renal deposition and excretion of mercury. *Br J Pharmacol, 31:* 560, 1967.
15. Clarkson, T.W. and Magos, L.: Effect of 2,4-dinitrophenol and other metabolic inhibitors on the renal deposition and excretion of mercury. *Biochem Pharmacol, 19:*3029, 1970.
16. Clarkson, T.W., Rothstein, A. and Sutherland, R.: The mechanism of action of mercurial diuretics in rats; the metabolism of ^{203}Hg-labelled chlormerodrin. *Br J Pharmacol, 24:*1, 1965.
17. Daniel, J.W., Gage, J.C. and Lefevre, P.A.: The metabolism of methoxyethylmercury salts. *Biochem J, 121:*411, 1971.
18. Dreisbach, R.H. and Taugner, R.: Renale "Stapelung" und Ausscheidung von ^{203}Hg-Sublimat bei der Ratte. *Nuclear Med, 5:*421, 1966.
19. Ellis, R.W. and Fang, S.C.: Elimination, tissue accumulation and cellular incorporation of mercury in rats receiving an oral dose of ^{203}Hg-labelled phenylmercuric acetate and mercuric acetate. *Toxicol Appl Pharmacol, 11:*104, 1967.
20. Eybl, V., Sykora, J. and Mertl, F.: Uber den Einfluss der Chelatbildner auf die Ausscheidung und Verhalten des Quecksilbers bei akuter Quecksilbervergiftung. *Arch Exp Path Pharmak, 252:*252, 1965.
21. Friberg, L.: Studies on the metabolism of mercuric chloride and methyl mercury dicyandiamide. *Arch Ind Health, 20:*42, 1959.
22. Gage, J.C.: Distribution and excretion of methyl and phenyl mercury salts. *Br J Ind Med, 21:*197, 1964.
23. Magos, L.: Mercury-blood interaction and mercury uptake by the brain after vapour exposure. *Environ Res, 1:*323, 1967.
24. Magos, L.: Transport of elemental mercury by blood. *Arch Pharmak Exp Ther, 259:*183, 1968.
25. Magos, L.: Effect of 2,3-dimercaptopropanol (BAL) on the urinary excretion and brain content of mercury. *Br J Ind Med, 25:*152, 1968.
26. Magos, L.: Uptake of mercury by the brain. *Br J Ind Med, 25:*315, 1968.
27. Magos, L. and Stoytchev, Ts.: Combined effect of sodium maleate and some thiol compounds on mercury excretion and redistribution in rats. *Br J Pharmacol, 35:*121, 1969.
28. Mambourg, A.M. and Raynaud, C.: Étude, a l'aide d'isotopes radioactifs du mécanisme de l'excrétion urinaire du mercure chez le lapin. *Rev Franc, Études Clin et Biol, 10:*414, 1965.
29. Nigrovic, V.: Der Einfluss von Chelatbildnern auf das Verhalten von Quecksilber im Organismus. *Arzneimittel-Forsch, 13:*787, 1963.
30. Norseth, T. and Clarkson, T.W.: Biotransformation of methylmercury salts in the rat studied by specific determination of inorganic mercury. *Biochem Pharmacol, 19:*2775, 1970.

31. Norseth, T. and Clarkson, T.W.: Studies on the biotransformation of ^{203}HG-labelled methyl mercury chloride in rats. Arch Environ Health 21:717, 1970.
32. Stoytchev, Ts., Magos, L. and Clarkson, T.W.: Studies on the mechanism of maleate action on the urinary excretion of mercury. Eur J Pharmacol, 8:253, 1969.
33. Suzuki, T., Miyama, T. and Katsunama, H.: Comparative study of the bodily distribution of mercury in mice after subcutaneous administration of methyl, ethyl and n-propyl mercury acetates. Jap J Exp Med, 33:277, 1963.
34. Swensson, A.K., Lundgren, K.D. and Lindstrom, O.: Distribution and excretion of mercury compounds after single injection. Arch Ind Health, 20:432, 1959.
35. Swensson, A. and Ulfvarson, U.: Experiment with different antidotes in acute poisoning by different mercury compounds. Int Arch Gewerbepath Gewerbehyg, 24:12, 1967.
36. Takeda, Y., Kunugi, T., Hoshino, O. and Ukita, T.: Distribution of inorganic, aryl, and alkyl mercury compounds in rats. Toxic Appl Pharmacol, 13:156, 1968.
37. Takeda, Y. and Ukita, T.: Metabolism of ethylmercuric chloride ^{203}Hg in rats. Toxicol Appl Pharmacol, 17:181, 1970.
38. Trojanowska, B., Piotrowski, J.K. and Szendzikowski, S.: The influence of thioacetamide on the excretion of mercury in rats. Toxicol Appl Pharmacol, 18:374, 1971.
39. Ulfvarson, U.: Distribution and excretion of some mercury compounds after long term exposure. Int Arch Gewerbepath Gewerbehyg, 19: 412, 1962.
40. Wisniewaka, J.M., Trojanowska, B., Piotrowski, J. and Jakubowski, M.: Binding of mercury in the rat kidney by metallothionein. Toxicol Appl Pharmacol, 16:754, 1970.
41. Wockel, W., Stegner, H.E. and Janisch, W.: Zum topochemischen Quecksilbernachweiss in der Niere bei experimenteller Sublimatvergiftung. Virchows Arch Path Anat Physiol, 334:503, 1961.

DISCUSSION

Goldwater: First I'd like to compliment Dr. Magos for having presented a great wealth of material very concisely. I think we could spend almost the whole day digesting what he presented in a very short time. I have a question or two and maybe a comment. You mentioned, Dr. Magos, a sort of protective action of ascorbic acid. Yesterday, mention was made of spironolactone as having had a protective action. If one looks at the literature one will find that pretreatment with mercuric chloride will protect rats against mercuric chloride. This seems to me as something we might speculate about, whether this is a different type of protective mechanism or whether this is a nonspecific sort of thing that might be produced by who knows what kind of compounds. What this is all about, I think, is a fascinating subject and I'd like to have your thoughts on this. And one other point: you mentioned the difference in retention between mercuric chloride which is sometimes called "corrosive sublimate" and mercury cysteine. Is part of this difference due actually to local tissue trauma from a highly corrosive compound which may in some way impede it's being taken up because the circulation is destroyed or something similar?

Magos: Doses used in these experiments were high: 2, 4 and 8 mg/kg. With these doses I was able to show that with the increased retention at the injection site there was an increase in the severity of edema in the control rats compared with those pretreated with ascorbic acid. This certainly was due to the fact that there was an irritation there. In the case of ascorbic acid, capillary permeability was decreased or non-irritative complexes were formed and that's why edema was less pronounced. When mercury was given with excess cysteine, cysteine neutralized by competition the effect of local thiol groups on the retention of mercury.

In analyzing the protective effect of mercuric chloride against mercuric chloride two possibilities are worthwhile for consideration. Firstly, slight tubular damage caused by the first dose might decrease the reabsorption of mercury from the tubular urine or the transport of mercury from the peritubular capillary space. Secondly, the first dose might increase the glomerular filtration of plasma proteins due to glomerular damage resulting in a subsequent increase of protein uptake by tubular cells. Such a protective effect was observed by frequent injections of hemoglobin.*

Goldwater: I'd like to carry this a little further because I think it's of extreme importance. First of all we know that there are three widely different compounds, all of which increase resistance or protect the animal. It's established that feeding small amounts of something can produce tolerance. This may be a factor in today's problems in which people are eating foods which presumably have dangerous amounts of this or that chemical.

If they've been doing this in small doses over a long period of time probably or possibly they develop some sort of protection in themselves, an immunity. So to relate the acute animal experiments to what happens to human beings who probably have been ingesting small amounts of things all of their lives I think can result in some misleading conclusions.

Piotrowski: I may have an answer to one of the questions Dr. Goldwater asked: namely, why mercuric chloride may cause higher resistance against mercuric chloride. I will show in my paper that mercury is being bound in the kidney to a great extent by metallothionein. When the rat is pretreated with mercury there occurs a stimulation of the biosynthesis of this chelating protein in the kidney resulting in a considerable increase of the binding capacity.

* Havill et al.: *J Exp Med*, 55:627, 1932.

A question to Dr. Magos: Can you offer an explanation for the mobilizing effect of maleic acid?

Magos: My theory is based on two observations. One is that a non-efficient dose of penicillamine is made efficient by sodium maleate.

The second observation is that sodium maleate did not decrease the concentration of protein thiol groups. Consequently it seems reasonable to suppose that maleate reacts with a non-thiol binding site which we can call X. The bond between maleate and X must be ionic because the effect of maleate is abolished by BAL.

We believe that mercury with one valence is attached to an SH and with one valence to X. Sodium maleate has a higher affinity to X than has mercury and when mercury is liberated from this bond by maleate it's more accessible to penicillamine. Actually maleate should decrease the effect of penicillamine because by reaction with cysteine it decreases the concentration of the diffusible thiol compounds. But the opposite effect is obtained. The effect lasts for more than 24 hours; during this time maleate concentration in the kidneys equals the concentration of the strong binding sites. It might be that the strong binding sites are strong because they have a nearby X, to which mercury is able to be linked by its second valence. In the case of weak binding sites, there is no nearby X but other groups and the bond between these other groups and mercury is very weak. Consequently penicillamine, cysteine or any diffusable thiol compound is able to react with the second valence of mercury without the help of the maleate.

Chapter 11

THE UPTAKE AND DISTRIBUTION OF METHYLMERCURY IN THE BRAIN OF *SAIMIRI SCIUREUS* IN RELATION TO BEHAVIORAL AND MORPHOLOGICAL CHANGES

M. Berlin, G. Nordberg and J. Hellberg

ABSTRACT

Scintillation counting, autoradiography, and a visual discrimination test were used to study the effects on 22 squirrel monkeys of MeHg given *per os* or intravenously. After ingestion of labelled MeHg, the blood concentration of mercury rose sharply for 24 hours, decreased rapidly until the fourth or fifth day, and thereafter declined more slowly with a half-life of 50 to 60 days. During an observation period of three weeks, the concentration of mercury in the blood was linearly related to dose and body burden up to a level of 1500 ng/g, and more than 95 percent of the ingested MeHg was absorbed. After 85 days, about 50 percent of the body burden was found in the fur. There was a latent period between peak of body burden and development of behavioral signs which decreased as the absorbed dose increased. Cortical damage was more widespread with a slow accumulation of MeHg than with a rapid increase in body burden, and additional behavioral signs were observed. A sudden visual disturbance occurred with subacute exposure, while a gradual constriction of the visual field along with impaired motor coordination and possible sensory disturbances were found with prolonged exposure.

REPORT

Introduction

At the International Conference of Occupational Medicine in Tokyo, 1969 (6), we reported that squirrel monkeys show signs similar to man at intoxication due to methylmercury (MeHg) exposure. This finding and the fact that these monkeys have a large brain in comparison to body size make them very suitable for studies of the uptake, retention and distribution of mercury in the brain. We have conducted such studies and combined them with psychological tests to ascertain if the toxic effects in the brain cause changes in behavior that can be correlated with the observed morphological changes and with the distribution of mercury in the brain (see Chap. 17). Dr. Grant will report on the morphological changes in Chapter 17.

Thus, the purpose of our studies has been as follows:

1. To elucidate the relation between the dose of MeHg absorbed and the concentration of mercury in the body and blood.

2. To study the kinetics of mercury distribution in the brain at ingestion of MeHg.

3. To relate changes in the behavior of the monkeys observed in standardized test situations to the duration of MeHg exposure and to the body burden of MeHg measured as concentration in the blood.

4. To study the relation between morphological changes, body burden of MeHg, and behavioral changes.

Material and Methods

Material

The investigations were made on squirrel monkeys of both sexes that weighed between 500 and 1000 grams. Methylmercury^{203} nitrate with a specific activity of 2 C/g or less was prepared according to procedures described earlier (1). MeHg was given *per os* mixed in food or by stomach tube. Blood samples were taken in the tail vein or vena saphena. Scintillation counting was used to determine the radioactivity of blood and the whole body. Measurements were made over a collinated sodium iodide crystal with a constant geometry and compared with a phantom standard that contained a known amount of MeHg of the same specific activity.

Chemical Analysis of Mercury

Chemical analyses of mercury in the blood and organs were performed according to the method of Schütz (7) with combustion in an oxygen atmosphere, absorption of mercury in potassium permanganate and reduction with stannous chloride followed by the determination with an atomic absorption spectrometer of mercury vapor in gas phase.

Autoradiography

Autoradiography was performed after sectioning of the frozen brain at $-12°C$ with a Leitz sleigh microtome. Sections were collected on tape according to the method of Ullberg (8) and apposed on x-ray film as described earlier (2). Kodax Structurix and Kodirex x-ray film were used.

Psychological Testing

Seven squirrel monkeys were trained to discriminate visually between two stimulus objects that differed in color, form and size. A modified Wisconsin General Test Apparatus was used which was mounted on the front of the living cage and contained two sliding trays that held the stimulus objects. A response consisted of pulling a tray within reach by means of an attached chain. A raisin reward was hidden on the tray with the correct stimulus object.

A new pair of stimulus objects was presented on each test occasion, a procedure which resulted in the formation of a learning set. Each stimulus pair was presented for 18 trials with an intertrial interval of about 20 seconds. The "correct" object was selected at random in advance.

After preliminary training, daily doses of MeHg were administered *per os* to four of the seven monkeys. The remaining three monkeys served as matched controls. Testing was conducted three times weekly until observable symptoms developed when daily tests were made.

Results were analyzed with respect to number of correct choices, response latency (time to select a chain), and response duration (time to pull in the tray). Notes were made on the motor coordination of each subject as manifested in the ability to grasp and pull the chain and to pick up the raisin. Motion picture films of the test procedure were taken both of control subjects and of experimental subjects at various stages of MeHg intoxication.

Pathological-anatomical Investigations on the Brain

After perfusion with phormol calcium, the brains were removed and parted at the midline. One half of each brain was frozen and sectioned sagitally for autoradiography. The other half was sectioned transversally in blocks 2 to 3 mm thick which were embedded in paraffin, sectioned, and deparaffinized for phase-contrast microscopy or for staining and light microscopy. Some transverse sections were also taken for autoradiography.

Results

The Relation Between Dose and Concentration of Mercury in Blood, Whole Body and Head

The relation between blood concentration and body burden of MeHg was studied in five monkeys given labelled MeHg by stomach tube every fourth day to build up the blood concentration to a toxic level. In two other monkeys given repeated doses of labelled MeHg, the concentration in blood and whole body was followed for 85 days after exposure ceased. The results of these experiments indicate that the blood concentration of mercury rises rapidly after ingestion and reaches a maximum within 24 hours. There is a phase of rapid decline over the next three days, followed by a slower decline that gives a half-life of about 50 to 60 days. Figure 11-1 shows the blood concentration in five monkeys during the six days after ingestion of the first dose. It can be seen that the period of rapid decline ends on the fourth day. The slow phase is illustrated in Figure 11-2, which shows that the blood concentration in two monkeys observed for 85 days after final exposure decreases with a half-life of 50 to 60 days, while the body burden decreases with a half-life of more than 150 days. The slower decline in the body burden is explained by the fact that half of the whole body concentration is found in the fur after 85 days.

In Figure 11-3, the relation between cumulative dose and blood concentration on the fourth day after every injection is

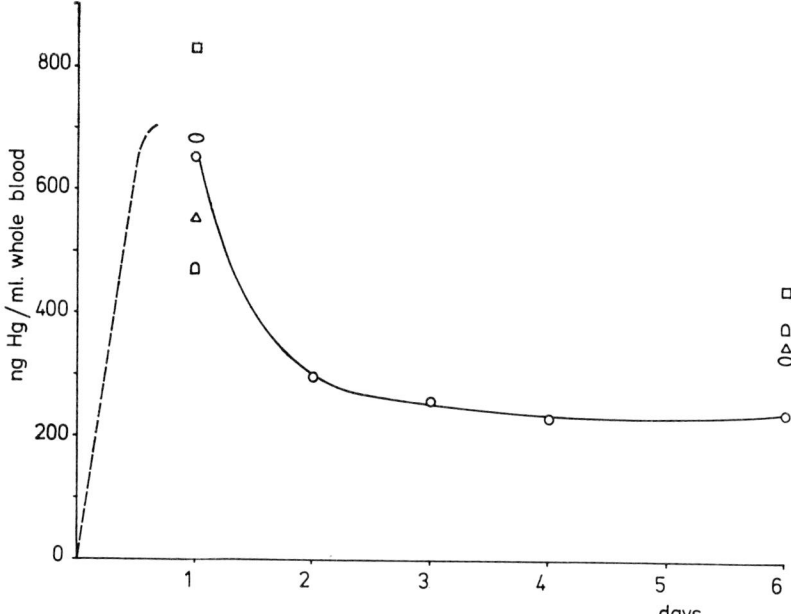

Figure 11-1. Blood concentration over six days after ingestion of a single dose of labelled MeHg in four monkeys sampled twice and one monkey sampled daily.

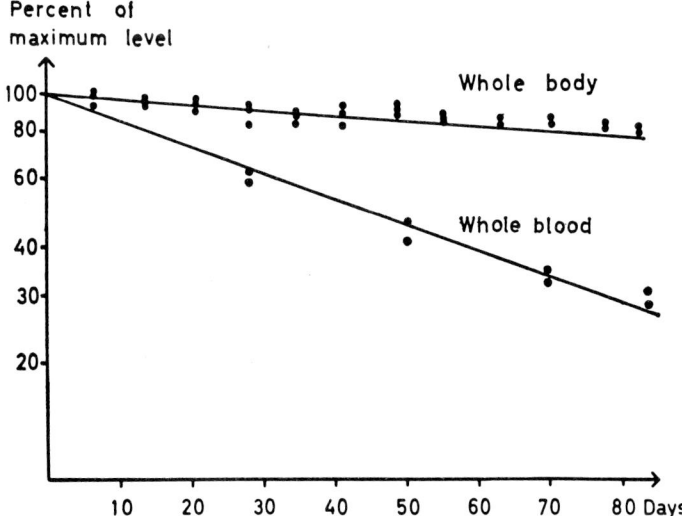

Figure 11-2. Decline in wholebody and blood concentration of mercury followed for 85 days after final exposure.

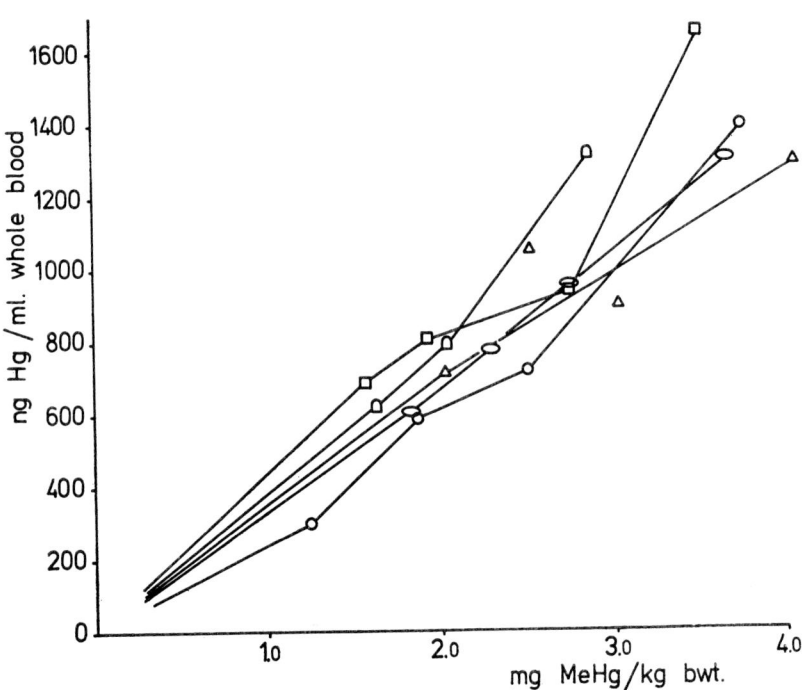

Figure 11-3. Blood concentration of mercury in five monkeys after repeated ingestion sampled every fourth day as a function of cumulative dose.

illustrated. It indicates that there is a linear correlation between blood concentration and body burden.

The whole body and head content of mercury were measured by scintillation counting in the five monkeys given consecutive doses. These are shown in relation to cumulative dose in Figures 11-4 and 11-5, respectively. It can be seen that both the whole body and head content of mercury increase linearly with dose.

The Relation Between Dose and Latent Period for Development of Signs

In early pilot experiments, it was observed that the latent period between cessation of exposure and development of signs varied from experiment to experiment. To determine what factors influence this latency, seven monkeys were given daily doses of

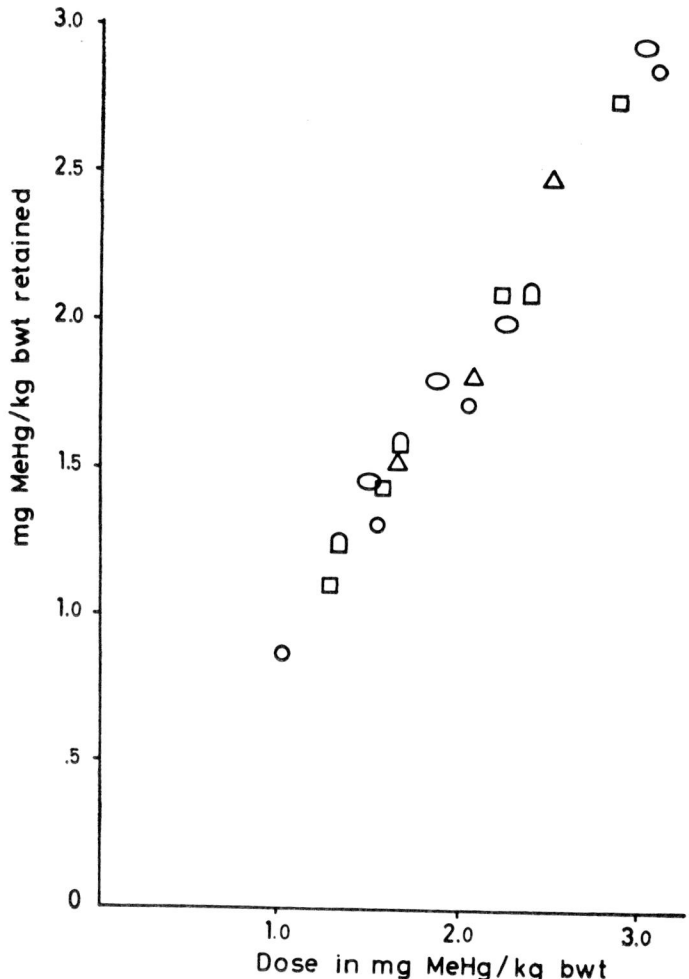

Figure 11-4. Wholebody concentration of mercury in five monkeys after repeated ingestion sampled every fourth day as a function of cumulative dose.

MeHg labelled with mercury[203] *per os* for 3 to 5 weeks. Exposure was controlled such that the blood concentration of mercury attained levels that varied from 0.5 to 2 ppm between subjects. Table 11-I gives the maximum concentration reached. It can be seen from Figure 11-6 that the latent period decreased as the dose of MeHg increased. The animal with the lowest dose

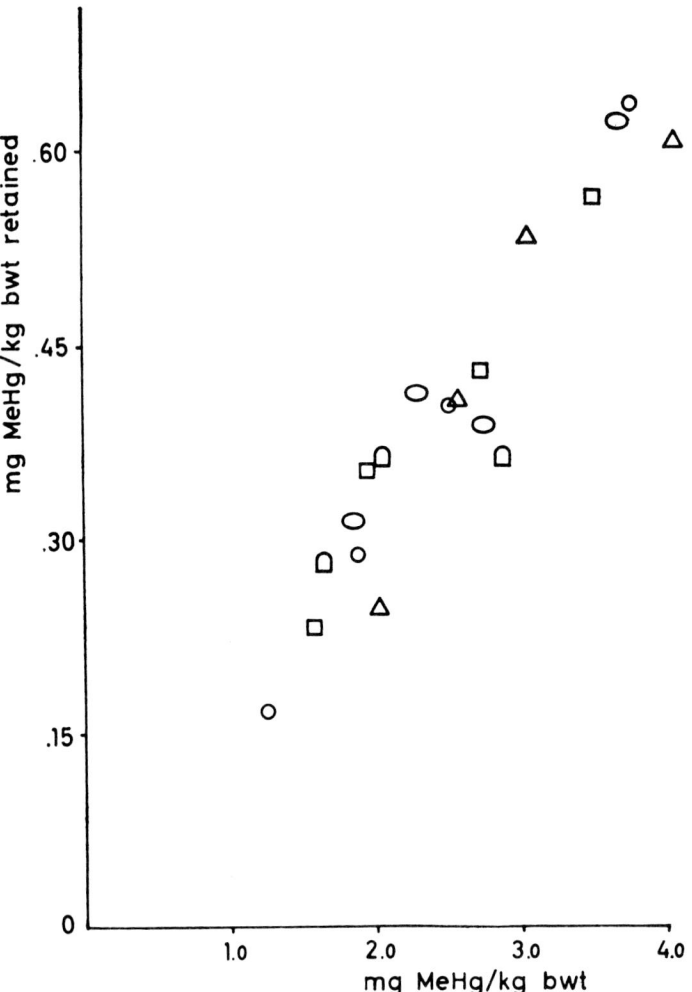

Figure 11–5. Mercury concentration in the heads of five monkeys after repeated ingestion sampled every fourth day as a function of cumulative dose.

(maximum 0.75µg/g whole blood) did not show any clinical or morphological signs after 15 weeks of observation.

The Distribution of Mercury in the Brain in Relation to Time

Two monkeys were given labelled MeHg (specific gravity, 2 C/g) in a single I.V. injection of about 0.5 mC. After eight

TABLE 11-I
RELATION BETWEEN DOSE, BLOOD CONCENTRATIONS, LATENT PERIOD AND SIGNS

Monkey No.	Max. Blood Conc. of MeHg (µg/g)	Max. Retained Dose (mg/kg)	Latent Period* (days)	Clinical Signs	Morphological Changes
3	1.80	5.5	0	Blindness	Severe damage in calcarine cortex
04	1.80	—	12	Blindness	
2	1.75	5.0	0	Blindness	Moderate damage in calcarine cortex
1	1.60	5.5	6	Blindness	Severe damage in calcarine cortex
4	1.40	4.3	37	Slowly progressing visual disturbance	Most severe damage in entire occipital cortex
5	1.20	3.6	85	No signs	Slight but evident changes limited to calcarine cortex
6	0.75	2.8	85	No signs	No changes

* Time between appearance of symptoms and dosing with mercury.

days, the brains were perfused with phormol calcium and dissected out for autoradiography. As the autoradiogram in Figure 11-7 shows, high concentrations of mercury were found in the cerebral cortex, cerebellar cortex and basal ganglia as compared to white matter. Apart from this general difference between white and gray matter, there are no significant concentration differences between different regions of the cerebrum, cerebellum or brain stem.

Six animals were given daily doses *per os* of labelled MeHg for 3 to 5 weeks. Four were sacrificed when clinical signs of MeHg intoxication became apparent (Table 11-I, Monkeys 1-4). Two animals given a lower dose were sacrificed 12 weeks after termination of exposure (Table 11-I, Monkeys 5 and 6). Figure 11-7 presents an autoradiogram from a sagittal brain section of a monkey that was exposed for four weeks (Monkey 1, Fig. 11-6), and showed obvious clinical signs of intoxication. A high concentration of mercury was found subcortically in the cerebrum. The concentration was most pronounced in the occipital region (Fig. 11-8), particularly around the calcarine sulcus. By means of an isotopic standard staircase, the concentration of mercury in the subcortical layer of the calcarine cortex was estimated to differ from other parts of the brain by a factor of about 16.

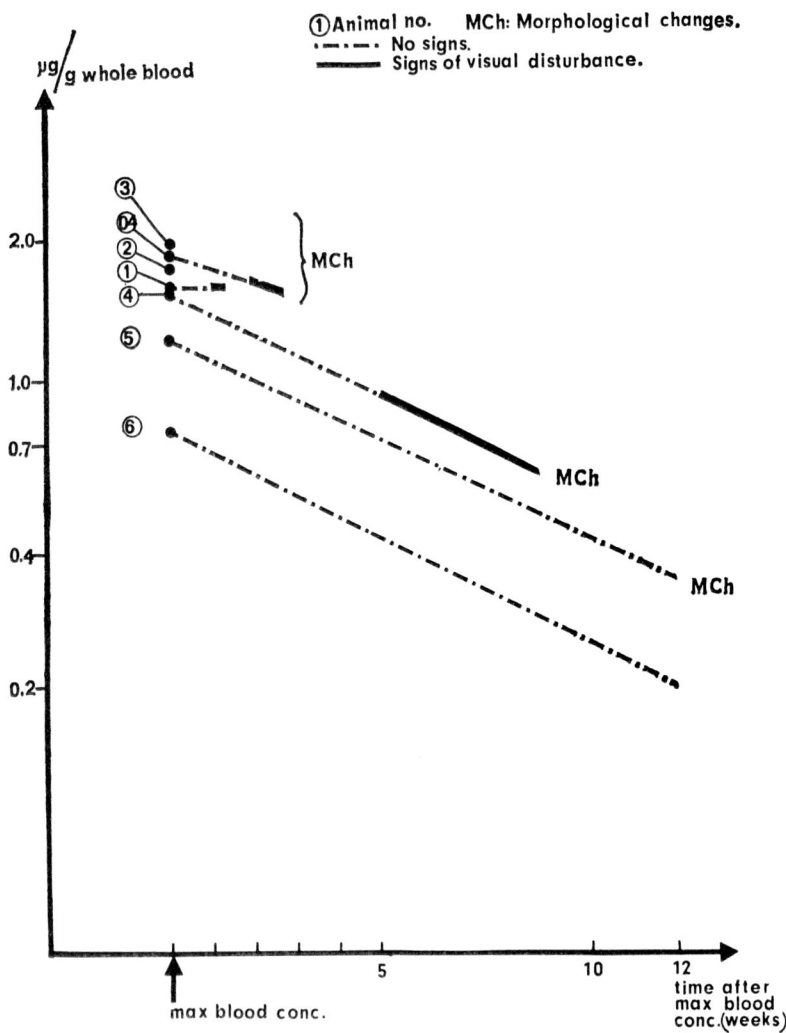

Figure 11-6. Latent period before development of signs as a function of blood concentration of mercury in seven monkeys.

The Relation Between Observed Behavior, Clinical Signs and Type of Exposure to MeHg

In earlier experiments with subacute exposure, a sudden visual impairment was observed which appeared after a certain latent period following the end of exposure. This impairment was in-

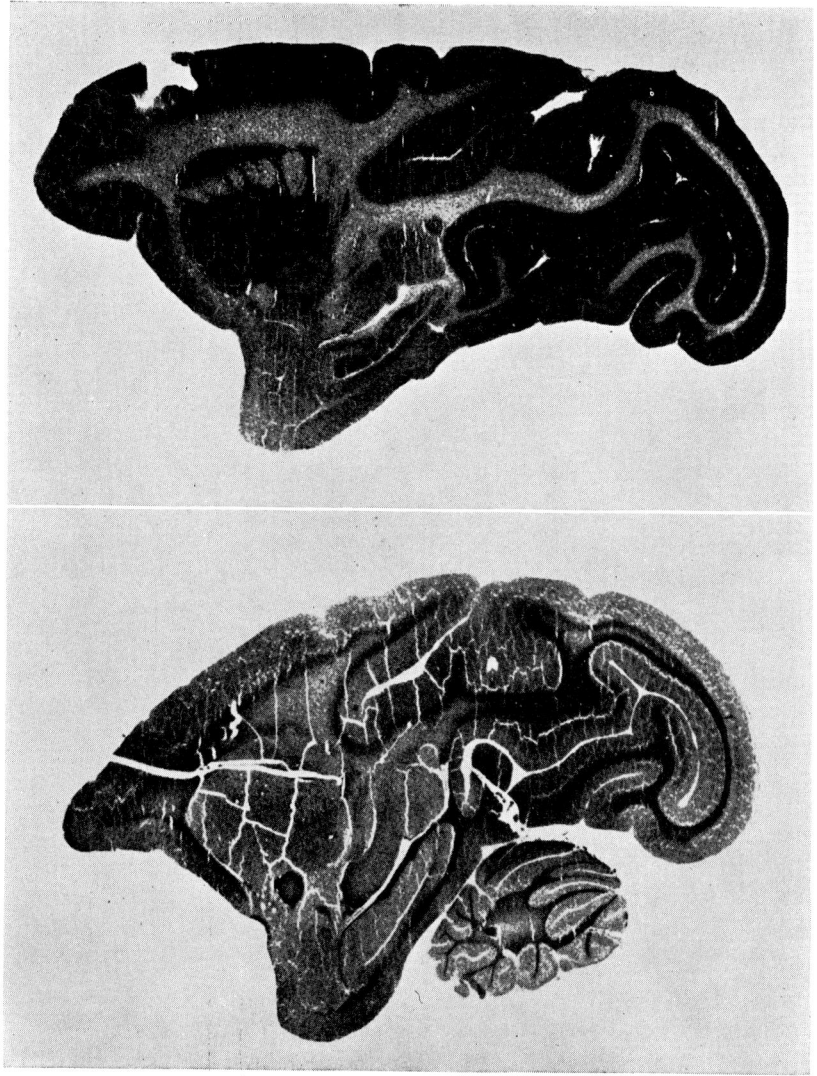

Figure 11–7. Autoradiograms of sagittal brain sections from monkeys given labelled MeHg. *Upper,* Monkey sacrificed eight days after a single I.V. injection. *Lower,* Monkey sacrificed after four weeks of daily exposure.

terpreted as being complete blindness because the monkeys did not respond to light or to sudden movement. In other monkeys with more prolonged exposure, vision was affected more gradually,

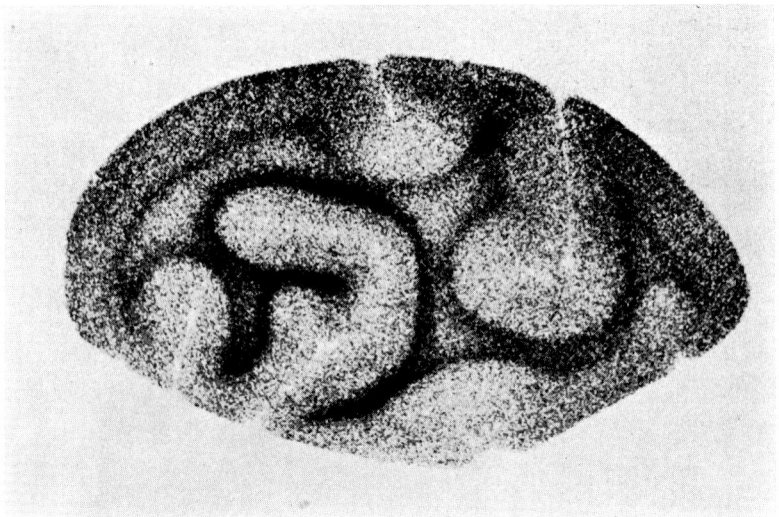

Figure 11-8. Autoradiogram of a transverse section from the paraffin-embedded occipital lobe of the monkey shown in Figure 11-7, *lower*.

and there were signs of defective motor coordination. The visual discrimination experiment was conducted to test these observations more objectively.

Two of the monkeys were exposed to MeHg in a subacute way, while the other two were given a more prolonged exposure. Figure 11-9 shows the rise in blood concentration in these animals and the time at which obvious clinical signs appeared.

During the period of mercury accumulation, visual discrimination performance was stable and did not differentiate the experimental monkeys from the controls. The average response latency, duration, and number of correct choices over the 25 or more test occasions before the first appearance of signs is given for each monkey in Figure 11-10 along with the standard deviation. The average score and variation for all monkeys on each of these measures is given in Table 11-II, while Table 11-III presents the average difference between paired experimental and control monkeys and the variation of the differences.

Test results did not differentiate the two monkeys with suba-

Behavioral and Morphological Changes

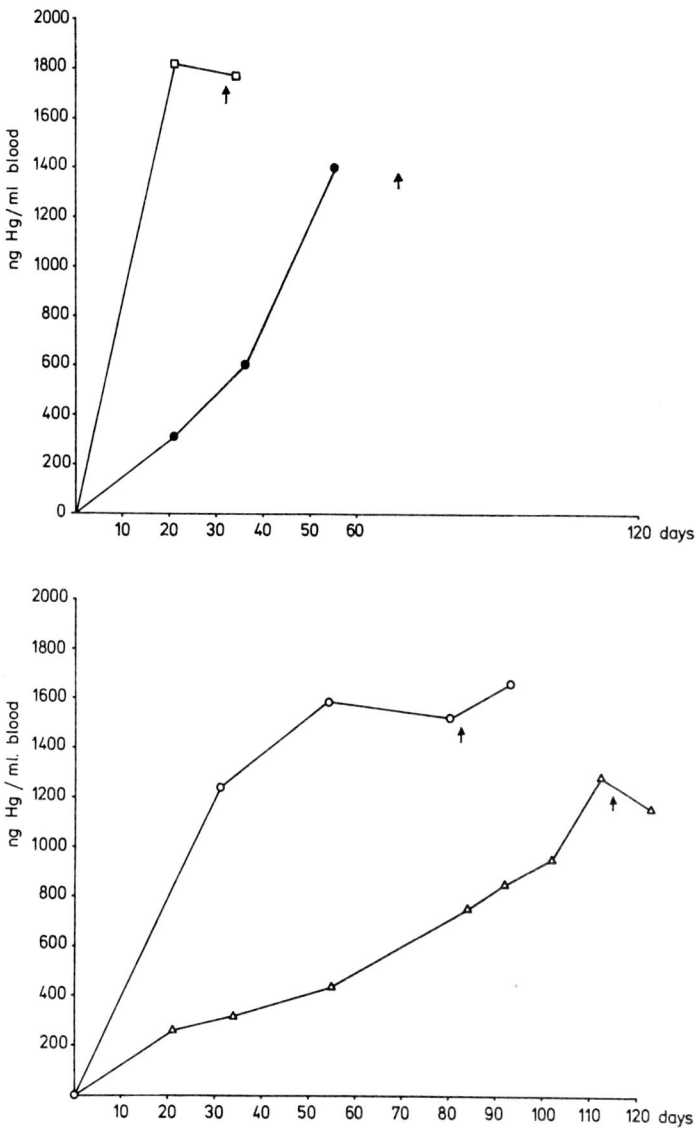

Figure 11-9. Blood concentration in monkeys with subacute and prolonged exposure to MeHg (top and bottom, respectively). Arrows indicate the onset of obvious clinical signs.

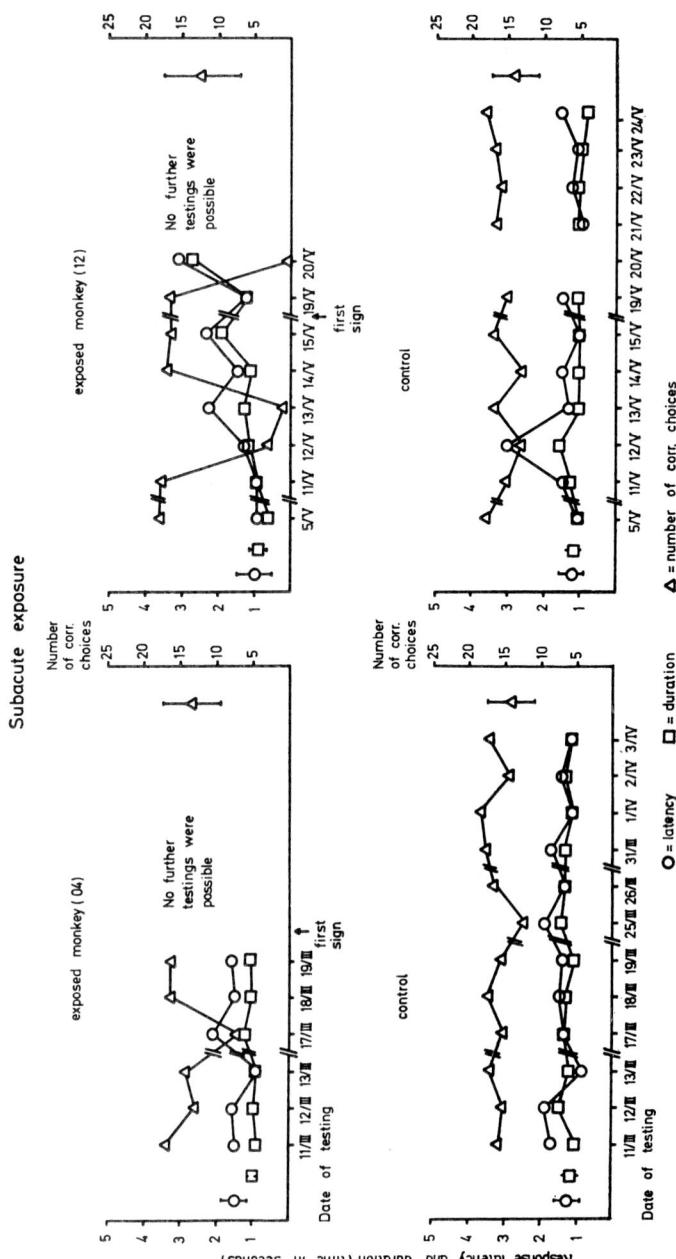

Behavioral and Morphological Changes 201

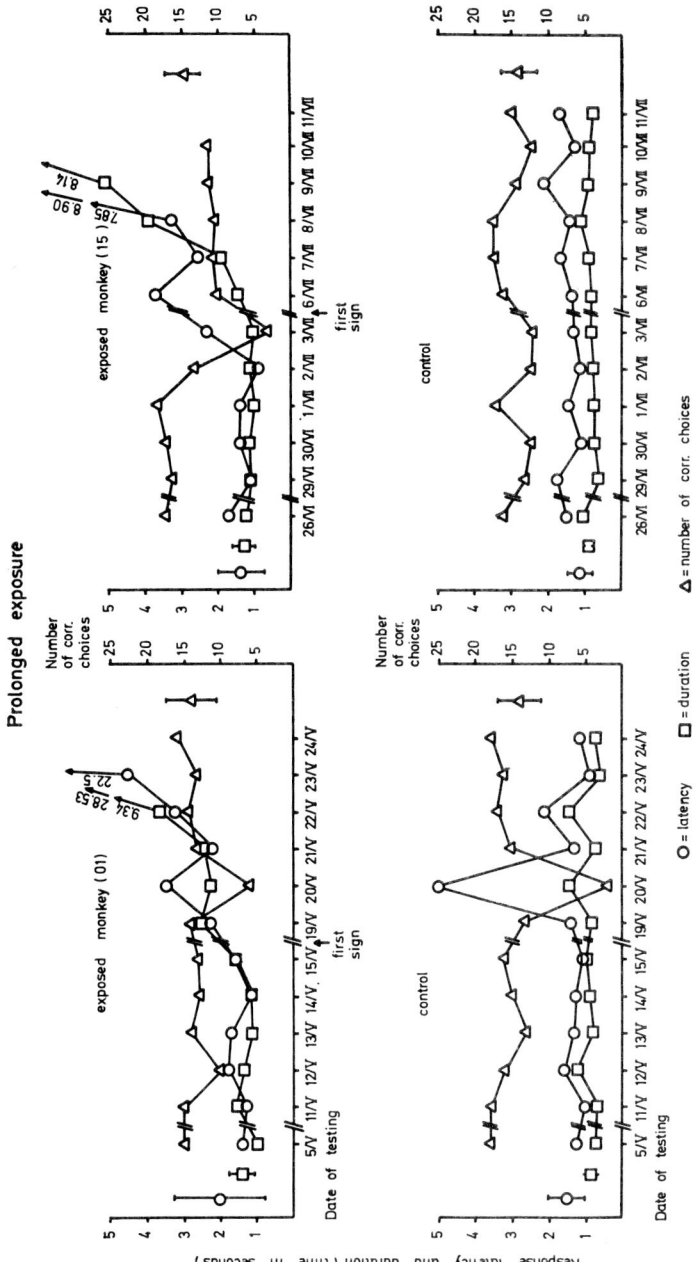

Figure 11–10. Test scores on three performance measures of visual discrimination (latency, duration, and number of correct choices) for four experimental and three control monkeys after the onset of signs and for the six prior test days. Average test score and standard deviation for each monkey before the onset of signs is indicated close to the relevant axis.

TABLE 11-II

MEAN AND COMPONENTS OF VARIANCE FOR SEVEN MONKEYS DURING LATENT PERIOD ON THREE MEASURES OF VISUAL DISCRIMINATION (N = 7)

Test Measure	Response Latency	Response Duration	Number of Correct Choices
Mean	1.42	1.12	13.9
Variance between Ss*	.08	.04	.4
Variance within Ss	.39	.05	11.8
Total variance	.57	.09	12.2

* Ss is "samples."

cute exposure from the controls until the sudden onset of severe visual impairment.

The first behavioral signs in the two monkeys with more prolonged exposure was an increase in response latency and in time to pull in the stimulus tray (see Fig. 11–10). This coincided with an observable disturbance in the fine movement of the fingers and in the visual-spatial coordination. These monkeys had obvious difficulties in locating a fixed object in space or to grasp it when found.

In the week following the appearance of the first symptom, the monkeys developed a severe visual impairment, a gradual constriction of the visual field, which prevented further testing. It could be observed that the increase in response latency was due to the fact that the monkeys required time to get the stimuli into the visual field. A certain clumsiness, possibly caused by sensory disturbances in the fingers, was also seen in the way the monkeys handled the chain and picked up the reward. In spite of these difficulties in performing the learned test procedure,

TABLE 11-III

MEAN AND COMPONENTS OF VARIANCE FOR DIFFERENCES BETWEEN PAIRED EXPERIMENTAL AND CONTROL MONKEYS DURING LATENT PERIOD ON THREE MEASURES OF VISUAL DISCRIMINATION (N = 3 PAIRS)

Test Measure	Response Latency	Response Duration	Number of Correct Choices
Mean of differences	.33	.22	.02
Variance of differences between Ss*	.04	.08	.3
Variance of differences within Ss	.50	.09	7.5
Total variance	.55	.17	7.8

* Ss is "samples."

the monkeys continued to try until the last day. In one monkey, the number of correct choices remained above chance, indicating that the learning set itself had not deteriorated. The monkey with the poor choice score was observed to concentrate on getting the end of the chain within view and probably was not able to see the stimulus objects at all.

Discussion

The experiments relating dose of MeHg to blood concentration and body burden of mercury show that the mercury concentration in blood rises rapidly for 24 hours after a single dose *per os*, followed by a rather rapid decrease during a phase of distribution. This rapid decrease diminishes after four to five days, and the blood concentration continues to decline with a rate parallel to that of the whole body. This decrease corresponds to a half-life of about 50 to 60 days in comparison to the half-life of the body burden which is more than 150 days. The difference is explained by the large amount of MeHg, 50 percent of the body burden, found in the fur at the end of the 85-day observation period.

It is obvious that a linear correlation exists between absorbed dose of MeHg and concentration of mercury in blood after the first phase of distribution which lasts four days. This linearity holds for blood concentrations of mercury greater than 1000 ng/g. On the basis of these findings, we have preferred to use blood concentration to index the body burden of mercury, especially since in a later state of prolonged exposure a rather large amount of the body burden is located in the fur, an uninteresting location from a toxicological point of view. It can be assumed that the concentration in blood reflects concentration in tissues and organs for better than the total body burden. Also, from a practical point of view, monkeys tend to handle their food before ingestion and to spread it around. Hence, a more accurate index of absorbed dose will be obtained by following blood concentration, providing that sampling during the rapid changes of the first four days is avoided.

The congruence between dose administered by stomach tube

and body burden measured with wholebody counting shows that the absorption is very efficient and the excretion, small. Our calculations indicate that more than 95 percent of the administered dose is absorbed.

The finding that about 15 percent of the body burden is to be found in the head is consistent with the finding in man. No delay in mercury uptake in the brain was observed during the three weeks of observation in our experiments. This is in contrast to what has been found in experiments with mice (3) and in tracer studies in man (4). However, it is possible that an increase in the proportion of mercury found in the brain in relation to the rest of the body might have been observed had the observation period been longer than three weeks. This will be investigated in further experiments.

From the distribution of mercury in the brain seen in the autoradiograms, it can be concluded that there is a redistribution of mercury requiring at least a month. This redistribution seems to be correlated with the development of toxic effects. However, it cannot be determined from this limited material whether the redistribution causes the toxic effect or results from it. Investigations are being conducted to clarify this question, as well as to localize the subcortical accumulation of mercury on a cellular level.

The purpose of the behavioral study was to test the ability of the monkeys to learn a visual discrimination based on form, color, and size, and to test their motor ability, coordination, and capacity to learn. The test also provided a standardized occasion on which to observe and to record changes in behavior. From the results it is evident that the behavioral symptomology differs, depending on whether intoxication was subacute or more prolonged. Ten monkeys with subacute exposure observed in our laboratories have shown similar symptoms with very little variation. With a more prolonged exposure and a slow accumulation of mercury in the body, the symptoms appeared more gradually and were more similar to those reported in man. These signs include sensibility disturbances and loss of coordination in peripheral extremities, followed at a later stage by constriction of the visual field (7). In our monkeys, as well as in man, brain injury seemed to be limited to visual cortex, primary sensory and motor areas, while higher functions were unaffected. The monkeys remained motivated to perform in spite of their handicaps, and the learning set persisted in one instance.

In the earlier stages of intoxication, the visual disturbance seemed to involve only the peripheral vision. The foveal vision evidently functioned until which time the monkeys could no longer perform the test. Autoradiogram results indicate that the turnover of mercury in visual cortex, where the most pronounced morphological changes were also observed, was faster than in other parts of the brain. It is possible that a slow accumulation of mercury affects a larger part of the brain and therefore involves more function. In fact, morphological changes were more widespread in monkeys with prolonged exposure than with subacute exposure, and these monkeys showed additional symptoms in the area of motor ability and coordination. However, motivation and learning capacity were apparently unaffected.

REFERENCES

1. Berlin, M.H.: Renal uptake, retention and excretion of mercury. II. A study in the rabbit during infusion of methyl- and phenylmercury compounds. *Arch Environ Health*, 6:626, 1963.
2. Berlin, M.H. and Ullberg, S.: Accumulation and retention of mercury in the mouse. I. An autoradiographic study after a single intravenous injection of mercury chloride. *Arch Environ Health*, 6:589, 1963.
3. Berlin, M.H. and Ullberg, S.: Accumulation and retention of mercury in the mouse. III. An autoradiographic comparison of methylmercuric dicyandiamide with inorganic mercury. *Arch Environ Health*, 6:610, 1963.
4. Miettinen, J.K.: Personal communication, 1969.
5. Nordberg, G.F., Berlin, M.H. and Grant, C.A.: Methylmercury in the monkey—autoradiographical distribution and neurotoxicity. Proc. 16th Int. Congr. Occup. Health, Tokyo, in press, 1970.
6. Report from an Expert group: Methylmercury in fish: A toxicologic—epidemiologic evaluation of risks, *Nord Hyg Tidskr*, (suppl 4) 1971.
7. Schütz, A.: Metod för bestämning av små mängder kvicksilver i blod, urin och annat biologiskt material. Rapport 691020 från yrkesmed klin vid Lunds lasarett, 1969.

8. Ullberg, S.: Studies on the distribution and fate of S^{35}—labelled benzylpenicillin in the body. *Acta Radiol,* (suppl.) 118, 1954.

DISCUSSION

Gage: You said that the half-life of mercury was about 60 days, but at that time 50 percent was in the fur. Does that mean the fur is the major route of excretion?

Berlin: The half-life in blood was about 50 days; they had 50 percent in the fur at the 85th day.

Kazantzis: I wonder what is the neurological lesion involved in relation to the difficulty which the monkey had at getting at the raisin. There were difficulties in getting to the raisin and in picking it up; this leaves us with the possibilty that there could be a posterior column lesion in the spinal cord causing loss of position sense, a cerebellar lesion causing incoordination or a parietal cortical lesion in which the monkey was unable to put sensory impressions together. Do you think this is primarily due to posterior column loss, damage to the cerebellum or to damage to the parietal center? If we want to investigate further, it would not be too difficult to develop a test in which the monkey discriminates objects placed in its hands. It would be important to define the functional lesion accurately, as this may precede any visible histological change.

Berlin: Dr. Grant will report later about specific lesions. As for a cortical lesion, we are trying to develop more specific tests since this is a distinct possibility.

Kazantzis: I would be a little surprised to find a parietal cortical lesion because the monkey had no difficulty in recognizing the raisin and knew what to do with it once it got it; what seemed to be the difficulty was getting the raisin in the first place.

Berlin: Sometimes the monkey took the raisin or believed it had the raisin in the hand and put the hand to the mouth even where there were no raisins in the hand. I feel that maybe the touch sense was affected.

Miettinen: Dr. Berlin, you mentioned that about 12 percent of the body burden was in the brains.

Berlin: In the head.

Miettinen: Do you know the percentage of body burden in the brain?

Berlin: We do not yet have that data available.

Xintaras: What has your laboratory done in relation to looking at cortico electroencephalographic (EEG) signals and correlating them with brain tissue levels of methylmercury?

Berlin: We haven't, so far, done anything. There are some reports in the literature (Niigata Report, "Report on the cases of mercury poisoning in Niigata," Ministry of Health and Welfare, Tokyo. Stencils, 1967. Jap; Sw translation). It seems that rather unspecific changes appear but again if you have the animal as his own control, I feel rather certain that this could be used to observe changes in some areas, e.g. in visual cortex. I therefore feel this may be a good approach but of course then you have to interfere with the monkey—you have to train it to have electrodes in the skull for at least five months.

Xintaras: Recently I had developed for behavioral toxicity studies two microminiature solid state beta probe detectors for picking up *in vivo* tissue levels of beta label methylmercury.

I would like to suggest that if your laboratory is interested in on-line testing in relation to your behavioral and chemical studies I would make available one of these probes. This would permit continuous *in vivo* monitoring. If your interest is in mathematically modeling the uptake of methylmercury and in also relating the tissue level with behavioral changes this would be a very good system to work with.

Berlin: We tested that idea some years ago. We made such a miniature probe. However, the irradiation from the mercury-203 is both beta and gamma rays and we did at that time come to the conclusion that the background irradiation of gamma rays from the rest of the body would interfere so much that there would be a question as to whether you were getting the local concentration, but if you use two probes— one shielded for beta and one not shielded—then you can subtract the effect of gamma rays.

Xintaras: Carbon-14-methylmercury would permit microlocalization within a half millimeter of the tissue.

Berlin: Yes, that would be a possibility.

Magos: What is your opinion whether metabolism of methylmercury depends on the dose: the higher the dose the less methylmercury is metabolized in monkeys per unit of time?

Berlin: There are obvious species differences in respect to biotransformation of methylmercury. The different half-lifes between species may be explained by a different rate of breakage of the bond but so far, we have no indication of whether there is any considerable breakage of the bond in this animal.

Grant: Dr. Miettinen, what have you seen with your EEG recordings?

Miettinen: We have not used EEG for the methylmercury-intoxicated animals as of today but it would be most interesting to correlate the neurochemistry with what has to be an electrophysiological change with the eventual corresponding behavioral change.

Kurland: Would peripheral nerve velocity measurements be useful to distinguish peripheral nerve from central involvement?

Berlin: So far we have no indications of any involvement of the peripheral nerves, but it may be a question of time of exposure.

Norseth: Dr. Berlin, do you know the form of mercury in the fur of these monkeys? Is it inorganic or is it still methylmercury? Is this evidence of the breakdown in the body of the monkey?

Berlin: We have no data whatever on that.

Jernelöv: I might point out regarding the discussion of the form of methylmercury in fur, we found that 75 to 80 percent of the mercury excreted in fur is as methylmercury.

Miettinen: Östlund's experiment* reported dose response for the half-time; when he had a high dose it was 12 days and the higher the intoxication of the animals the longer the half-time. We have determined in fish this same dose response and found that it is the other way around.

When you approach the toxic doses the half-time is radically reduced to about half of what it is in the low-dose area.

Magos: I can confirm Dr. Jernelöv's findings; in the skin and fur of rats it was about 80 to 90 percent of the total mercury as methylmercury. So there was little difference in the percentage of the methylmercury if the whole body or fur plus skin was assayed. Dr. Jernelöv can you comment on half-life studies in fish?

Jernelöv: Yes, but it's complicated. When you calculate your fish content and biological half-life from concentrations you run into a problem because the excretion in fish is a small factor generally compared to dilution through growth. The growth rate is more important than excretion. But actually the growth by dilution should decrease the concentration and not increase it.

Wood: Dr. Berlin, did you observe any excessive fighting in the monkeys?

Berlin: No. They were not aggressive. They don't like to be caught and they do fight a little if you try to catch them, but these are friendly animals, especially if they are used to your handling them.

Fassett: Is there any possibility that the fur level actually results from dust contamination rather than the diet?

Berlin: Not really. It doesn't make any difference if you give it in this form or in a form bound to protein. It obviously results from metabolic inclusion of methylmercury at the formation of the pile.

* Östlund, K: *Acta Pharmacol et Toxicol,* 27 (suppl): 1, 1969.

Chapter 12

METABOLIC FATE OF ETHYLMERCURY SALTS IN MAN AND ANIMAL

Tsuguyoshi Suzuki, Tai-ichiro Takemoto,
Hiroshi Kashiwazaki and Tomoyo Miyama

ABSTRACT

By using methods for estimating the inorganic and total mercury content of biological specimens, the metabolism of ethylmercury salts was studied in man and animals. The C-Hg bond of ethylmercury salts has been shown to break very rapidly and to a great extent in men, who were patients and were transfused with a commercial product of human plasma with 0.01% sodium ethylmercurithiosalicylate, and also in mice injected subcutaneously or intravenously with ethylmercury chloride or sodium ethylmercurithiosalicylate solution.

The increasing level of inorganic mercury and its percentage to total mercury content in the brain were quite distinguishable with post-injection time in mice, which resulted in a longer biological half-life time of total mercury than that reported for methylmercury injection.

The mercury in red cells from patients was mainly organic; in contrast that of the plasma and urine was inorganic. Only the former showed a rapid decrease with time.

REPORT

Introduction

The toxic nature of ethylmercury has been considered to be fairly similar to that of methylmercury salts (2). In the recommendation of an international committee on Maximum Allowable Concentration for mercury and its compounds, ethyl-

The authors should like to express their thanks to the staffs of the Clinics of Pediatrics and of Surgery, and the Department of Blood Supply in the University Hospital and the Department of Pathology in the Faculty for their kind cooperation, and Prof. Katsunuma and Assoc. Prof Koizumi for their kind reviewing of the manuscript.

mercury was grouped with methylmercury salts (15). Reports on human intoxication with ethylmercury salts have usually reported symptoms similar to those of methylmercury poisoning, which is accentuated by the typical neurological symptoms (8, 9, 12, 14, 31, 37), although there have been a few reports that noted slightly different symptoms from the typical features of methylmercury poisoning (10, 12).

In acute experiments on animals, ethylmercury has an LD_{50} similar to that of methylmercury salts (32) and a high neurotoxicity similar to that of methylmercury and n-propylmercury salts (3, 22). In relation to neurotoxicity expressed as mercury content in the brain when neurological symptoms appear, the toxic dose of ethylmercury was suggested to be higher than that of methylmercury salts (27, 40).

In comparative studies on bodily distribution of mercury after administration of alkylmercury salts in mice or rats, ethylmercury showed a slightly different pattern from methylmercury and n-propylmercury salts; for instance, there was a relatively larger retention of mercury in the kidney and/or in the liver, and a smaller one in the brain with ethylmercury as compared to methylmercury and n-propylmercury salts (24, 26, 38). A difference was also found in rats between methylmercury and ethylmercury salts in the distribution of mercury in the blood; the ratio of red cells: plasma was 30:1 for ethylmercury salts, but about 200:1 for methylmercury salts at 24 hours after a single subcutaneous injection (11).

As to the biotransformation of methylmercury salts, controversial results on the *in vivo* release of inorganic mercury have been reported (1, 6, 19). But, chemical analyses estimating total and organic mercury, or total and inorganic mercury, have revealed the existence of inorganic mercury in organs, tissues and excreta (6, 19).

The controversy is mainly about the chemical form of mercury in the brain after methylmercury administration. Norseth and Clarkson (19) noted that 2.8 percent of the total mercury in the brain was in the inorganic form at one to ten days after injection of a single dose of 1.0 mg Hg/kg in rats. Hitherto, Miller *et al.* (17), Sadakane (21), Takeda *et al* (35), Nose (20) and Kitamura *et al* (13) investigated the metabolism of ethylmercury salts in experimental animals. With the exception of Takeda *et al.* who extracted all mercury in the liver, kidney and excreta with dithizone CCl_4 and separated different forms of mercury with chromatography (33, 34), the organic mercury was separated by solvent extraction and the amount estimated by gas-chromatography or colorimetry using dithizone. About 11 to 46 percent of the total mercury in the rat brain was found as inorganic mercury after intraperitoneal injection of ethylmercury phosphate and about 22 to 38 percent after a 190-day feeding period (21); ethylmercury constituted between 25 and 50 percent of the total mercury in the brain for 72 hours after intravenous injection of ethylmercury chloride or Thimerosal (ethylmercurithiosalicylate) (13). The observed percentage of inorganic to total mercury in the brain are noteworthy in comparison to the value obtained (2.8%) in the case of methylmercury injection.

Our intention in this report is to describe the metabolic fate of ethylmercury salts by using a method which estimates inorganic mercury in the presence of organic mercury. We had to study this problem after the unfortunate episode in the Hospital of the University of Tokyo, where "Human Plasma" with sodium ethylmercurithiosalycilate added as a preservative had been used occasionally for the treatment of patients; autopsy has, however, revealed large amounts of retained mercury in all organs and tissues. Additional chemical analyses from one patient autopsy, follow-up studies on patients administered the "Human Plasma," and experiments with animals administered ethylmercury salts have been carried out.

Methods for Analysis of Mercury

Magos and Cernik's method (16), using a chemical reaction by stannous chloride at high pH and an ultraviolet mercury vapor photometry, was used for estimation of inorganic mercury in the undigested urine, red cells, plasma, tissue homogenate and hair extract. For the total mercury content, all specimens were oxidized with a mixed solution of concentrated H_2SO_4 and HNO_3 or solutions of concentrated H_2SO_4 and 3% $KMnO_4$ in a reflux condenser. By these two analytical procedures estimation of inorganic and total mercury, nonoccupational exposure (presumably to methylmercury in food) and occupational exposure to mercury have been separately estimated in workers exposed to mercury vapor (29), and the values for oxidized red cells or whole blood, and oxidized extract of hair coincided well with those obtained with activation analysis (36). Gage and Warren (7) modified slightly Magos and Cernik's method; they used the lability of organic mercurials in the presence of acid cysteine and reported estimations for distinguishing between organic and inorganic mercury in urine, feces, blood and kidney tissue.

Inorganic and total mercury were estimated by the method described above for mice exposed to mercury vapor. Table 12-I shows results on the blood and brain of mice exposed to a very high concentration of mercury vapor for four hours. Two kinds of measurement yield nearly identical values for the blood and brain. In the repeated exposure to mercury vapor, similar results were obtained (Table 12-II).

TABLE 12–I
ESTIMATIONS FOR INORGANIC AND TOTAL MERCURY
IN MICE EXPOSED TO MERCURY VAPOR
(SINGLE EXPOSURE)

Exposure	Time	Organ		(A) Inorganic Hg* ($\mu g/g$)	(B) Total Hg ($\mu g/g$)	(A)/(B) ×100
8 mg/m³ 4 hrs	20 min	Brain	1.	1.74	1.79	97.2
			2.	2.08	1.91	109.0
		Red cells	1.	1.46	1.54	94.8
			2.	4.20	4.48	93.8
		Plasma	1.	0.83	0.90	92.2
			2.	1.26	1.34	94.0
6.5 mg/m³ 4 hrs	48 hr	Cerebrum	1.	0.15	0.12	125.0
			2.	0.17	0.12	142.0
			3.	0.13	0.11	118.0
		Cerebellum	1.	0.33	0.33	100.0
			2.	0.29	0.41	70.7
			3.	0.28	0.26	108.0
		Brain stem	1.	0.22	0.18	122.0
			2.	0.29	0.25	116.0
			3.	0.21	0.32	65.6

* Tissue homogenates used for estimation are about 10% (W/V) in phosphate buffer solution (pH 7.8).

Metabolic Fate of Ethylmercury Salts 213

TABLE 12-II
ESTIMATIONS FOR INORGANIC AND TOTAL MERCURY
IN MICE EXPOSED TO MERCURY VAPOR
(REPEATED EXPOSURE)

Exposure	Organ		(A) Inorganic Hg ($\mu g/g$)	(B) Total Hg ($\mu g/g$)	(A)/(B) $\times 100$
1 to 2 mg/m³ for	Kidney	1.	10.90	10.00	109.0
2 hrs in a day,		2.	10.30	12.00	85.8
5 times in a week	Liver	1.	0.30	0.27	111.0
and for 7 weeks		2.	0.33	0.30	110.0
	Cerebrum	1.	0.92	0.73	126.0
		2.	0.76	0.65	117.0
	Cerebellum	1.	1.70	2.42	70.2
		2.	1.41	1.96	71.9

Extractive and Nonextractive Mercury with 1 N HCl in Tissues from Mice Exposed to Mercury Vapor or Ethylmercury Chloride

Mercury in the hair of workers exposed to mercury vapor and of people without any occupational exposure to mercurials, and that in the placental tissue obtained after spontaneous delivery of healthy pregnant women was almost completely extracted with 1 N HCl solution (29, 30). Whether or not the extraction with 1 N HCl would be applicable to other tissues in different conditions of contact with various mercurials was tested.

One gram of each organ was homogenized with 5 ml of 1 N HCl solution in a glass homogenizer. The homogenate was washed by adding 5 ml of 1 N HCl solution, and centrifuged at 3,000 rpm for about ten minutes. The supernatant was separated, and 10 ml of 1 N HCl solution was added to the precipitate, after which it was stirred with a small glass rod and separated again by centrifugation. An aliquot of combined supernatant was oxidized by solutions of concentrated H_2SO_4 and 3% $KMnO_4$ at 60° C for ten minutes in a reflux condenser.

The percentage of extracted to total amount of mercury is shown in Table 12-III. A striking difference was observed between mice exposed to mercury vapor and those injected with ethylmercury chloride. Hydrochloric acid did not extract the mercury retained in the organs and tissues, especially the brain tissues, after vapor exposure, but did efficiently extract the mercury after ethylmercury injection. The extent of extraction varied in organs with both mercurials, and its order was quite opposite.

TABLE 12-III
EXTRACTION OF MERCURY WITH 1 N HCl SOLUTION

Compound	Exposure or Dose	Organ	Time* (hr)	Extracted Amount of Hg %†	Range
Mercury vapor	1 to 2 mg/m³ for 2 hrs per day, 5 times in a week and for 7 weeks	Liver	24	29.3‡	25.9– 32.7
		Kidney	24	5.9‡	5.0– 6.7
	2 to 3 mg/m³ for 2 hrs per day, 5 times in a week and for 3 weeks	Liver	24	10.9	3.4– 25.0
		Kidney	24	11.8	9.2– 13.2
		Cerebrum	24	3.8	0.8– 5.9
		Cerebellum	24	0.9	0.7– 1.3
		Brain stem	24	0.5	0– 0.9
Ethyl Hg Cl	2.4 mg Hg/kg (I.V.)	Liver	4	65.6	52.5– 74.5
			144	52.9	37.8– 62.8
		Kidney	4	68.3	59.4– 78.8
			144	68.1	63.5– 72.5
		Brain	4	87.3	80.0– 92.9
			144	95.7	79.3–105.0

* Time after cease of final exposure to mercury or after injection.
† Values are the average of data obtained from three mice or two.‡

Distributions of Mercury in the Dead Patient Who Received Repeated Transfusion of "Human Plasma" Including Ethylmercury Salt

The patient was a 13-year-old boy having suffered from protein-losing enteropathy since four years of age. He was repeatedly hospitalized with recurrent worsening of his illness, and the final admission was at about one year before death. Transfusion of whole blood or plasma had been indispensable in supplementing the loss of protein through the intestinal membrane. When fresh blood was available, the blood was transfused, but the blood type of the boy was unfortunately AB, and fresh blood of the AB type was not readily obtainable for such a repeated and long-term need. The commercial product of human plasma ("Human Plasma"), in which less than 0.01 percent of sodium ethylmercurithiosalicylate (Merthiolate Na) was added as a preservative, was therefore purchased and transfused to the boy.

Total transfused volume of "Human Plasma" had amounted to 9,000 ml from November 16, 1969 to February 13, 1970 (death occurred on February 18, 1970). The usual daily dose was 200 or 400 ml and the transfusion was carried out at one- to seven-day intervals.

Contents of total and inorganic mercury in "Human Plasma" were determined as 31, 5 and 0.20µg/ml, respectively, from a commercial sample. So, the total amount of mercury administered was calculated at being 283.5 mg based on a determined content or 450 mg by the highest content stated by the producer.

Total, inorganic and HCl-extractive mercury were estimated from frozen specimens taken at autopsy. Blood and urine were sampled at three or four days before death, when the physician in charge first asked us to estimate mercury content. Figure 12–1 shows the results for total and inorganic mercury contents. The highest content was noted on the proximal one third of the hair. Then, the order from high to low contents in the tissues was as

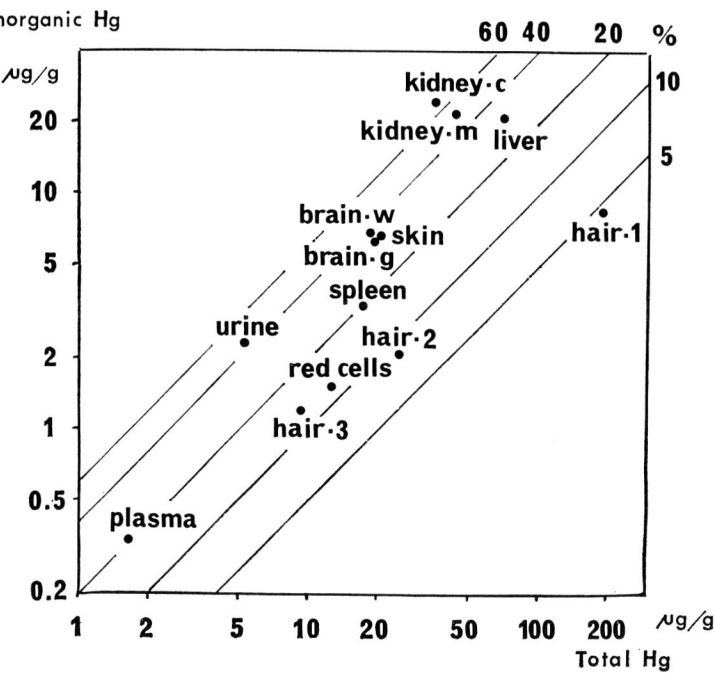

Figure 12–1. Percentages of inorganic to total mercury from a dead patient who prior to dying had received long-term and repeated transfusions of commercial human plasma containing 0.01% sodium ethylmercurithiosalicylate for his protein-losing condition. Kidney-c and kidney-m mean the cortex and medulla of the kidney. Brain-w and brain-g; white and grey matter. Hair 1, 2 and 3 are each one third from proximal to distal parts.

TABLE 12-IV
TOTAL MERCURY, INORGANIC MERCURY AND HCl-EXTRACTIVE MERCURY IN ORGANS AND TISSUES (FROZEN SPECIMENS) FROM A DECEASED PATIENT

Organ and Tissue	(A) Total Hg (μg/g)	(B) Inorganic Hg [μg/g (B/A × 100)]	HCl-Extractive Hg	
			(C) Total Hg [μg/g (C/A × 100)]	(D) Inorganic Hg [μg/g (D/C × 100)]
Cerebrum (Frontal lobe)				
Grey matter	19.4	6.3 (32)	17.0 (88)	1.0 (6)
White matter	18.8	6.8 (36)	20.8 (111)	0.7 (3)
Skin (Head)	20.3	6.7 (33)	13.1 (65)	0.3 (3)
Kidney				
Cortex	35.3	24.3 (69)	13.9–21.1 (39–60)	1.7–4.0 (12–19)
Medulla	43.2	21.9 (51)	12.5–16.5 (29–38)	1.8–1.9 (12–14)
Liver	69.4	21.4 (31)	43.3 (62)	7.1 (17)
Spleen	17.3	3.4 (20)	9.4 (54)	1.1 (12)
Hair 1 (Proximal)	—	—	186.7	8.7 (5)
2 (Middle)	—	—	24.7	2.1 (9)
3 (Distal)	—	—	9.3	1.2 (13)
Red cells	12.5	1.5 (12)	—	—
Plasma	1.7	0.3 (20)	—	—
Urine	5.3	2.3 (44)	—	—

follows; the liver, the kidney, the middle part of the hair, the skin, the brain, the spleen and others. The largest percentage of inorganic to total mercury was in the cortex of the kidney. Both white and grey matter of the brain had similar contents of total mercury and percentages of inorganic to total mercury.

The percentage of inorganic to total mercury was compared with two methods; one was our regular procedure in which total mercury was estimated from oxidized tissue and the inorganic mercury from an homogenate in phosphate buffer; the other procedure yielded the total mercury from the oxidized HCl-extract and the inorganic mercury from the unoxidized one. Results are shown in Table 12-IV with the HCl-extractable mercury, the inorganic mercury was less both in the actual level and in the percentage to the total as compared to the mercury estimated by the regular procedure. As observed in animals injected with ethylmercury chloride (Table 12-III), the HCl solution extracted most of the mercury in the brain.

The distribution of mercury in various parts of the nervous system was studied by a combined procedure of HCl extraction, oxidation and application to mercury vapor photometry (Tables 12-V and 12-VI). Except for a slightly higher content in the cerebellar cortex and a marked reduced content in the lumbar spinal ganglion and *Nervus ischiadicus*, no significant difference was observed, even between grey and white matter of the three lobes of cerebral cortex.

According to two possible calculations regarding the total administered dose of ethylmercury, the percentage of mercury distributed in each organ or tissue varies considerably (Table

TABLE 12-V
COMPARISON OF MERCURY CONTENTS* BETWEEN GREY
AND WHITE MATTER OF CEREBRUM

		Grey Matter ($\mu g/g$)	White Matter ($\mu g/g$)
Parietal lobe		13.9	22.7
Occipital lobe	1	18.5	17.8
	2	18.3	16.2
Frontal lobe		17.0	19.8

* In Tables 12-V and 12-VI, mercury contents were measured by a combination of HCl-extraction, oxidation of extract and application to mercury vapor photometry, and the specimens were fixed with ten percent Formalin solution for about one month.

TABLE 12-VI
MERCURY CONTENTS IN VARIOUS PARTS OF THE NERVOUS SYSTEM

Part	Hg Content ($\mu g/g$)
Cerebellar cortex	23.9
Mesencephalon	
Substantia nigra	13.2
Corpora quadrigemina	18.3
Tectum opticum	18.2
Nucleus ruber	20.3
Spinal cord (the uppermost part)	
Grey matter	18.1
White matter	16.1–19.3
Lumbar Spinal ganglion	9.3
Nervus ischiadicus	9.2

12-VII). The retained percentage in the brain, being important in relation to the neurotoxicity of ethylmerucry salts, was calculated to be 5.4 to 8.6 percent.

TABLE 12-VII
PERCENTAGE DISTRIBUTION OF MERCURY TO TOTAL ADMINISTERED DOSE IN A DECEASED PATIENT

Organ	Weight (kg)	Hg Content ($\mu g/g$)	Amount of Hg (mg)	% (1)*	% (2)†
Brain	1.28	19.0	24.3	8.6	5.4
Liver	1.10	69.4	76.3	26.9	17.0
Kidney	0.24	40.0	9.6	3.4	2.1
Spleen	0.08	17.0	1.4	0.5	0.3
Blood	1.80‡	7.0	12.6	4.4	2.8
Skin	2.00‡	20.0	40.0	14.1	8.9
Sum	6.5	—	164.2	57.9	36.5

* The value calculated from the administered dose of 283.5 mg Hg.
† The value calculated from the administered dose of 450 mg Hg.
‡ Estimated from the body weight (23 kg) at autopsy.

Follow-up Observations on Four Patients Treated Surgically and Administered with "Human Plasma"

The death of the patient who received a long-term administration of "Human Plasma" and had shown symptoms of poisoning with alkylmercury salts, alerted medical personnel whose patients had received a transfusion of "Human Plasma" after surgery. The mercury contents in red cells, plasma and urine from four patients is shown in Table 12-VIII. During the follow-up observations, they did not show any suspectable neurological symptoms, and

Metabolic Fate of Ethylmercury Salts

TABLE 12-VIII
PATIENTS EXAMINED FOR MERCURY CONTENT IN RED CELLS, PLASMA AND URINE AFTER THE ADMINISTRATION OF "HUMAN PLASMA"

Patient	Sex	Age	Body Weight (kg)	Dose of "Human Plasma" (ml)	Dose of Ethyl Hg as Hg (mg)	Frequency of Transfusion	Illness
H.S.	m.	65	43.0	6600	208–330	18	Gas gangrene of legs
M.T.	f.	61	45.0	800	25.2–40	4	Cancer in the pancreas
T.S.	m.	51	36.5	700	22.1–35	3	Cancer in the pancreas
Y.K.	m.	79	69.0	100	3.15–5	1	Ileus

one of them (T.S.) succumbed from a malignant tumor in the pancreas.

A decrease of mercury with time after the final administration was observed only in red cells, in which the level of inorganic mercury was the lowest among the three kinds of specimens (Fig. 12–2). The biological half-life time of mercury in red cells was calculated to be about one week. Mercury content in the plasma stayed almost constant except for one patient who received a single transfusion of 100 ml of "Human Plasma." Plasma mercury contents and daily excreted amounts of mercury in the urine were relatively similar in both total and inorganic mercury (Fig. 12–3).

Metabolic Fate of Ethylmercury Injected Subcutaneously or Intravenously in Mice

Two experiments were carried out. In experiment 1, aqueous solutions of ethylmercury chloride (EMC) and of sodium ethylmercurithiosalicylate (Merthiolate Na) were injected subcutaneously at a dose of 2 mg Hg/kg in female mice (about 25g) of ICR-JCL strain. After this single injection, three mice from each EMC and Merthiolate Na group were sacrificed (by guillotine) at 4, 48, 96 and 168 hours; the brain, liver and kidney were removed for analyses of mercury. The total mercury and inorganic mercury were estimated by the regular procedure described above.

In experiment 2, the EMC solution was injected intravenously

220 Mercury, Mercurials And Mercaptans

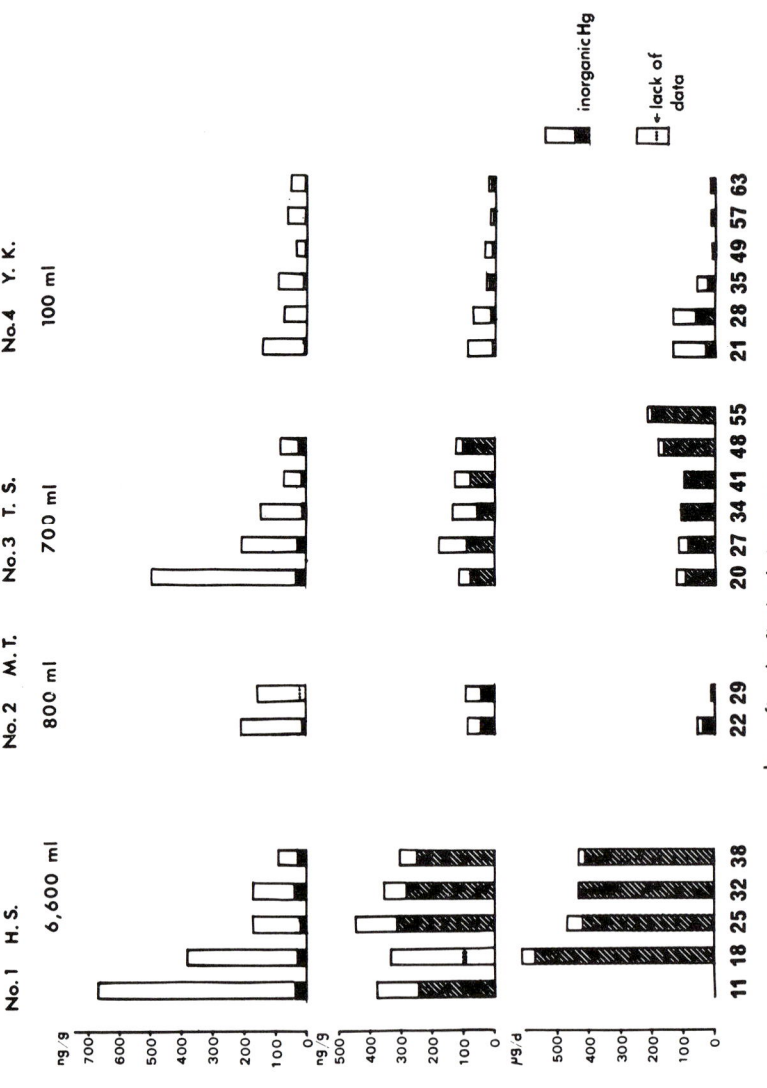

Figure 12–2. Levels of inorganic and total mercury in red cells, plasma and urine in four patients who received the transfusion of human plasma after surgery. The number below the patient's name (e.g. 6,600 ml) indicates the transfused volume of human plasma.

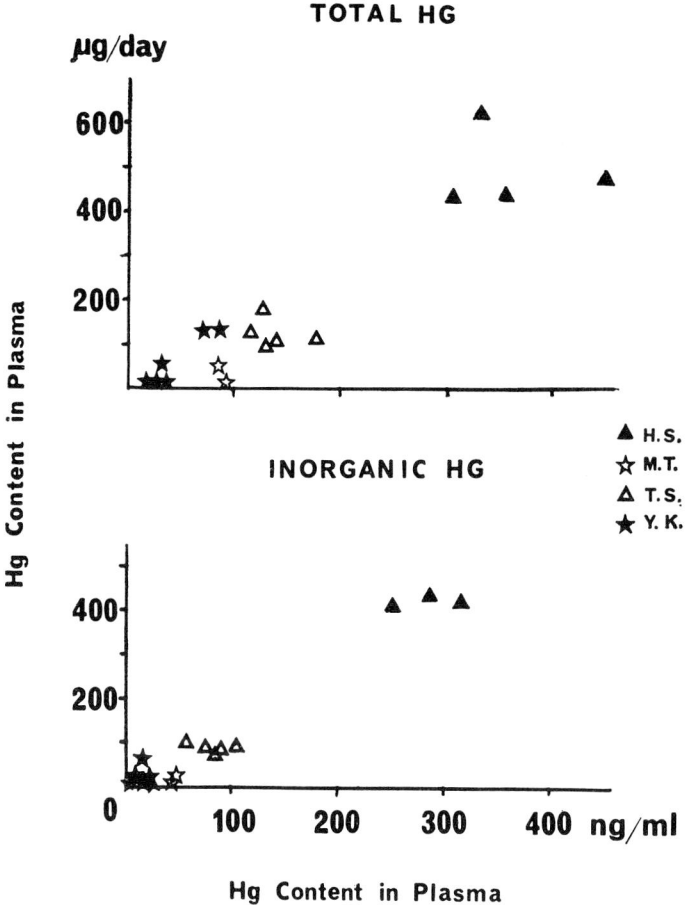

Figure 12–3. Relationships of total mercury or inorganic mercury content in plasma to each amount excreted daily. The various symbols represent different patients.

to male mice (about 40g) of ICR-JCL strain. After a single injection of a dose of 2.4 mg Hg/kg, three mice were sacrificed (by guillotine) at 4, 24, 96, 144, 216 and 312 hours. The estimation of mercury was carried out as above.

Percentage Retention of Mercury in Organs

The results of experiments 1 and 2 are shown in Figures 12–4 to 12–6. Percentages (means and ranges) of total mercury in organs to the injected dose are arranged with the time course.

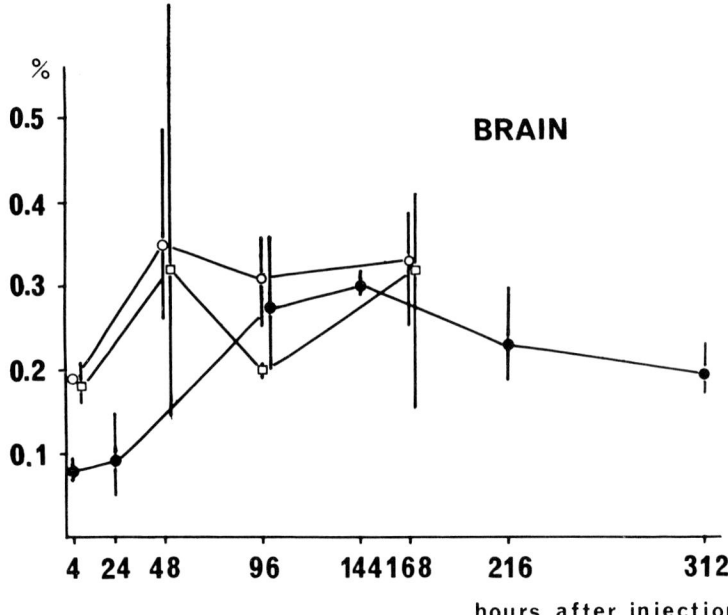

Figure 12–4. Change in the percentage distribution of total mercury in the brain after a single injection of ethylmercury salts to mice. Solid circle: intravenous injection, ethylmercury chloride, 2.4 mg Hg/kg; open circle: subcutaneous injection, ethylmercury chloride, 2.0 mg Hg/kg; and open square: subcutaneous injection, sodium ethylmercurithiosalicylate, 2.0 mg Hg/kg. Each dot and bar means the average and the range of the data obtained from three mice.

There was no great difference between the EMC and Merthiolate Na groups in the subcutaneously injected series. The intravenous injection gave a different percentage retention among three organs; e.g. much more retention in the kidney and less in the brain. The observation of less retention in the brain after intravenous injection relative to that retained from subcutaneous injection was similar to that observed with inorganic mercury salts (18). With inorganic mercury salts, the intravenously injected mercury was retained less in the kidney and more in the liver as compared to similar subcutaneously injected; the percentage of retention in each organ after subcutaneous injection of ethylmercury compared well to those of previous reports (24, 26, 38).

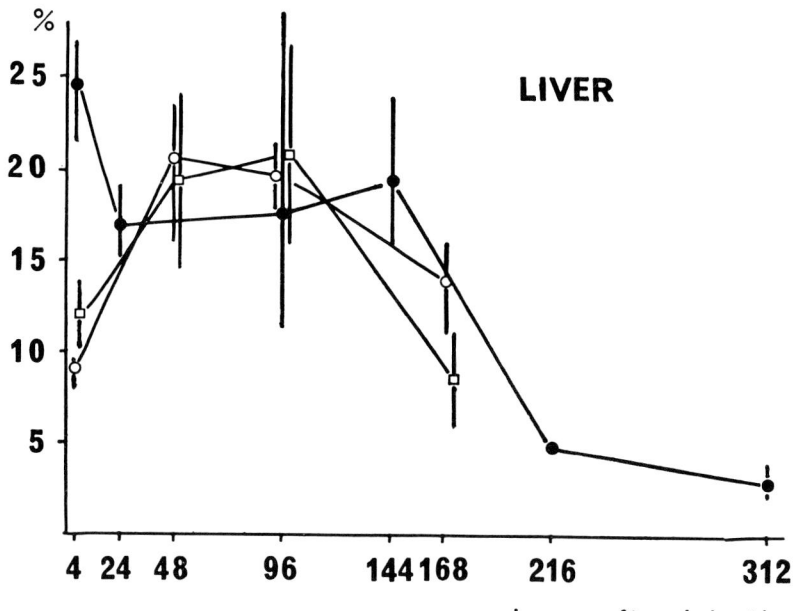

Figure 12–5. Change in the percentage distribution of total mercury in the liver after a single injection of ethylmercury salts in mice. Symbols are the same as those in Figure 12–4.

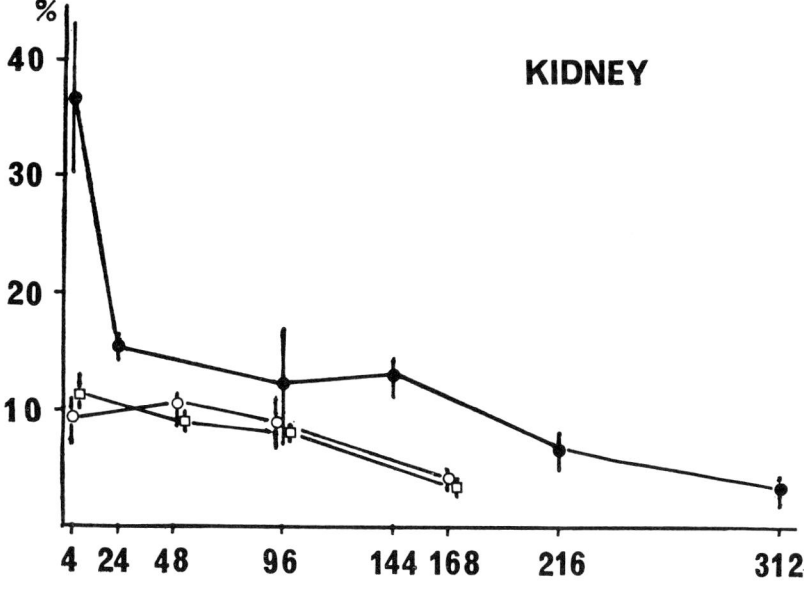

Figure 12–6. Change in the percentage distribution of total mercury in the kidney after a single injection of ethylmercury salts in mice. Symbols are the same as those in Figure 12–4.

Inorganic Mercury in Each Organ

Percentages of inorganic to total mercury in the organs showed gradual increases with post-injection time with two ethylmercury salts with two routes of administration (Tables 12-IX–12-XI).

TABLE 12-IX
INORGANIC MERCURY IN THE BRAIN, LIVER AND KIDNEY AFTER A SINGLE SUBCUTANEOUS INJECTION OF ETHYLMERCURY CHLORIDE (2 mg Hg/kg) IN MICE*

Time (hrs)	Brain % (range)	Liver % (range)	Kidney % (range)
4	4.7 (3– 6)	3.6 (2– 6)	3.8 (3– 5)
48	8.2 (6–10)	8.6 (7–10)	9.2 (8–10)
96	12.2 (11–13)	12.0 (10–14)	14.0 (10–19)
168	27.4 (19–41)	9.4 (8–11)	19.3 (13–25)
48†	—	6.3 (5– 8)	17.5 (14–20)
192†	—	6.2 (6– 7)	34.1 (28–41)

* Three mice per sample.
† Data were obtained from three rats injected with a dose of 20 mg Hg/kg by Takeda et al. (35).

TABLE 12-X
INORGANIC MERCURY IN THE BRAIN, LIVER AND KIDNEYS AFTER A SINGLE SUBCUTANEOUS INJECTION OF MERTHIOLATE Na (2 mg Hg/kg) IN MICE*

Time (hrs)	Brain % (range)	Liver % (range)	Kidney % (range)
4	4.5 (3– 8)	6.1 (5– 7)	5.1 (4– 7)
48	7.3 (5–10)	9.6 (8–11)	9.5 (7–12)
96	18.9 (17–22)	12.0 (6–18)	14.9 (11–18)
168	21.5 (19–25)	13.8 (12–16)	19.5 (17–21)

* Three mice per sample.

TABLE 12-XI
INORGANIC MERCURY IN THE BRAIN, LIVER AND KIDNEY AFTER A SINGLE INTRAVENOUS INJECTION OF ETHYLMERCURY CHLORIDE (2.4 mg Hg/kg) IN MICE*

Time (hrs)	Brain % (range)	Liver % (range)	Kidney % (range)
4	8.6 (– †)	7.5 (7– 8)	7.5 (7– 8)
24	12.6 (10–16)	7.5 (6– 9)	10.9 (10–11)
96	13.9 (12–18)	8.9 (8–10)	19.3 (14–25)
144	32.5 (30–35)	11.2 (9–15)	16.0 (13–19)
216	29.5 (24–36)	24.3 (21–30)	29.4 (22–30)
312	50.0 (47–52)	29.9 (28–32)	52.2 (42–70)

* Three mice per sample.
† Data from two mice were lost.

Noticeably, the percentage in the brain became the highest or almost similar to that in the kidney. The lowest value in all cases was in the liver.

Actual levels of organic mercury, calculated from the values of total and inorganic mercury, and of inorganic mercury are shown in Figures 12–7 to 12–10. A sharp rise in the level of inorganic mercury in the brain was observed in two experiments. In Figures

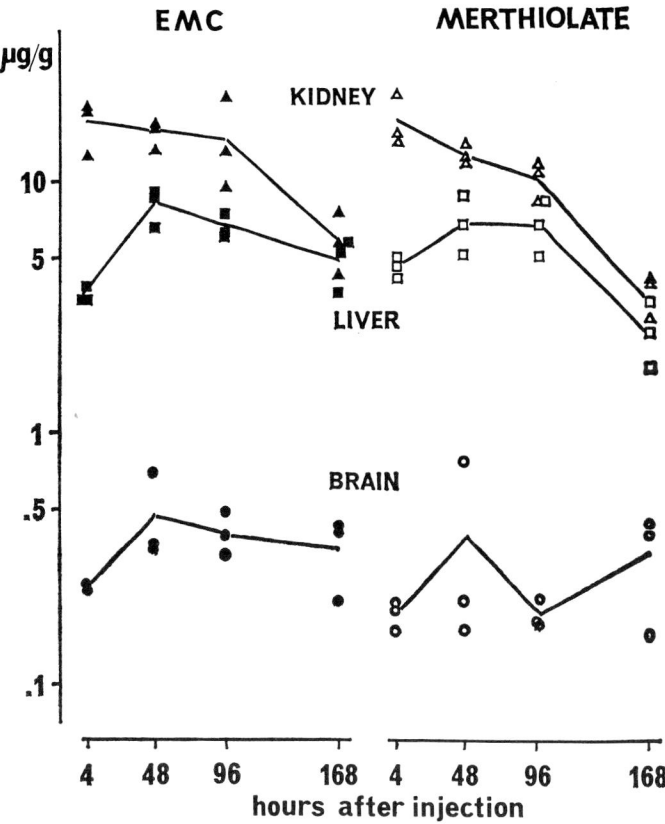

Figure 12–7. Changes in the content of organic mercury in three organs (kidney, liver and brain) after a subcutaneous injection of ethylmercury salts (2.0 mg Hg/kg) in mice. EMC (ethylmercury chloride); Merthiolate Na (sodium ethylmercurithiosalicylate).

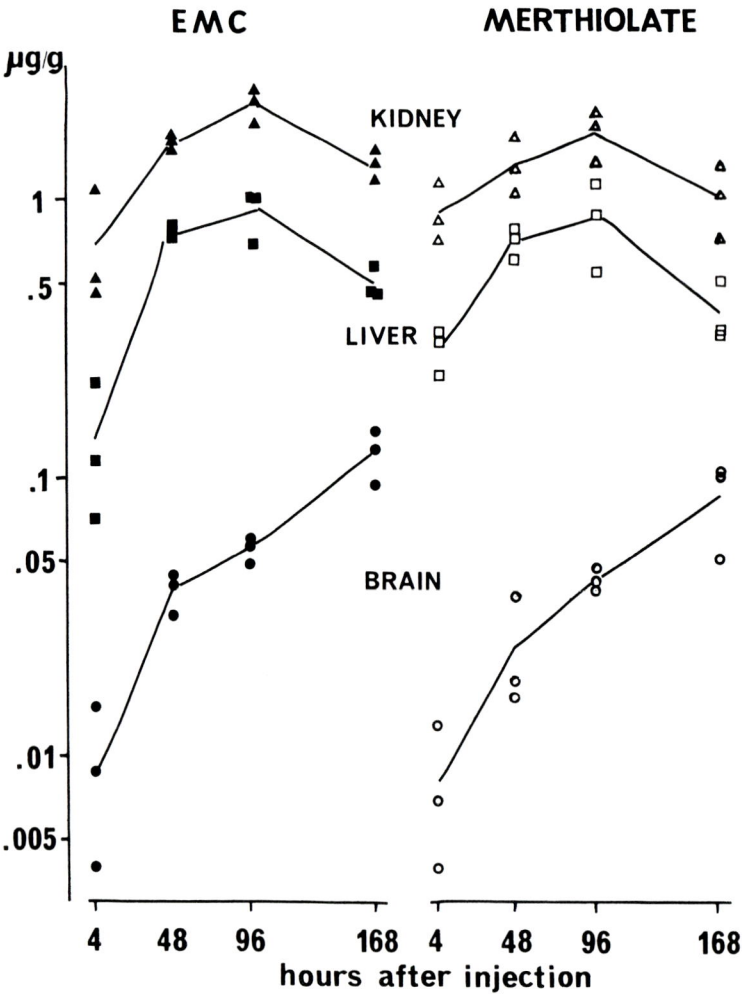

Figure 12–8. Changes in the content of inorganic mercury in three organs (kidney, liver and brain) after a subcutaneous injection of ethylmercury salts (2.0 mg Hg/kg) in mice. Abbreviations are the same as those in Figure 12–7.

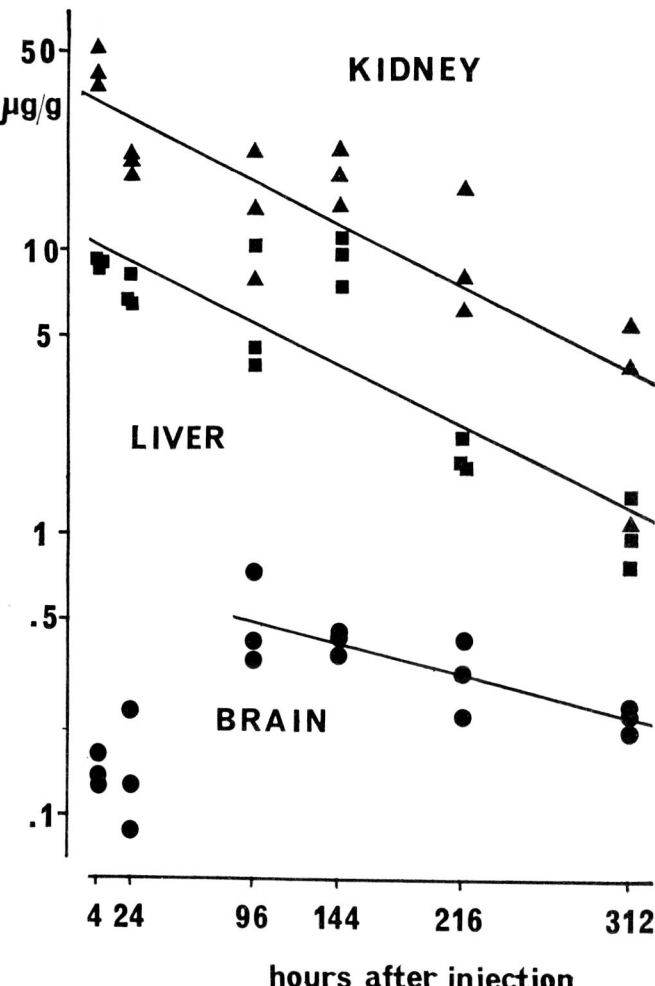

Figure 12-9. Changes in the content of organic mercury in three organs (kidney, liver and brain) after an intravenous injection of ethylmercury chloride (2.4 mg Hg/kg) in mice.

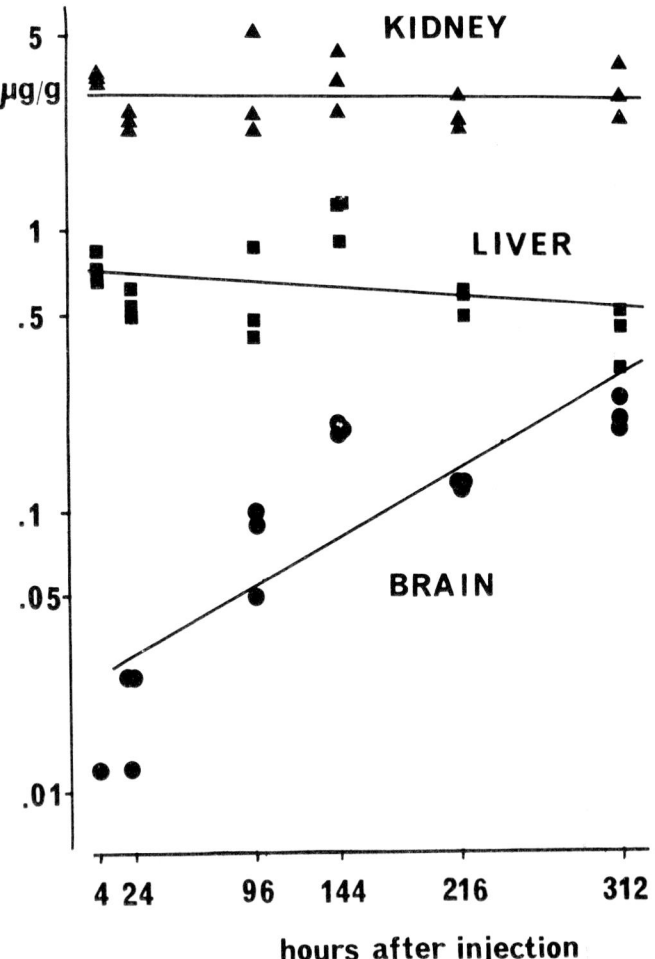

Figure 12–10. Changes in the content of inorganic mercury in three organs (kidney, liver and brain) after an intravenous injection of ethylmercury chloride (2.4 mg Hg/kg) in mice.

12–9 and 12–10, regression equations are fitted to the data. The organic mercury in the kidney and liver decrease similarly; the biological half-time is about 100 hours (4.2 days). Organic mercury in the brain decreases at a slightly slower rate; its biological half-life time is 197 hours (8.2 days).

Discussion

The release of inorganic mercury from ethylmercury as compared to methylmercury, although we do not know exactly where and how it occurs, can be said to occur more rapidly in the body.

Among the organs and tissues studied, the brain displayed a constant increase of inorganic mercury after injection. The increasing inorganic mercury in the brain may come from the plasma, which has been assumed to play the role of a temporary depot for inorganic mercury from the results with four human cases studied after the transfusion of "Human Plasma"; with this mechanism, some damage in the blood-brain barrier due to ethyl- or inorganic mercury which may effect the permeability of substances (5, 23) may have to be taken into consideration. But, it is necessary to keep in mind that possibly breakage of C-Hg bonds in the brain may occur.

The behavior of inorganic mercury in the brain is considered to be different from that observed with methylmercury; i.e. inorganic mercury is retained in the brain (39) and is not easily replaced by subsequent administration of the same mercury salt (4, 25). The biological half-life times of mercury in the whole brain or each compartment of mice after methylmercury and phenylmercury injections are reported to be six to seven days and 14 to 20 days, respectively (28). The calculated total mercury in the brain after intravenous injection of ethylmercury chloride was 21.4 days in the present experiment; thus ethylmercury is quite similar to phenylmercury. Also, the differences in extractability of mercury between mercury vapor exposure and ethylmercury injection with 1 N HCl solution suggests that the difference may depend upon the chemical binding of mercury to protein, or the nature of the subcellular distribution.

The metabolism of ethylmercury has some characteristics similar to the metabolism of inorganic mercury. As to urinary excretion, the release of inorganic mercury from ethylmercury was the main observation with the four patients injected with "Human Plasma." With the ethylmercury retained in the kidney and liver, the breakage of C-Hg bonds, the excretion to urine and bile, and the return to the bloodstream of ethyl- or inorganic mercury should occur simultaneously in short half-life times of organic mercury in the liver and in the kidney and should be a reflection of these three processes.

In relation to neurotoxicity and a toxic dose of ethylmercury, the distinction between the toxic effects of ethylmercury itself and of inorganic mercury is still to be made. This distinction is, I feel, related to the brain "site" and its associated mercury or ethylmercury contents. The "site" in the brain is the next target in the toxicology of mercury.

REFERENCES

1. Berglund, F. and Berlin, M.: In M. W. Miller and G. G. Berg (Eds.): Chemical Fallout. Springfield, Thomas, 1969, p. 269.
2. Bidstrup, P.L.: *Toxicity of Mercury and its Compounds.* Amsterdam, Elsevier, 1964.
3. Brown, W.J. and Yoshida, N.: *Shinkei-Kenkyu-no-Shimpo*, 9:34, 1965.
4. Friberg, L.: *Acta Pharmacol Toxicol*, 12:411, 1956.
5. Fukuhara, N., Oguchi, K. and Tsubaki, T.: *Shinkei-Kenkyu-no-Shimpo*, 14:313, 1970, (in Japanese).
6. Gage, J.C.: *Br J Ind Med*, 21:197, 1964
7. Gage, J.C. and Warren, J.M.: *Ann Occup Hyg*, 13:115, 1970.
8. Haq, I.U., Brit. Med J., 1 (1963) 1579.
9. Höök, O. and Lundgren, K.D., and Swensson, A.: *Acta Med. Scand*, 50:131, 1954.
10. Jalili, M.A. and Abbasi, A.H.: *Br J Ind Med*, 18:303,1961.
11. Katsunuma, H., Suzuki, T. and Miyama, T.: *Jap J Med Progress*, 48: 373, 1961 (in Japanese).
12. Katsunuma, H., Suzuki, T., Nishii, S. and Kasheima, T.; *Rep Inst Sci Labor*, (no. 61) (1963), 33.
13. Kitamura, S. and Sumino, K., Hayakawa, K., and Hirano, I.: *Jap J Publ Health*, 17:768, 1970, (in Japanese).
14. Koelsch, F.: *Arch Gewerbepathol Gewerbehyg*, 8:113, 1937.
15. Maximum allowable concentrations of mercury compounds. *Arch Environ Health*, 19:891, 1969.
16. Magos, L. and Cernik, A.A.: *Br J Ind Med*, 26:144, 1969.
17. Miller, V.L., Klavano, P.A. and Csonska, E.: *Toxicol Appl Pharmacol*, 3:459, 1961.
18. Miyama, T., Murakami, M., Suzuki, T. and Katsunuma, H.: *Ind Health*, 6:107, 1968.
19. Norseth, T. and Clarkson, T.W.: *Biochem Pharmacol*, 19:2775, 1970.
20. Nose, K.: *Jap J Hyg*, 24:359, 1969.
21. Sadakane, T.: *Jap J Ind Health, 6:489, 1964*, (in Japanese).

22. Sebe, E. and Itsuno, Y.: *Jap J Med Progress,* 49:607, 1962, (in Japanese).
23. Steinwall, O. and Olsson, Y.: *Acta Neurol Scand,* 45:351, 1969.
24. Suzuki, T., Miyama, T. and Katsunuma, H.: *Jap J Med Progress,* 48:717, 1961 (in Japanese).
25. Suzuki, T. and Miyama, T., Katsunuma, H.: *Jap J Med Progress,* 49: 754, 1962, (in Japanese).
26. Suzuki, T., Miyama, T. and Katsunuma, H.: *Jap J Exp Med,* 33:277, 1963.
27. Suzuki, T.: *Nihon Ishikai Zasshi (J Jap Med Assoc),* 61:1051, 1969 (in Japanese).
28. Suzuki, T., Miyama, T. and Katsunuma, H.: Proc XVI *Int Cong Occup Health,* 563, 1969.
29. Suzuki, T., Miyama, T. and Katsunuma, H.: *Ind Health,* 8:39, 1970.
30. Suzuki, T., Miyama, T. and Katsunuma, H.: *Bull Environ Contam Toxicol,* 5:502, 1971.
31. Swensson, A.: *Acta Med Scand,* 143:365, 1952.
32. Swensson, A. and Ulfvarson, U.: *Occup Health Rev,* 15:5, 1963.
33. Takeda, Y., Kunugi, T., Hoshino, O. and Ukita, T.: *Toxicol Appl Pharmacol,* 13:156, 1968.
34. Takeda, Y., Kunugi, T., Tefao, T. and Ukita, T.: *Toxicol Appl Pharmacol,* 13:165, 1968.
35. Takeda, Y. and Ukita, T.: *Toxicol Appl Pharmacol,* 17:181, 1970.
36. Takemoto, T., Suzuki, T., Miyama, T. and Nagao, H.: *Jap J Ind Health,* 13:46, 1971.
37. Togi, H., Yazaki, Y., Yoshino, Y., Mozai, T. and Tsukagosh, H.: *Rinsho-Shinkeigaku (Clin Neurol),* 6:69, 1966, (in Japanese).
38. Ulfvarson, U.: *Int Arch Gewerbepathol Gewerbehyg,* 19:412, 1962.
39. Watanabe, S.: Proc XVI Int Cong Occup Health, 553, 1969.
40. Yoshino, Y.: *Rinsho-Shinkeigaku (Clin Neurol),* 5:130, 1965.

DISCUSSION

Magos: Have you considered that when you release inorganic mercury from different homogenates, the rate of release depends on the concentration and kind of homogenate? So your peaks won't be the same with 5 percent liver homogenate as with 0.5 percent kidney homogenate.

Suzuki: About the homogenates, I agree with you that the shape of the peak-change depends on the kind of tissues used and so I must say there is some difficulty in estimating precisely without an inner standard. But in the case of oxidized samples the shape of peak was quite consistent in any case.

Goldwater: We have to be careful here about how deep we're going, whether we're just talking about the epithelial layers or the corium or just which part and certainly we have to consider this.

Miettinen: As far as I remember the mercury content in skin on average is about the same as in muscle or in blood but of course in hair it is about 300 to 1,000 times higher than in blood. Heavy metals in general are quite effectively concentrated into hair. The weight of hair tissue is small compared with those of skin, muscle and blood, however.

Chapter 13

ABSORPTION AND ELIMINATION OF DIETARY MERCURY (Hg^{2+}) and METHYLMERCURY IN MAN

JORMA K. MIETTINEN

ABSTRACT

Investigations were carried out to determine the percentage of absorption and the rate of elimination of protein-bound mercury (Hg) and methylmercury (MeHg) using ^{203}Hg-labeled compounds and the whole-body counting technique.

Fifteen volunteers were fed $2\mu Ci$ Me ^{203}Hg in fish muscle protein, eight subjects $4\mu Ci$ to $8\mu Ci$ ^{203}Hg in calf liver protein and two subjects $4\mu Ci$ and $14\mu Ci$ ^{203}Hg$(NO_3)_2$ in water solution. Urine, feces and blood samples were collected and counted regularly. Whole-body counting was performed in a steel room using the 43-cm Argonne chair geometry and one 4" x 8" NaI(Tl)-crystal. Within the first 4 to 5 days about six percent of the administered Me ^{203}Hg and 85 percent of the ^{203}Hg were excreted, mostly in the feces. The biological half-time of the remaining activity was 76 ± 3 days for MeHg and 42 ± 3 days for Hg.

The activity ratio of red blood cells/plasma was about 10 for MeHg, and about 0.4 for $Hg^{2\pm}$. For both mercury compounds, males showed a slightly longer half-time than females.

Radioactivity of the red blood cells decreased with a biological half-time of 50 days in the case of MeHg and 16 days in the case of Hg, ie. in both cases considerably faster than in the whole-body. For both compounds, MeHg and Hg, the elimination rate is the same whether the compound is administered as proteinate or in an ionic form.

REPORT

Introduction

An understanding of the kinetics of dietary mercury and methylmercury in man has become of vital importance recently because of the presence of these compounds in various

This study was financed by Contract No. 702/RB of IAEA and a grant from Nordforsk.

foods, especially fish and shell fish. The retention and biological half-time of MeHg were recently determined in three males, who swallowed a few microcuries of MeHg $(NO_3)_2$ in a water solution. A retention of 95 percent and a half-time of 70 to 74 days for the retained activity was determined (1).

For inorganic mercury, the ICRP states a half-time of ten days (6), the percentage of absorption not being known exactly.

In fish flesh, however, these compounds (mercury and methylmercury) are not free but protein-bound, and it may be assumed that the proteinate has a different retention and elimination rate than the ionic form. This problem was studied, therefore, by giving radio-labeled compounds to 10 and 15 volunteers, collecting all excreta for a few days determining the body activity by whole-body counting, and sample counting the blood activity (plasma and RBC).

Methods

Fifteen volunteers (nine males, six females; all scientists) were each given orally about $2\mu Ci$ of Me^{203}Hg bound to fish muscle protein. The fish (a 600 g burbot, *Lota vulgaris*) was labeled by feeding it with a suspension of calf liver protein, which contained firmly bound, labeled MeHg. After ten days, the fish was killed, cleaned, cooked and divided into sections of about 10 g each, which were counted and then eaten by the subjects. The inorganic mercury was given to two subjects as ionic^{203}Hg$(NO_3)_2$ in water and to eight subjects as ^{203}Hg bound to calf liver protein (liver pastry). In both cases, it was shown by extraction according to Westöö (7) and thin-layer chromatography that neither demethylation nor methylation, respectively, took place in the fish or the liver suspension.

During the first four days, the subjects collected daily in plastic containers their urine and in plastic bags their feces. Later, 24-hour samples were collected at wider intervals. Blood samples (50 to 80 ml) were occasionally collected from about half of the subjects. All of the samples were counted by a pulse height analyzer and corrected for decay. After counting, the red blood cells and plasma were separated by centrifugation and counted again.

The whole-body counting was performed in a steel room using the 43-cm Argonne chair geometry and one 4" × 8" NaI(Tl)-crystal (2). Counting times from 5 to 60 minutes were used.

Results

During the first four days after administration of Me^{203}Hg, about six percent was eliminated, mainly in the feces (Fig. 13–1).

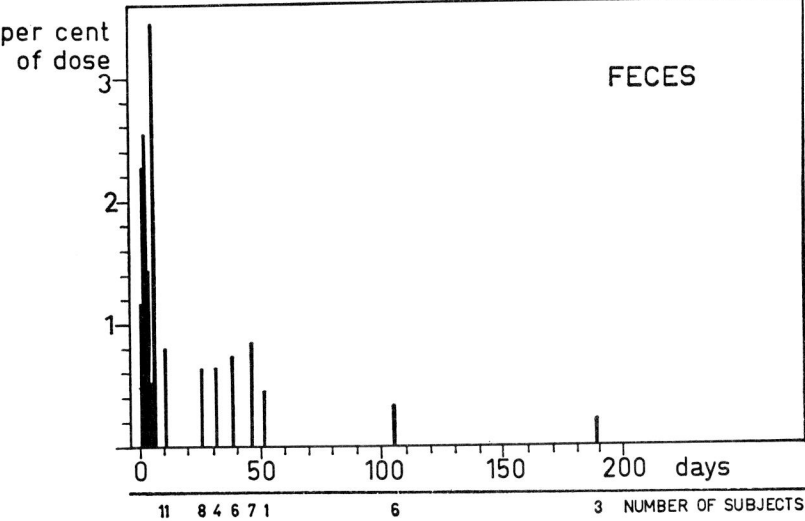

Figure 13–1. Daily excretion in feces of ^{203}Hg after oral administration of ^{203}Hg-methylmercury.

Excretion in urine was at first negligible, but then increased slowly (Fig. 13–2). After 100 days, about 20 percent of the daily excretion was in the urine. Elimination of the absorbed activity followed an exponential function very accurately (Fig. 13–3). For the fifteen subjects, a mean biological half-time of 76±3 days was obtained (Table 13-I). The mean value for the six women was 71±6 days and for the nine men 79±3 days.

Regarding blood activity, about 90 percent was in the red blood cells (RBC) and the biological half-time for this fraction was only about 50±7 days (Fig. 13–4). After two weeks, the whole-blood contained about 7 percent of the whole-body ac-

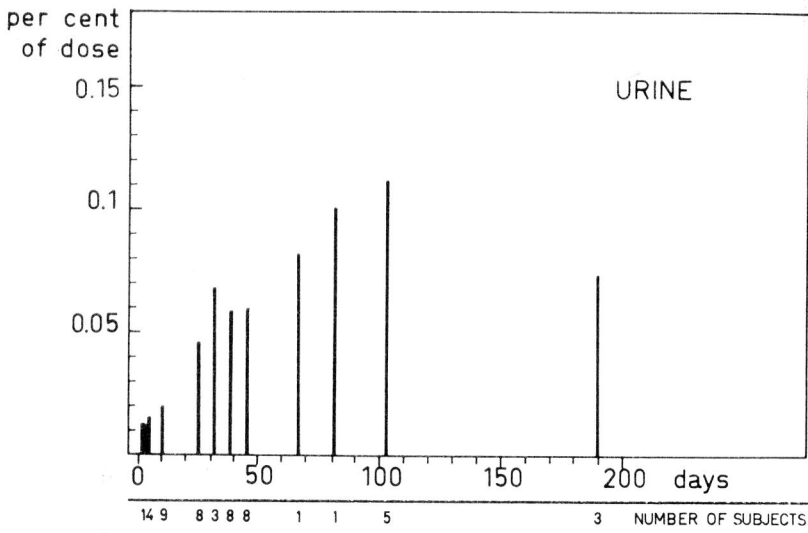

Figure 13-2. Daily excretion in urine of ^{203}Hg after oral administration of ^{203}Hg-methylmercury.

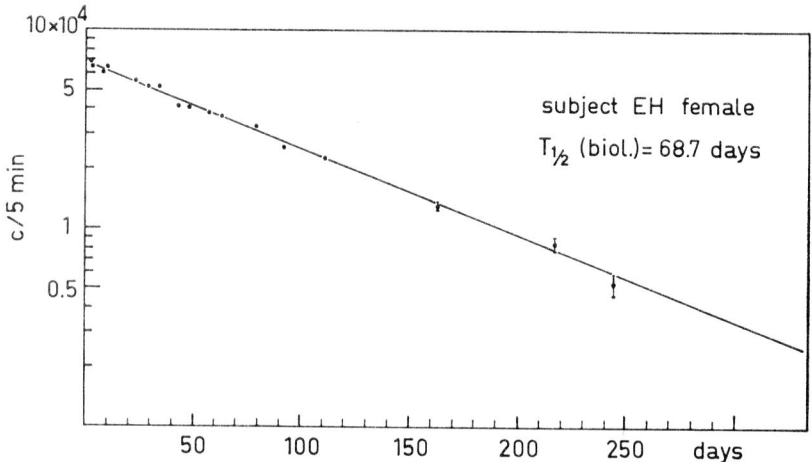

Figure 13-3. Retention curve of ^{203}Hg in the whole body after oral administration of ^{203}Hg-methylmercury.

Absorption and Elimination of Dietary Mercury

TABLE 13-I
BIOLOGICAL HALF-TIME OF ^{203}Hg-ACTIVITY IN THE WHOLE BODY AFTER ORAL ADMINISTRATION OF PROTEIN-BOUND CH$_3$^{203}Hg (ERROR MARKED AS ONE STANDARD DEVIATION OF THE MEAN)

Subject	$T_{1/2}$ biol (days)	Subject	$T_{1/2}$ biol (days)
Female		Male	
A.E.	52	R.E.	72
E.H.	69	A.H.	88
L.J.	73	P.K.	87
A.N.	88	J.K.	74
K.R.	87	M.M.	78
L.U.	56	J.M.	70
		V.M.	78
		R.P.	93
		H.T.	74
mean	71 ± 6	mean	79 ± 3

The whole group, mean 76 ± 3

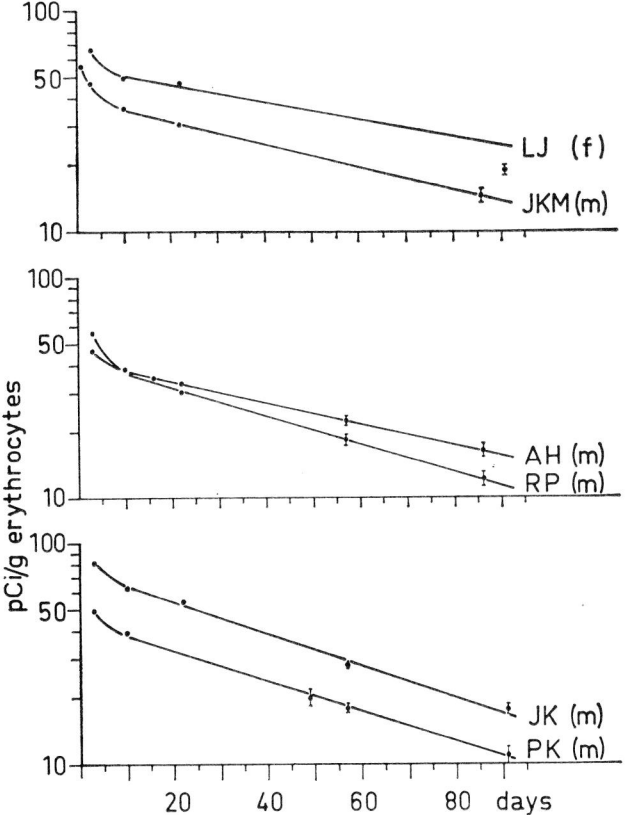

Figure 13-4. Retention curve of ^{203}Hg in red blood cells after oral administration of ^{203}Hg-methylmercury.

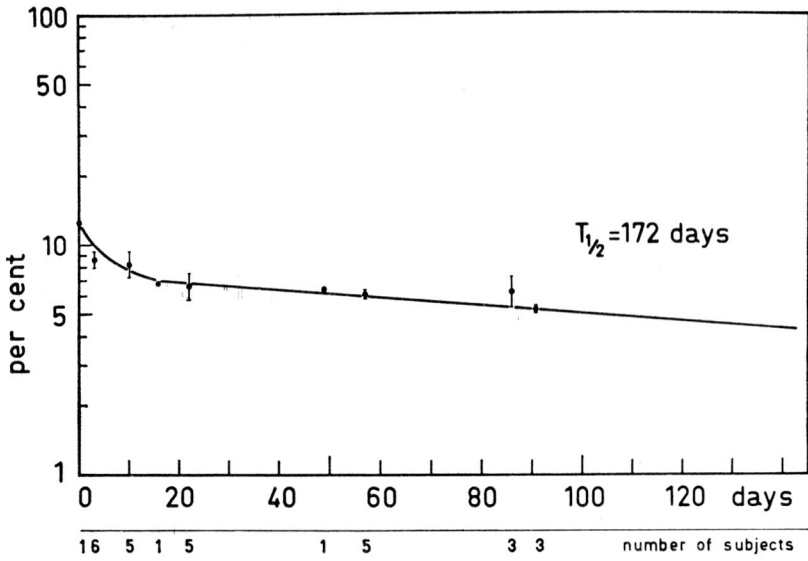

Figure 13-5. ^{203}Hg activity in whole blood as percent of that in the whole body after oral administration of ^{203}Hg-methylmercury.

tivity, but this percentage was reduced with a "half-time" of about 170 days (Fig. 13-5). The distribution for five subjects was measured in Stockholm by Falk (3), who found 10 to 15 percent of the activity in the head region.

Absorption of inorganic mercury was very low: only about 15 percent was retained (Table 13-II). During the first 4 to 5 days, about 85 percent of the administered dose was excreted in the feces and only about 0.17 percent in the urine. The blood concentration also remained rather low. Only about 0.3 percent of the dose was found in the whole-blood after 24 hours, which is less than one percent of the whole-body value. Furthermore, the ratio of activities in red blood cells/plasma was only 0.4 for inorganic mercury during the first 50 days, while this ratio was 10 for MeHg. The mean biological half-time was 37 days for women and 48 days for men, while the mean of the whole group was 42±3 days (Table 13-III). For five of these subjects, the half-time of MeHg was also determined, but there is no clear correlation between the excretion rates of the two compounds.

TABLE 13-II
ADMINISTERED AMOUNT OF ^{203}Hg, PERCENTAGES ^{203}Hg ELIMINATED VIA FECES AND URINE DURING THE FIRST FOUR TO FIVE DAYS AND PERCENTAGE OF DOSE IN WHOLE BLOOD 24 HOURS AFTER ADMINISTRATION

Subject	Sex	Administered ^{203}Hg Amount μCi	Percentage ^{203}Hg Excreted via Feces During the First 4-5 Days	Percentage ^{203}Hg Excreted via Urine During the First 4-5 Days	Percentage of Dose in Whole Blood 24 h After Administration
AE	female	8.2	82	0.11	< 0.07
MG*	female	14		0.11	
AN	female	8.0		0.21	< 0.18
TP	female	6.8			0.18
KR*	female	4.2	75	0.38	0.48
AH	male	6.5	91	0.06	0.15
PK	male	7.6	75	0.15	0.42
JM	male	3.8	88	0.05	0.44
MR	male	4.5	91	0.21	0.25
IV	male	7.8	92	0.23	0.22
average			85	0.17	0.27

* ionic mercury-Hg-203.

TABLE 13-III
BIOLOGICAL HALF-TIME OF INORGANIC MERCURY IN THE WHOLE BODY FOR THE GROUP OF TEN SUBJECTS AND THE CORRESPONDING HALF-TIME OF METHYLMERCURY IN FIVE SUBJECTS

Subject	Sex	$T_{1/2}$ biol Inorganic Hg (days)	$T_{1/2}$ biol methyl Hg (days)
AE	female	29	52
MG	female	38*	
AN	female	36	
TP	female	39	
KR	female	41*	87
average		37 ± 3	
AH	male	50	88
PK	male	60	87
JM	male	51	70
MR	male	45	
IV	male	32	
average		48 ± 5	
average 10 subjects		42 ± 3	
average 15 subjects			76 ± 3

* Administered in ionic form.

Conclusions

The human biological half-time of inorganic mercury, 42±3 days is considerably longer than the value of ten days as stated by ICRP (2). It is about half that of MeHg, 76 days. As the ratio of Hg in red blood cells/plasma is quite different for these two compounds (MeHg 10, Hg 0.4), it is possible to calculate their relative concentrations in man by determining this ratio. For more accurate analyses, the time lapse since intoxication must be known.

It is quite possible that in some populations both Hg and MeHg may be present simultaneously. With the above information, accurate prognoses can be made. More detailed reports of these investigations are published elsewhere (4,5).

REFERENCES

1. Åberg, B., Ekman, L., Falk, R., Greitz, U., Persson, G. and Snihs, J.O.: Metabolism of methly mercury (^{203}Hg) compounds in man (excretion and distribution). *Arch Environ Health*, 19:478, 1969.
2. *Directory of Whole-Body Radioactivity Monitors.* Vienna, IAEA, 1970. Fi 1.2.
3. Falk, R.: Whole-body counting of ^{203}Hg-labelled protein-bound methylmercury given orally to volunteer subjects. Report SSL:1970–001 (in Swedish).
4. Miettinen, J.K., Rahola, T., Hattula, T., Rissanen, K. and Tillander, M.: Elimination of ^{203}Hg-methylmercury in man. *Ann Clin Res*, 3:116, 1971.
5. Rahola, T., Hattula, T., Korolainen, A. and Miettinen, J.K.: Absorption and elimination of dietary mercury (Hg^{2+}) in man. *Ann Clin Res*, 3:116, 1971.
6. *Recommendations of the International Commission on Radiological Protection.* ICRP publications *2* and *10*, Oxford Pergamon Press, 1959 and 1968, resp.
7. Westöö, G.: Determination of methylmercury compounds in food stuffs. II. Determination of methylmercury in fish, egg, meat and liver. *Acta Chem Scand*, 21:1790, 1967.

DISCUSSION

McDuffie: Did I understand it correctly that in the red blood cells of your experimental patients, the half-life for methylmercury was about 40 or 50 days?

Miettinen: Yes, for methylmercury it's about 50 days, for inorganic about 15 days, always somewhat shorter than in the whole body which is 76 days for methylmercury and 42 for the inorganic mercury.

Vostal: Do you have information about the distribution in the body?

Miettinen: We cannot count persons sectionally but Falk* found 10 to 15 percent in the head region. Then, of course, there is the question what is actually in the brain of that in the head. All the half-times were measured during the excretion phase, not during accumulation.

Foulkes: One of the volunteers was Dr. Miettinen himself and I think he deserves our respect for that. But I have certain reservations about the use of urinary excretion values especially in a case of a highly diffusable compound like methylmercury. The volume of urine flow, just as in the case of urea, is going to affect critically the clearance of the compound. So I wonder whether these values in themselves are really very informative unless we have some idea of the total urine flow. If you increase the urine flow let us say to three liters per day you would probably significantly increase the clearance.

Miettinen: Yes, but it's only after heavy beer drinking or something like that. The more usual flow is 1½ liters per person per day.

Berlin: With inorganic mercury certainly we have a multicompartment system; the brain may be the slowest compartment. With methylmercury it is still open whether turnover in the brain is slower than that of the rest of the body. Do you think that the intake of methylmercury in food every day could influence the observed half-life by dilution in the enterohepatic circulation?

Miettinen: There's no difference. The half times obtained by Åberg et al.† were 70 to 74 days for three subjects, ours was 76 days for 15 subjects. And the retention was the same, we had 95 percent—Åberg 95 percent. In our experiment methylmercury was given protein-bound and diluted by food, while Åberg gave it in ionic form in a water solution.

Grant: We now have three known cases where the individuals concerned have retained mercury to which they were exposed, in the form of mercury vapor, in the brain for a period of about ten years in the case of the two Japanese miners and 13 years in the case of our plumber.‡ The Japanese miners were exposed of course for a period of several years but the plumber, as far as we know, on one single occasion. The values obtained at autopsy after presumably mercury-free intervals of ten and 13 years were of the order of six to ten ppm. And that's quite a high figure.

Vostal: Dr. Miettinen, do you have any data concerning the retention of inorganic mercury?

Miettinen: Yes, about 85 percent was eliminated in the first three or four days; so the retention was some 15 percent.

Kazantzis: Dr. Foulkes mentioned the volume of urine excreted as a variable which might need to be taken into account in considering inorganic

---

* Falk, R.: Report SSL:1970–001 (In Swedish).

† Åberg et al.: Arch Environ Health, 19:478, 1969.

‡ Methyl mercury in fish. A toxicologic-epidemiologic evaluation of risks. *Nordisk Hygienisk Tidskrift*, (suppl 2) 1971, p. 153.

mercury half-life in the body. Another variable which perhaps ought to be considered is dietary intake and in such studies maybe some form of standardized diet ought to be devised.

Hammer: In studies both in hair growth and also metal uptake in animal hair using irradiated compounds, one technique that's used is to shave one half of the animal and leave the hair on the other half.*

McDuffie: Did you exclude the hair in your whole body counts?

Miettinen: We did not exclude the hair but we have a chair geometry in which the detector, a large NaI crystal, does not "see" the hair very well. The chair geometry is unfavorable for the whole head region; it gives only the average of total body.

Foulkes: If you shave half an animal you may alter radically the blood flow to the skin so that it may be difficult to interpret these results.

Herman: Has an attempt been made to relate the mercury content not to fresh weight but to the weight of the protein content? Hair is almost pure protein. It doesn't contain water and the protein content may be as much as 50 times as high as, for instance, the protein content of the brain.

Goldwater: There are three possibilities for mercury getting into the hair. One, it could be excreted as the hair is formed in the hair follicles. Two, fallout onto the hair from outside or the use of hair tonics and other cosmetics which frequently have mercury in them as preservatives. And three, incorporation of ambient air mercury binding with proteins in the hair.

McDuffie: Is mercury excreted to any extent in sweat?

Goldwater: Yes, and in saliva too.

Berlin: With regard to incorporation of mercury in hair. Different parts of the hair from Niigata patients were analyzed. The data indicated it was built in and by analyzing different parts you have a kind of calendar.

Nechay: Does anyone know the distribution of mercury in the fish?

Miettinen: Yes. A lot of work has been carried out on this subject. Roughly two-thirds of whole body mercury in fish, in chronic cases, is in the musculature, so unfortunately if you eat the fish flesh, you get the mercury.

Fassett: If there's so much mercury in the hair why isn't the body load mainly in the skin? I was thinking from the point of view of SH binding. The skin has an enormous amount of SH groups.

Goldwater: The relative richness of circulation to the skin compared to brain or liver would be, I think, quite small. So the mercury is not being presented to the skin in the same concentrations as it might be elsewhere. I see Dr. Vostal is shaking his head.

Vostal: I think such large differences between skin and hair concentrations cannot be related solely to circulation. But I admit it's very difficult to explain.

Kazantzis: The fact that we are able to get a time profile of mercury

---

* Jaworowski et al.: *Int J Radiat Biol*, 11:563, 1966.

exposure by analyzing successive segments of hair suggests a high blood supply to the skin otherwise we wouldn't be able to get such a profile. The histology and physiology of the skin supports this, with an abundant vascular plexus both superficial and deep to the fatty insulating layer, which has an important thermoregulatory function. I imagine that the blood supply to the hair roots is in fact extremely rich.

SESSION IV

BIO-COMPLEXES AND CHELATES OF MERCURY

George Kazantzis, *Chairman*

Chapter 14

FURTHER INVESTIGATIONS ON BINDING AND RELEASE OF MERCURY IN THE RAT

Jerzy K. Piotrowski, Barbara Trojanowska,
Justyna M. Wisniewska-Knypl and Wanda Bolanowska

ABSTRACT

Some data are given on the role of metallothionein in the systemic storage and release of mercury in the rat. It has been revealed that—as in the kidneys—the main fraction of mercury in the liver is identical to that of the Hg-metallothionein complex. Lesser deposition of mercury in the liver than that in the kidneys may result from a lower concentration of metallothionein in the former organ.

It was found that in the kidneys of rats injected with several doses of mercuric chloride, the bulk of mercury was present in the form of a metallothionein complex. This was accompanied by an increase of the metallothionein level: the induced biosynthesis of this protein could be confirmed also by an increased *in vivo* rate of incorporation of ^{14}C-cysteine into this fraction. However, neither the level of metallothionein nor the rate of ^{14}C-cysteine was increased in the liver.

Stimulation of mercury release by thioacetamide results in increased urinary excretion of high molecular weight mercury-containing proteins. 2,3-dimercaptopropanosulphonate was shown to release mercury to the urine partly as a low molecular weight mercury complex but the bulk of urinary mercury remained in the high molecular weight fraction. None of both chemicals influenced the relative distribution pattern of mercury among protein fractions of the kidney. It is tentatively concluded that a nonspecific action of both compounds on the kidney membranes may have been responsible for the release of mercury, together with some parallel signs of damage to the kidney.

These studies were performed under the Polish-American Agreement 05–002–3 with Bureau of Occupational Safety and Health, PHS, USA.
The assistance of Mrs. Janina Jabłońska and Mr. A. Sapota in the experimental work is appreciated.

REPORT

Introduction

Apart from the general belief that in a living organism mercury is being bound by different proteins, little has been known about the chemical forms in which this element, when administered to animals, is retained in their tissues. In our recent studies (3,7), using the technique of gel filtration, it has been shown that in the kidney and liver of rat, where the bulk of mercury is stored, the element exist mainly in the form of a protein complex of molecular weight of approximately 10.000. Led by the data of Vallee and co-workers (4,5,11) we assumed that the binding substance present in these tissues may be identical with metallothionein. These authors (11) have shown that, apart from cadmium, human metallothionein contained also mercury. Our identification studies performed with respect to mercury present in the kidney of rats to which mercuric chloride had been administered further substantiated this assumption (15).

Following the hypothesis that binding of mercury by metallothionein may play an essential role in the systemic behavior and detoxication of mercury, our further studies have concentrated on this problem; some of our recently obtained observations are presented in this report.

Identification Studies of Hg-metallothionein in the Liver of the Rat

As indicated in our previous reports (3,7), mercury derived from mercuric chloride and also from phenylmercury acetate administered to rats is found in the liver in the form of complexes with proteins of various molecular weights. However, the 10,000 molecular weight fraction has an essential role because more than half of the mercury found in this organ is present in such a form.

The identification studies of this fraction have been performed according to the same scheme as described for detection of Hg-metallothionein in the kidney (15). From the group of rats given mercuric chloride (^{203}Hg), the livers were removed, homogenized, and subjected to deproteinization, dialysis and subsequent chromatography on Sepadex gel G-75 and DEAE cellulose. The identification was based on molecular weight determination, absorption spectra of the eluate, observations of ^{203}Hg and cadmium behavior after chromatography on Sepadex gel G-25 in alkaline and acid media, and the determination of the -SH groups. Three series of experiments were performed on rats: (1) to which cadmium had been previously administered for a longer period (total dose about 20 mg/kg); (2) to which HgCl$_2$ was administered only; (3) which were pretreated with a single dose of cadmium one day before administration of ^{203}HgCl$_2$.

In the first series all of the data indicated that the fraction in question and Hg-metallothionein were the same. It is known, however, that prolonged administration of cadmium induces biosynthesis of this protein in the liver (9,16). Therefore, further studies seemed necessary on rats not pretreated with cadmium. In series (b) all the test were positive except for that of Cd behavior after chromatography at various pH (too low Cd content for chemical determinations). In the third series this protein was traced with cadmium *in vivo* to an extent satisfactory for chemical determinations. The results obtained confirmed the identity of the fraction in question with metallothionein.

The Level and Biosynthesis of Metallothionein in Tissues

The discovery of metallothionein was linked with its high content of cadmium. It has often been assumed that this protein could not be of importance in animals not exposed previously to cadmium. Contrary to this view our opinion was that the only difference between "intact" and "pretreated" animals with respect to the presence of metallothionein and its capacity for binding metals could be expressed by differences in the levels of this protein in the tissue.

In order to test the above hypothesis, we have developed a simple quantitative method for determination of metallothionein in tissues (9). The method is based on tracing of metallothionein *in vitro* with ^{203}Hg, separation of the proper fraction and counting of the tracer. Internal standards for metallothionein are obtained from the equine renal cortex as described by Kägi and Vallee (4,5).

It may be seen from Table 14-I that the normal level of metallothionein is much higher in the kidney than in the liver. In rats pretreated with prolonged administration of cadmium the levels of metallothionein rose by almost one and two orders of magnitude in the kidneys and liver, respectively.

TABLE 14-I
THE LEVEL OF METALLOTHIONEIN IN THE KIDNEY AND LIVER OF RATS, WITH AND WITHOUT EXPOSURE TO METALS

Exposure to Metals	Metallothionein, mg/g of Tissue			
	No. of Rats	Kidney	No. of Rats	Liver
None	9	0.39 ± 0.08 (0.26-0.52)	5	0.12 ± 0.05 0.07-0.20
Cadmium 7 months	3	3.1 (2.6-4.8)	4	4.3 (2.6-9.0)
Mercury 3 weeks	6	2.5 2.1-2.7	5	0.17 ± 0.06 (0.12-0.26)

Mean values and standard deviations; high and low values in parentheses.

We expected that a similar situation might be obtained with respect to the rate of biosynthesis of this protein. Studies on the biosynthesis were performed using *in vivo* incorporation of ^{14}C-cysteine in the relevant protein-fraction (Fig. 14-1, Fig. 14-2). The incorporation was easy to study in rats previously exposed to cadmium, whereas in intact rats only incorporation at the threshold of detection seemed to occur. However, from the quantative point of view these results are not contrary to the presence of metallothionein biosynthesis found also in the intact animals. A higher incorporation rate could not be expected if one takes into consideration differences in absolute levels of metallothionein in both cases.

Hg-metallothionein Complexes in Prolonged Exposure to Mercury

All experiments referred hitherto were accomplished by administering a single dose of mercury. However, our recent studies have indicated that the importance of metallothionein with respect to mercury is also reduced in prolonged exposure to this element.

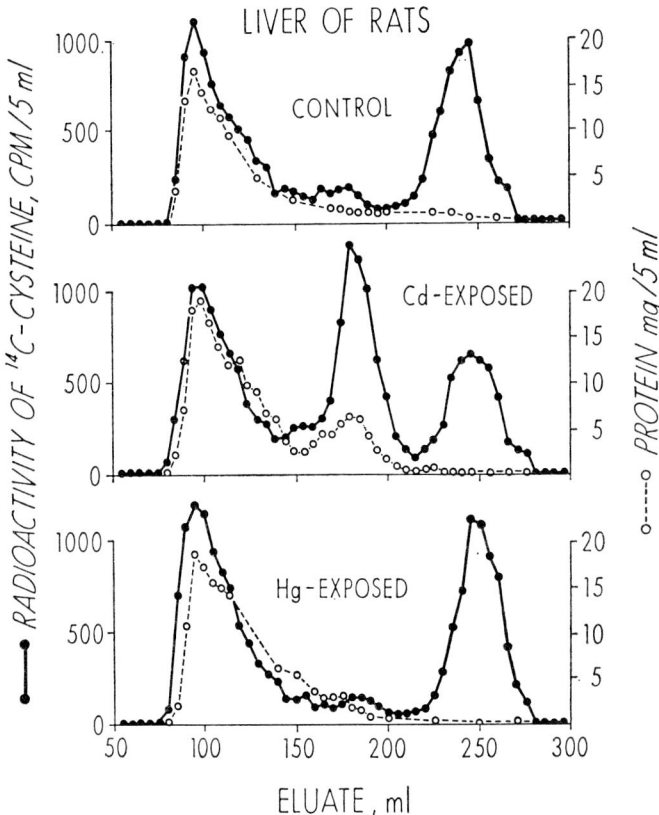

Figure 14-1. The chromatography of the liver homogenates of rats administered *in vivo* with ^{14}C-cysteine: Sephadex G–75, formate buffer, pH 8. Exposure: cadmium, 24 weeks, total dose 27 mg/kg; mercury, 9 weeks, total dose 12.5 mg Hg/kg.

When rats were injected every second day with mercuric chloride (0.5 mg Hg/kg per dose) the bulk of mercury in the kidney was found again in the 10,000 MW fraction. The distribution of ^{203}Hg among different protein fractions was the same when comparing the rats given all doses of ^{203}Hg and those given all doses of stable mercury, followed by a single dose with the tracer (Fig. 14–3). This seems to point to the fact that at least over several days the pattern of distribution of Hg among the proteins of the kidney has not changed substantially. Taking into account the limited pool of metallothionein in the kidney,

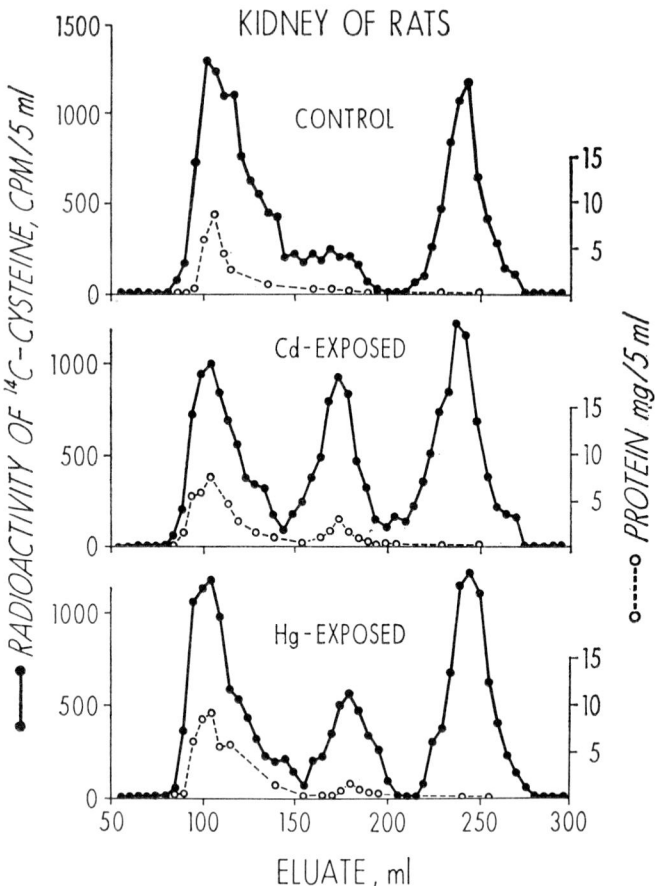

Figure 14–2. The chromatography of the kidney homogenates of rats administered *in vivo* with ^{14}C-cysteine. For other details see Figure 14–1.

this may point to an increase of the content of these proteins in the course of exposure to mercury.

In order to confirm the above hypothesis we determined the levels of metallothionein using the method mentioned above, and we also followed the rate of incorporation of ^{14}C-cysteine into the protein-fraction under study. Our preliminary findings seem to point to the stimulating effect of Hg on the biosynthesis of metallothionein in the kidney. The concentrations of this protein in the kidneys of rats increased from about 0.4 mg/gm of tissue in intact animals to about 2.5 mg/gm in rats given

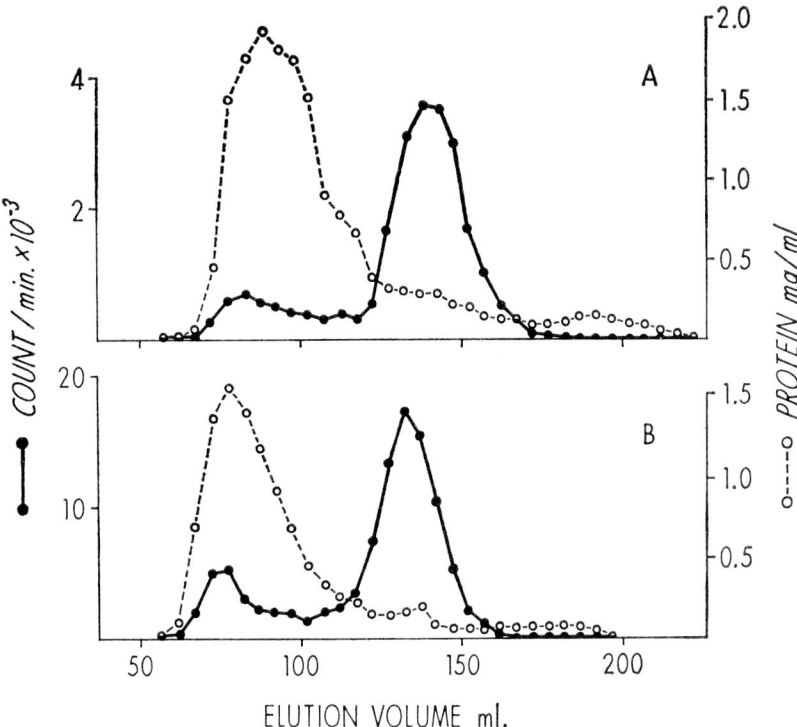

Figure 14–3. The chromatography of the kidney homogenates of rats, exposed to several doses of mercury: (a) nine doses of stable Hg followed by a single dose of ^{203}Hg; (b) nine doses labeled with ^{203}Hg. Exposure: subcutaneous injection of $HgCl_2$, dose 0.5 mg/kg, three times a week, over three weeks. Sephadex G-75, formate buffer, pH 8.

9 to 12 doses of $HgCl_2$. Also the rate of incorporation seemed to be evidently higher, as compared to the control rats (Fig. 14–2).

The above does not apply to the liver. In rats given several doses of mercuric chloride the total deposition of Hg in the liver does not increase and the one existing is composed in minor part only of the Hg-metallothionein complex (Fig. 14–4). In the course of prolonged exposure the level of metallothionein did not change considerably (between 0.1 and 0.2 mg/g of tissue in both groups, the differences are not statistically significant) and no signs of increased biosynthesis could be found in this organ (Fig. 14–1).

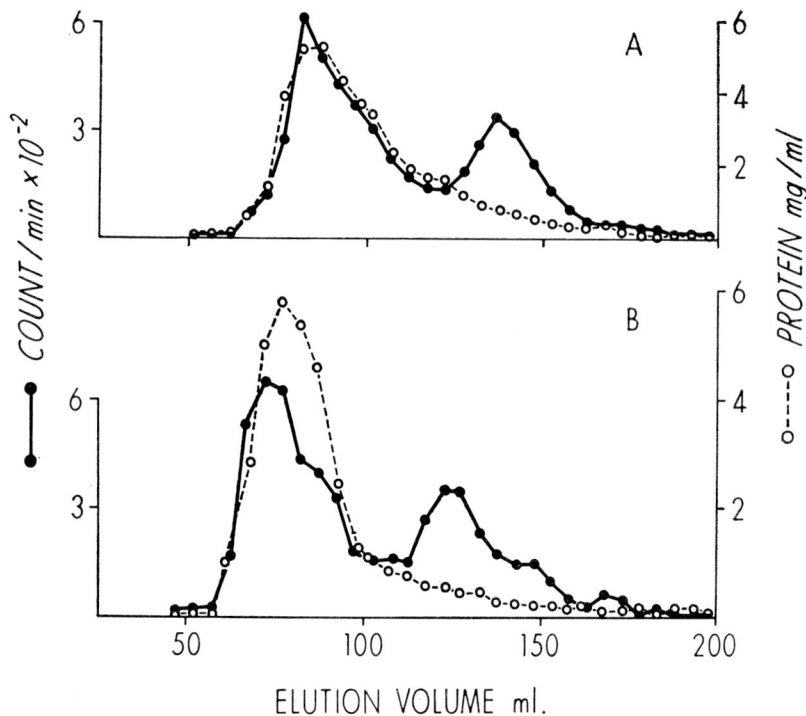

Figure 14-4. The chromatography of the liver homogenates of rats exposed to several doses of mercury: (a) 11 doses of stable Hg followed by a single dose of ^{203}Hg; (b) 11 doses labeled with ^{203}Hg. For other details see Figure 14-3.

Behavior of Hg-metallothionein during Stimulated Release of Hg from the Body

Two chemicals have been investigated in our laboratory for their ability and mechanism of releasing mercury from the organism. Details on the action of thioacetamide (TAA) have been given in one of our recent reports (13) where it was shown that this compound releases mercury almost exclusively from the kidney without any change in the mercury levels in other body compartments. The mechanism did not seem to depend upon formation of specific low molecular weight complexes with mercury and it has been suggested that the release of mercury may be due to some form of kidney damage.

In further studies we tried to identify more closely the mechanism of action of thioacetamide, and special attention was given to the behavior of the Hg-metallothionein complex in the kidney. Simple calculations showed that when one stimulates a release of about 80 percent of Hg present in the kidney, the release must occur at the cost of the Hg-metallothionein complex, which accounts for about 60 to 75 percent of the total mercury in the kidney. Sephadex-filtration studies performed in rats in different periods of the TAA treatment did not reveal any specific change in the distribution pattern of mercury among different proteins of the kidney. This points to the fact that *in vivo* the mercury is subject to equillibration processes among different categories of binding compounds and that the equilibrium obtained 2 to 3 days after injection of mercury (3) is further maintained when one also stimulates prompt release of the element with thioacetamide.

The urine of rats, however, treated with TAA, did not contain detectable amounts of the Hg-metallothionein complex and nearly all the Hg present could be found in the fraction of proteins of the higher molecular weight (Fig. 14–5). One of the possible explanations, that TAA inhibits biosynthesis of metallothionein, has been rejected since no decrease in the level of this protein in the kidneys of the TAA treated rats could be found. Thus, a possible mechanism of mobilization could be dependent upon a simple increase in permeability of membranes against the high molecular weight Hg-proteins, a process that would be followed by redistribution of Hg present in a complex with metallothionein, the metal moving into the fraction of high molecular weight proteins. The only reservation that has to be kept in mind when discussing the lack of the Hg-metallothionein in the urine is that the behavior of this complex when added to the urine has not been yet studied *in vitro* and we do not know whether or not this compound would be subject to polymerization or destruction, thus indicating that a lack of the fraction in question might have been an artifact.

The other chemical studied was the Russian drug Unithiol (2,3-dimercaptopropanosulphonate) on which basic data are available from Ashbel (1). Later, Dutkiewicz and Ogiński (2)

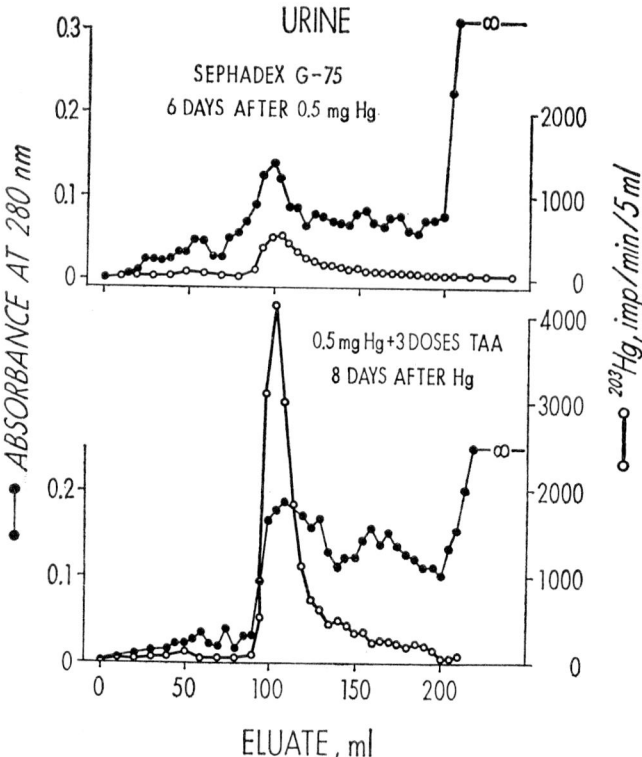

Figure 14-5. The chromatography of urine, six and eight days after administration of Hg at a dose of 0.5 mg Hg/kg. Upper graph: control; lower graph: after TAA treatment.

also proved the effectiveness of this drug in releasing the mercury-deposits. It also appeared effective in our studies on rats given a single low dose of ^{203}HgCl$_2$ (0.2 mg Hg/kg).

It might have been expected beforehand that in this case a classical chelation of mercury by this drug would take place. This was confirmed by the relatively high content of dialysable mercury in the urine (45% as compared with 10% in control). Also, when applying the gel filtration technique, a low molecular weight fraction was separated from the urine (Fig. 14-6) suggesting that simple chelation was taking place and was, at least partly, responsible for the increased rate of release from the body. However, the share of the low molecular weight

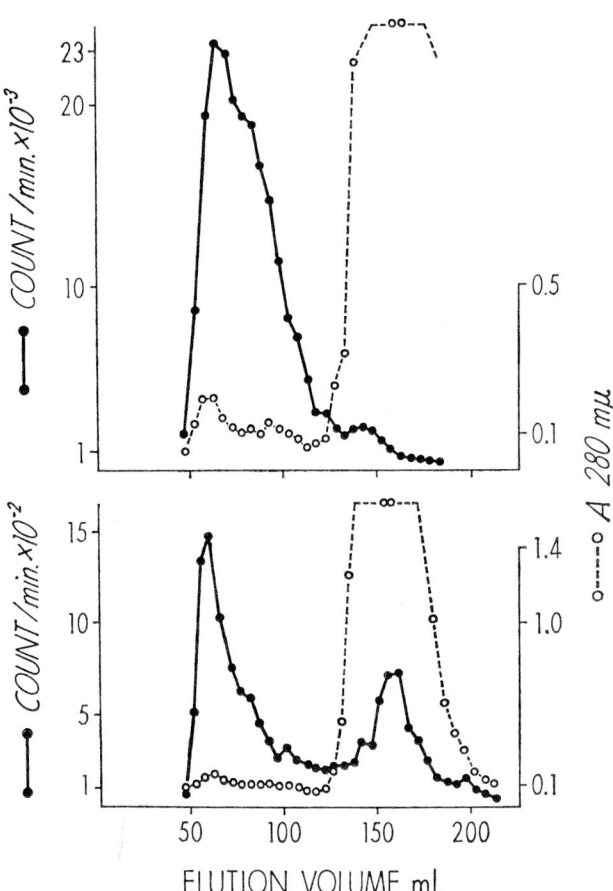

Figure 14-6. The chromatography of urine on Sephadex G-75. Upper graph: control; lower graph: after Unithiol treatment.

fraction was far below that expected if one accounted for the increased excretion. This points to the fact that in this case also simple chelation does not represent the only mechanism involved. Signs of tissue damage are observed when the drug is applied following a relatively high dose of mercury. Also in the case of Unithiol the distribution pattern of mercury among the protein fractions of the kidneys remained unchanged. Thus, it is difficult to estimate which kind of complexes present in the kidney was first subject to the release of mercury.

Discussion

From the data presented in this report it is evident that metallothionein represents the protein responsible for binding and storage of mercury in the body, especially in the kidney.

When comparing the behavior of two elements, cadmium and mercury, with regard to their binding to this protein it is striking that, whereas mercury is bound in this form mostly in the kidney, cadmium prefers the liver. It seems probable that this fact is connected with the ability of mercury to stimulate the biosynthesis of metallothionein in the kidney only, and not in the liver as cadmium does.

Whether or not the binding of mercury by metallothionein *in vivo* represents merely a simple process of chelation by a compound rich in -SH groups is not clear at present. Comparing our observations with *in vivo* and *in vitro* conditions it is striking that the *in vivo* binding was much more efficient than could be reproduced in the *in vitro* tissues. In the latter case, the share of Hg-metallothionein in the kidney homogenate of the rat never exceeded 10 percent of the total, whereas *in vivo* it often exceeded 70 percent. Similar situations, only to a lesser extent, have been found with cadmium (16). Metallothionein is located in the cytoplasm of the cell. It seems likely that the more selective binding of metals on metallothionein *in vivo* is due to the role of the subcellular membranes.

From the levels of metallothionein in the kidney it follows that kidneys of a control rat may contain about 0.8 mg of this protein. The binding capacity for mercury, as determined in the calibration of the method, is approximately 1:10 (by weight); from this it follows that the total binding capacity of metallothionein in kidneys of a normal rat is about 80μg of Hg. *In vivo* the toxic effects start at a level of about 0.4 to 0.6 mg/kg (8,14) corresponding to the total dose of approximately 100μg of which half (about 50μg) is promptly located in the kidneys. When the dose of mercury exceeds this level, the relative retention of Hg in the kidneys diminishes considerably, accompanied by a prompt release of the excess metal via urine. Simultaneously, toxic effects become evident and more serious.

The near-equal amounts for saturated binding capacity and toxic levels tend to confirm our view (6) that binding with metallothionein represents a true detoxication process and that biochemically and morphologically recognizable toxic action of mercury start only when the binding capacity of the renal metallothionein is exceeded.

The binding capacity of metallothionein in the kidney increases with repeated exposure to mercury because of a stimulation of the biosynthesis of the protein. This fact provides an explanation for the almost linear increase of Hg in the kidney over several weeks of daily exposure to mercury (12). Lack of the stimulatory effect in the liver is consistent with the stable level of mercury in this organ despite repeated introduction of daily doses of this metal.

Our studies of the stimulated release of mercury from the kidney show that such a process occurs at the cost of the renal Hg-metallothionein complex. It is striking that no specific mechanism of this process could be found. The only explanation that seems consistent with observations made in our laboratory would involve a nonspecific action of TAA on the cellular membranes thus enabling simple exchange of mercury between metallothionein and competing high molecular weight proteins followed, with increased dosage, by damage to the kidney and prompt release of the high-molecular-weight mercury-proteins via urine. Despite the occurrence of a simple chemical chelation with the other chemical, Unithiol, the above discussed simultaneously occurring nonspecific process involving damage of the kidney seems to limit a practical application of this compound for the mobilization and elimination of mercury in human beings.

REFERENCES

1. Ashbel, S.I.: *Intoksikatsi Rtut-organitsheskimi jadokhimikatami.* Medicina, Moskwa, 1964.
2. Dutkiewicz, T. and Oginski, M.: *Int Arch Gewerbepath Gewerbehygiene,* 23:197, 1967.
3. Jakubowski, M., Piotrowski, J. and Trojanowska, B.: *Toxicol Appl Pharmacol,* 16:743, 1970.
4. Kagi, J.H.R. and Vallee, B.L.: *J Biol Chem,* 235:3460, 1960.

5. Kagi, J.H.R. and Vallee, B.L.: *J Biol Chem, 236*:2435, 1961.
6. Piotrowski, J.K.: *Proceedings of the Symposium on Occupational Health,* PL-480 Program, Dubrovnik, 1970, in press.
7. Piotrowski, J., Bolanowska, W., Trojanowska, B. and Szendzikowski, S.: *Med Pracy, 20*:589, 1969.
8. Piotrowski, J.K. and Bolanowska, W.: *Med Pracy, 21*:338, 1970.
9. Piotrowski, J.K., Bolanowska, W. and Sapota A.: IX Conference of the Polish Biochem. Soc., Katowice, 1971.
10. Piscator, M.: *Nord Hyg Tidskr, 45*:76, 1964.
11. Pulido, P., Kagi, J.H.R. and Vallee, B.L.: *Biochemistry, 5*:1768, 1966.
12. Trojanowska, B.: Dynamika rozmieszczania i wydalania rteci u szczurów. DSc thesis, Medical Academy in Gdańsk, 1969.
13. Trojanowska, B., Piotrowski, J. and Szendzikowski, S.: *Toxicol Appl Pharmacol, 18*:374, 1971.
14. Wisniewska, J. and Trojanowska, B.: *Med Pracy, 19*:405, 1968.
15. Wisniewska, J.M., Trojanowska, B., Piotrowski, J. and Jakubowski, M.: *Toxicol Appl Pharmacol, 16*:754, 1970.
16. Wisniewska-Knypl, J.M. and Jablonska, J.: *Bull Acad Polon Sc Serie Sc Biol, 18*:321, 1970.

DISCUSSION

Piscator: How pure was your metallothionein standard?

Piotrowski: There are no commonly accepted criteria for the purity of metallothionein. The one we used was obtained in the way described in the laboratory of Vallee.*

Klein: Has metallothionein been looked for in other tissues besides the kidneys, i.e. the central nervous system, liver, muscle?

Piotrowski: I cannot answer this question precisely. If one performs the determinations of metallothionein using our method, traces will be found in most media. However, the identification studies have been performed only with respect to liver and kidneys. On the other hand when searching for mercury complexes in the blood serum we did not succeed in finding the metal bound with metallothionein, so all the mercury was attached to proteins of higher molecular weight.

Klein: Did you observe any effect of alkylmercury compounds on metallothionein?

Piotrowski: No, only with inorganic mercury and phenylmercury acetate.†

Goldwater: What cellular elements in the kidney are the main sites for metallothionein? Is there any evidence as to whether you can produce an increase in metallothionein under the stress of either cadmium or

* Kagi and Vallee: *Biochemistry, 5*: 1768, 1966.
† Piotrowski, J. and Bolanowska, W.: *Med Pracy, 21*:338, 1970.

mercury or something else and then if you remove the stress is there a return to the base metallothionein in the cells or the tissues? And, is there any indication, since possibly this represents a detoxication or protective mechanism, that this may be a clue to some essential function for these metals and others as essential trace elements?

Piotrowski: The information available is based partly on experience gained in the preparative isolation of metallothionein and partly on the distribution of tracer metals which more or less selectively bind on this protein (cadmium, mercury). From these data it seems that metallothionein of the kidney is contained in the cortical and subcortical zone, and when applying the ultracentrifugation technique, it appears mostly in the cell cytoplasm.

Goldwater: Is it an epithelial cell or is it glomerular capsular cell or what part of the kidney?

Piotrowski: Such data could be eventually obtained by autoradiography but I do not think that such experiments have ever been done.

Observer: Can you return to a base line after an increase?

Piotrowski: Our experiments aimed at showing the possible stimulation of biosynthesis of metallothionein by mercury but no dynamic observations of the process have yet been made.

Fassett: In 1965 when we* first described the protective effect of the very tiny dose of cadmium against a subsequent higher dose, we did some studies to see how long this protective action would last and we were quite surprised that it lasted about a month.

The amount needed to produce the protective effect was as low as a few micrograms/kilogram. We interpret this protective effect to have been due to what we now know as the induction of metallothionein.

Fang: How long have you exposed mercury to the liver before you tried to separate the mercury metallothionein complex?

Piotrowski: In one of the series we exposed rats to cadmium for several weeks, aiming at inducing biosynthesis. However, the basic two series were performed either without pretreatment or with a single dose of cadmium aiming at tracing the existing metallothionein with cadmium for the identification purposes.

Vostal: You mentioned that you did not detect metallothionein either in plasma or in urine, but that metallothionein was readily accumulated in the kidney after administration of mercury. Does it mean that metallothionein synthetized in the kidney cannot diffuse or is firmly bound in the renal tissue?

In relation to your proposed detoxifying action of metallothionein for mercury it must be assumed that this function of metallothionein is limited; when the concentrations of mercury exceed the limit we may observe toxic symptoms. However, if the mercury-induced biosynthesis of metallothionein is stimulated in the kidney only and not in the liver,

* Terhaar, C.J. et al.: *Toxicol Appl Pharmacol,* 7:500, 1965.

why do we not see toxic symptomatology in the liver sooner than in the kidney?

Piotrowski: I would like to differentiate between two situations: when a single dose of mercury of various magnitude is given, and the other, when repeated doses of mercury are applied. In the first case the damage of the kidney appears soon after administration of mercury, and, after about three days, signs will disappear.* The damage seems to occur when the dose applied is too high to be chelated quickly and efficiently by metallothionein. The process of new biosynthesis of this protein probably needs some time to be developed and only in the later period may be of importance for the chelating of the excess mercury, increasing the relative amount of mercury bound with this compound.†

A different situation exists with repeated low doses of mercury. There is a rise in the level of metallothionein in the kidney which allows greater amounts of mercury to be stored successively without evident damage.

Vostal: Does your retoxification theory imply why we see the toxic effects of inorganic mercury mainly in the kidney and not anywhere else?

Piotrowski: The mechanisms of biosynthesis of metallothionein and of its induction have not been recognized to such an extent which would make a complete answer possible. I think it is a question of quantity because the uptake of inorganic mercury is rather slight in the liver as compared with the kidney.

Vostal: Would you comment on what could be the reason for it?

Piotrowski: This may be connected with processes of glomerular filtration and tubular transport which act in the kidney but not in the liver. I agree, however, that this question is not yet sufficiently clear.

Magos: Where is the metallothionein formed?

Piotrowski: I cannot answer this at present. As a working hypothesis we accept that this process may occur both in the liver and in the kidney.

Fassett: Dr. Piotrowski's experiments were made by injecting mercury intravenously. This would mean that in relatively small doses mercury would be carried to hit the kidney directly and the liver would be reached only secondarily. The results might have been different if one had given the dose orally in which case the liver would have then been exposed immediately to the mercury with the opportunity to be induced.

Foulkes: It seems to me that the comparison between cadmium and mercury is somewhat difficult. One administers cadmium for seven months, mercury for three weeks. To conclude that cadmium stimulates the liver whereas mercury does not would require that both metals be given for the same period of time. Couldn't it be that if you had measured cadmium for three weeks too you wouldn't have found anything in the liver either?

Piotrowski: From other sets of our experiments not quoted here, it

---

* Piotrowski, J. et al.: *Med Pracy,* 20:589, 1969.

† Jakubowski et al.: *Toxicol Appl Pharmacol,* 16:743, 1970.

follows that in the case of cadmium, stimulatory effects in the liver are already evident after a few days of exposure.

Piscator: In humans there seems to be a capacity for the renal metallothionein to bind mercury but still the amount of mercury in these kidneys or in the metallothionein fraction was relatively low compared with the amount of zinc and cadmium in these metal binding fractions.

Magos: I should like to point to the possibility that the increased metallothionein level in the kidney was not the result of induction in kidneys but metallothionein was partly formed elsewhere, e.g. in the liver and taken up by the kidney.

Piotrowski: This is not excluded but there is no proof for such hypothesis either. Therefore, we accept presently the hypothesis of biosynthesis occurring in both these organs independently which is much simpler and does not need additional assumptions.

Vostal: Would somebody comment on what could be the mechanism of metallothionein uptake by the kidney?

Magos: It may be glomerular filtration followed by uptake and storage by the tubular cells. The notion of glomerular filtration and tubular reabsorption of much larger molecules than metallothionein is now new. There have been published histological evidence for the reabsorption from the tubular urine or gelatine, gelatine polymers, serum albumin[*] and hemoglobin.[†]

[*] Oliver, J.: *Harvey Lect*, 40:102, 1944–45.
[†] Rather, J.L.: *J Exp Med*, 87:163, 1948.

Chapter 15

BILIARY COMPLEXES OF METHYLMERCURY: A POSSIBLE ROLE IN ORGAN DISTRIBUTION

Tor Norseth

ABSTRACT

The mechanisms of organ distribution and excretion of mercury after exposure to methylmercury salts are discussed with specific relation to *in vivo* complexes of mercury. The biotransformation of methylmercury salts releasing inorganic mercury is important both for excretion and organ distribution of mercury. Release of inorganic mercury with redistribution takes place in the rat and the mouse. About 50 percent of the excreted mercury is in the inorganic form. More than ten percent of the methylmercury chloride from a single injection to the rat passes through enterohepatic circulation during the first 24 hours. Also, mouse bile contains a high amount of mercury, which is mostly reabsorbed. Mercury is bound to different compounds in rat and mouse bile; this may be of importance for differences in organ distribution. Mercury is bound to some small molecular compound found in plasma from normal rats but is absent in rats when the bile duct is ligated or cannulated; this compound may be filtered in the kidneys and thus determine both kidney uptake and urinary excretion of mercury. The strong binding of methylmercury in rat red cells is also discussed in relation to biliary excretion and organ distribution. The importance of these results for the general understanding of the pharmacokinetics of of metals and organometals is underlined.

REPORT

The large hazard related to methylmercury exposure compared to most other mercurials is related to organ distribution with penetration into the brain, and to the slow excretion with accumulation of toxic doses from repeated exposure to small amounts. The mechanisms determining organ distribution and excretion of mercury after exposure to methylmercury salts have therefore attracted considerable interest in recent years.

Biotransformation and Organ Distribution

The biotransformation of methylmercury salts and its relation to organ distribution have been investigated by Norseth and Clarkson (6). It was found that except for a time dependent accumulation of mercury in the rat kidney and some inorganic mercury in the liver, most mercury in the organs was not inorganic mercury. Inorganic mercury seemed to play a part, however, in the excretion of mercury after exposure to methylmercury salts. About 50 percent of the mercury in the feces was found to be inorganic. As biliary excretion would account for only a limited amount of this mercury, and no other major excretory pathway for inorganic mercury to the intestinal tract could be found, bacterial degradation of the organomercurial in the gastrointestinal tract was postulated to be responsible. Studies on germ free rats by Norseth (4) failed to verify this theory, and the mechanisms of these reactions are at present unknown. *In vitro* incubations of intestinal content, intestinal cells or pieces of intact rat intestine have not been found to produce inorganic mercury in the amounts necessary to explain the excretion rate of inorganic mercury in feces.

Enterohepatic Circulation and Organ Distribution

Mercury transport in the intestinal tract is dependent upon the form of mercury present after exposure to methylmercury salts (7). On the other hand, biliary excretion of mercury is clearly higher than the fecal excretion. On the other hand, biliary excretion of inorganic mercury is lower than the fecal excretion of mercury in this form. It was also shown by Norseth and Clarkson (7) that there is a selective reabsorption of the intact organomercurial excreted in bile, while inorganic mercury is probably completely excreted in feces. These mechanisms are related to the binding of mercury in bile; inorganic mercury bound to some high molecular weight compound is excreted, while methylmercury, bound to a smaller molecule, is reabsorbed.

That reabsorption is dependent on the ligand has recently found application in a new method for treating experimental methylmercury poisoning. Clarkson, Small and Norseth (1) showed that a resin containing sulfhydryl groups and given orally to mice in order to bind methylmercury in an unabsorbable form increased the excretion of mercury. Mercaptodextran (MW 500.000) (2) has been tested with corresponding results in rats by Norseth and Jellum (unpublished results).

Further studies on the enterohepatic circulation of mercury after exposure to methylmercury indicated that accumulation of mercury in the kidney seemed to be specifically dependent upon this mechanism (Fig. 15-1). When the bile duct is ligated, the mercury content in blood in the liver and brain increases, but that of the kidney decreases (7). There are at present experimental evidence for three different mechanisms to explain these results, none of which are unequivocal. The mechanism may however, be a combination of the possibilities outlined below.

Rat kidneys from controls did not contain more inorganic mercury than those with ligated bile ducts. Inorganic mercury

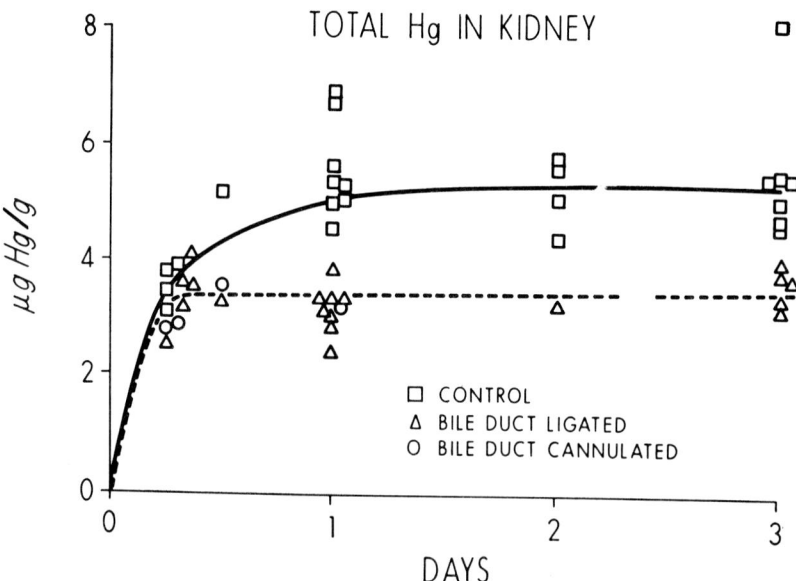

Figure 15-1. Total mercury in the kidney at different time intervals in rats with the bile duct ligated or cannulated compared to normal rats. Dose was 1 mg Hg/kg as methylmercury (170μg/rat).

in the kidney must be derived from methylmercury in the kidney, and from biotransformation in other organs. Biliary excretion may therefore be responsible for the selective distribution of mercury after enterohepatic circulation to the kidney, either because of a greater instability of the biliary compound after reabsorption or because of increased uptake by the kidney of the intact compound. About 20 percent of the biliary excretion the first day after a single injection should be accumulated by the kidney if the redistribution theory is correct, but in preliminary tests only 10 to 12 percent was found. The decrease in kidney content disappears, however, with injection of methylmercury bound to cysteine when compared to control rats injected with methylmercury chloride. This indicates a small sulfhydryl molecule to be of importance in this mechanism. Probably the small molecule in bile is also a sufhydryl. In a previous paper (7) this compound was postulated to be cysteine, but this has not been verified by recent experiments. Binding to sulfhydryl groups seems to decrease the stability of some organomercurials (9). Some evidence was found for a lower stability of the reabsorbed compound based on injection of the partially purified substance to otherwise untreated rats. About 25 percent of the inorganic mercury was found in the liver and 17 percent in the kidney as compared to less than 10 percent in both organs for control rats.

There was also a much lower retention of mercury in the blood after injection of the partially purified substance from bile. The red cell to plasma ratio was lower than in methylmercury chloride injected rats, 80 ± 33 and 271 ± 22, respectively. This result offers a possible explanation for the decrease in the mercury content of the kidney when the bile duct is ligated. The increase in mercury content in the liver may be a result of inhibited biliary excretion, and the increase in mercury content in the brain by the blood content of the organ. Thus, what seems to be a specific kidney mechanism may be the same for all organs and may be dependent upon the penetration of the reabsorbed complex into the red cells, but still related to biliary excretion.

If the enterohepatic circulation of mercury is of importance it should be possible to show differences in the plasma distribution of mercury in normal and bile ligated or cannulated rats. Figure 15–2 shows that a small amount of mercury is bound to a plasma compound which may be the same as that in bile as determined by Sephadex column chromatography. This peak corresponds to the major peak in rat bile. Filtration of this compound by the kidney, followed by reabsorption leading to retention in the kidney or to excretion, adds another aspect to biliary excretion and enterohepatic circulation of mercury. Rats with ligated bile ducts had a lower urinary excretion of mercury than control rats, but the mercury content in the kidney was also diminished in these rats. As these rats, however, had a higher plasma level of mercury, this suggested that excretion of mercury by the kidney is related to the mercury content of the kidney or to a specific compound absent from the blood. It was shown

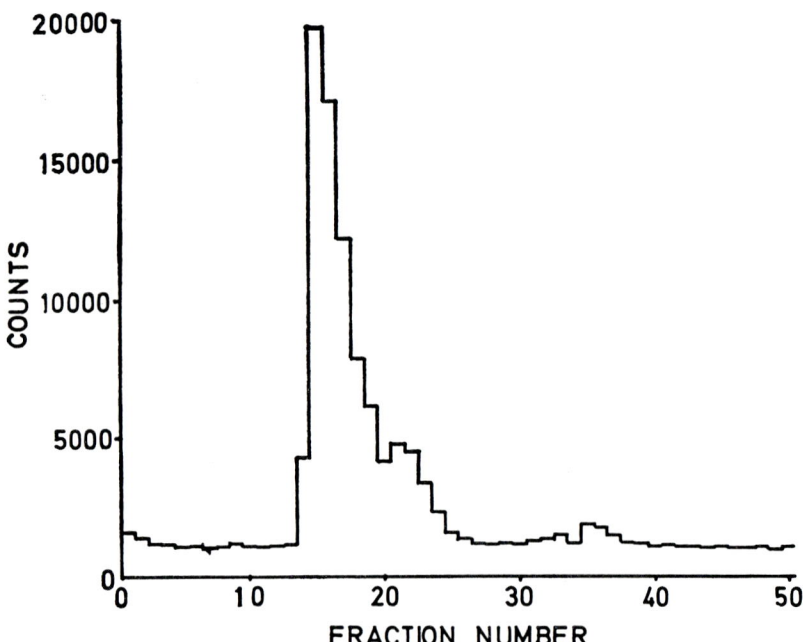

Figure 15–2. Column chromatography on Sephadex G-25 of rat plasma collected four hours after intravenous injection of 1.7 mg Hg/kg as methylmercury. Supporting and eluting solvent was chloroform saturated distilled water; 5 ml fractions were collected.

by Norseth and Clarkson (6) that a relationship exists between relative amounts of inorganic mercury in the kidney and in the urine, which supports the assumption that mercury in urine is derived from the kidney and not from blood. Thus, if the small molecular compound is filtered it should be reabsorbed and deposited in the kidneys.

Species Differences in Organ Distribution

The rat accumulates more mercury in the kidney relative to other organs than other animals tested, but the relative amount of the injected dose is about the same. This is because of the high amount of mercury in the rat blood. Table 15-I shows the

TABLE 15–I
MERCURY IN ORGANS FROM RAT AND MOUSE

Organ	1 day		8 days	
	Rat	Mouse	Rat	Mouse
Blood	39.0%*	16.0%	21.0%	7.0%
Liver	7.6%	10.0%	2.7%	7.0%
Kidney	7.0%	6.9%	6.8%	7.0%
Brain	0.2%	0.5%	0.3%	0.8%

* Values given as percent of dose. Dose 1 mg Hg/kg as methylmercury.

relative amount of the original dose found in different organs for the rat and the mouse after two different time periods. The strong binding of mercury in the rat blood must be borne in mind when species' differences in excretion and distribution of mercury are discussed. Differences in red cells between the two species may be important for mercury content of the kidney if reduced penetration of the biliary complexes into the red cells can explain the distribution of mercury in the rat as discussed earlier. Enterohepatic circulation may still be important as seen from previous results. Biliary excretion of mercury in the mouse was therefore tested (3). The total amount of mercury excreted by the mouse in the bile cannot be estimated since the amount of bile secreted per day is unknown. There are considerable species' differences in the amount of bile secreted even among animals with gallbladders, and the rat differs markedly by not having a gall bladder. However, the recent results by Clarkson, Small and Norseth (1) indicate the existence of a considerable enterohepatic circulation of mercury also in the mouse.

Species' differences in organ distribution of mercury may, however, be explained because methylmercury is bound to different compounds in the mouse and rat bile. Filtration in the kidneys without reabsorption may then explain the lower kidney concentration compared to that for liver. Urinary excretion is, relative to fecal excretion, higher in the mouse than in the rat, but the red cell to plasma ratio is lower (8). Figure 15–3 shows that mercury is bound to different ligands in the bile for the two species. A complicating factor is that binding of mercury in the mouse bile changes with time, the second peak of radioactivity being higher after 24 hours. There is also some time

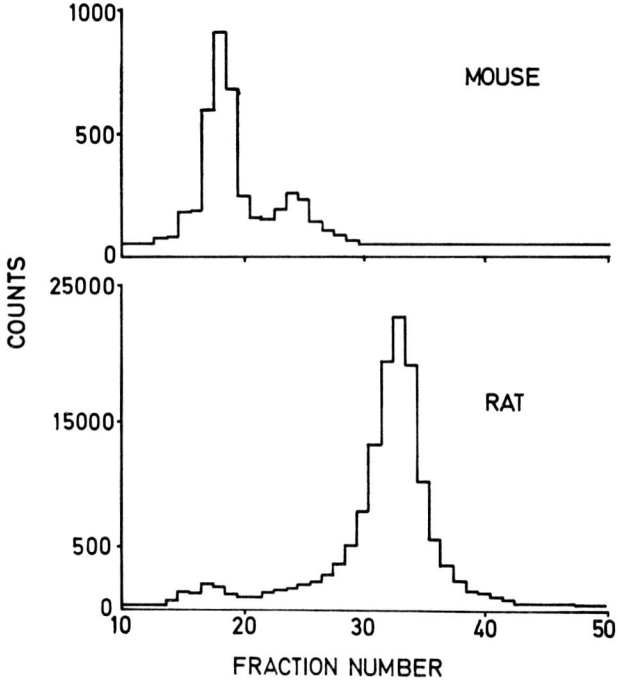

Figure 15–3. Column chromatography of mouse and rat bile in the same system as described in Figure 15–2. Dose was 1 to 2 mg Hg/kg as methylmercury. Animals were intraperitoneally injected and bile collected the first six hours (gall bladder removed after six hours for mice).

dependent changes in the rat, as the peaks six hours after intravenous injection are slightly different from those after 24 hours and following intraperitoneal injection. The fractions have been freeze-dried for thin layer chromatography, electrophoresis and reinjection. Final identification of the compounds has not been achieved. The dominating compound in mouse bile for the first six hours may be glutathione; the other compounds are unidentified. The dominating compound in rat bile is not cysteine as suggested in a previous paper (7), but mercury is released from this ligand by acid hydrolysis. Preliminary identification indicates that the compounds are peptides other than glutathione; cysteinyl-glycine is a possibility.

Intracellular distribution of mercury in mouse and rat liver has been compared (5). Preliminary results indicate that there are no major differences between rat and mouse liver. Differences in biliary excretion are thus unlikely to be explained on the basis of intracellular distribution.

Comments

Some aspects of mercury distribution and excretion after experimental exposure of rats to methylmercury salts have been outlined. The results and the discussion clearly indicate the need for further work on the mechanisms of mercury excretion and distribution. Our mercury pollution problem may not be solved along these lines, but the work is basic to our understanding of how the body treats foreign compounds. Our knowledge on the biotransformation and metabolism of metals and organometals is relatively limited, and I assume that the chemical hazards to the public's health from general metal pollution do not end with the solution of our present mercury problem which may hopefully soon be achieved.

REFERENCES

1. Clarkson, T.W., Small, H. and Norseth, T.: The effect of a thiol containing resin on the gastrointestinal absorption and fecal excretion of methylmercury compounds in experimental animals. *Fed Proc,* 1971, in press.

2. Jellum, E., Aaseth, J. and Eldjarn, L.: Mercaptodextran, A new thiol-protecting and metal chelating agent, 1971, in preparation.
3. Norseth, T.: Biotransformation of methylmercury salts in the mouse studied by specific determinations of inorganic mercury. *Acta Pharmacol toxicol,* 1971, in press.
4. Norseth, T.: Biotransformation of methylmercury salts in germ free rats. *Acta Pharmacol toxicol,* 1971, submitted.
5. Norseth, T. and Brendeford, M.: Intercellular distribution of inorganic and organic mercury in rat liver after exposure to methylmercury salts. *Biochem Pharmacol,* 1971, in press.
6. Norseth, T. and Clarkson, T.W.: Studies on the biotransformation of ^{203}Hg-labeled methyl mercury chloride in rats. *Arch Environ Health, 21:*717, 1971.
7. Norseth, T. and Clarkson, T.W.: Intestinal transport of ^{203}Hg-labeled methyl mercury chloride. Role of biotransformation in rats. *Arch environ Health, 22:*568, 1971.
8. Östlund, K.: Studies on the metabolism of methyl mercury and dimethyl mercury in mice. *Acta Pharmacol toxicol, 27* (suppl 1.): 1969.
9. Weiner, J.M., Levy, R.I. and Mudge, G.H.: Studies on mercurial diuresis. *J Pharmacol Exp Ther, 138:*96, 1962.

DISCUSSION

Kazantzis: You told us the intestinal sample taken from the lower small intestine was obtained through a colostomy; you also had some higher intestinal samples. How did you obtain those?

Norseth: I couldn't collect those until I sacrificed the rat and took out the sample from the intestinal tract.

Kazantzis: Is there any possibility that the peristaltic wave which goes through the rat the moment before it dies might alter the distribution of your samples in relation to the anatomical site?

Norseth: It's possible.

Kazantzis: I visualize the fall in mercury concentration of the kidney following bile duct ligation to be related to complexing with some substance normally excreted in the bile, whose renal threshold is then exceeded. If, for example, this mercury complexing agent behaved like bilirubin, following bile duct ligation, the concentration in the blood would rise. The renal threshold would be exceeded and this hypothetical substance would be excreted in the urine, mobilizing mercury from the kidney in the process. Do you see things at all like this?

Norseth: I'm really not certain what's happening.

Kazantzis: With the biliary duct cannulated did you observe any biliary holdup?

Norseth: I have higher rather than lower excretion volumes than reported by others;* 15.4 ml (SD = 0.22, n = 18) of bile from 150 to 200 gram rats in 24 hours.

Hook: What happens to renal function in general when you cannulate or ligate the bile duct?

Norseth: There must be differences in the kidney's renal blood flow and filtration. After six hours the kidney levels in the ligated animals and control animals are the same. If decreased filtration in the kidney was important a difference also appearing during the first six hours would have been probable.

Magos: In experiments done in collaboration with Dr. J.M. Barnes we gave 0.85 mg/kg Hg as methylmercury to rats into the stomach five times a week. Animals sacrificed after different doses showed that the brain and liver levels of mercury increased faster between 0 and 20 doses than after 20 doses. This decline in uptake was more pronounced in liver than in brain. Kidneys showed a quite different curve. Mercury concentration in the kidneys reached a peak level between 20 to 30 doses after which the mercury concentration was more pronounced than the decrease in total mercury pointing to some deterioration in the uptake of this organomercurial.

What I should like to propose in the knowledge of the experiments of Dr. Norseth is that it might be that after a time of exposure the coupling mechanism in liver is actually poisoned or inhibited by the methylmercury. The result is a decrease in the concentration of the complex which is more easily taken up by the kidney than any other form of methylmercury.

Norseth: I have tested biliary distribution of mercury at several time periods after one single injection, but I have never tested it after serial administration.

Vostal: It was interesting to see the sudden change in the red blood cell/plasma distribution of mercury in experiments where you administered methylmercury isolated from bile. Did you observe a similar change in blood distribution after the bile duct ligation?

Norseth: No, I checked for this but could not find it. I have also tested the red cell plasma/distribution with injected methylmercury cysteine and could not find a significant change, indicating again that the biliary compound probably is not methylmercury cysteine.

I have no results regarding the red blood cell plasma/distribution with the compound isolated from mouse bile. It's difficult to get sufficient amounts of mouse bile. The red cell plasma/distribution after ligation of the bile duct in the rat is more like the one found in mouse or in humans; they have also a smaller amount in the kidney, and a higher relative amount in the brain.

Vostal: Would it be possible to explain these results on the basis of species' differences in bile secretion between rat and mouse? There is no gallbladder in the rat and consequently rats have to excrete large volumes

---

* Altman, P.L.: *Blood and Other Body Fluids.* Washington, 1961, p. 412.

of bile per day compared to the mouse where the system of bile secretion and its volume approximate that of man.

Norseth: I have not found any references to the rate of bile secretion in the mouse, but some animals with gallbladders have a higher secretion than the rat.* Mercury coming out in the bile is not reabsorbed in the gallbladder.

Vostal: McMaster† and later Smith and Ivy‡ reported that the hepatic secretion of bile in mice is only 0.6 ml/gm liver and 24 hours as compared to 1.25 ml/gm liver and 24 hours in the rat. In the character of small bile output per gram of liver tissue and presence of gallbladder with large concentrating capacity the mouse is more comparable with similar conditions in man.§ I wanted to mention the species differences in bile secretion only in relation to the fact that your results showed lower concentrations of methylmercury in the rat liver than in mice.

Norseth: The explanation may be that a greater part of the mercury is found in the red cells in the rat. There's also a different compound in the bile of the rat; in the mouse it's probably methylmercury glutathione, in the rats some other compound. I don't think the gallbladder bears any relation to this problem.

Jernelöv: What is known about the excreted methylmercury compound in the intestine that is not being absorbed again?

Norseth: It was a small amount of mercury compared to the amount coming out in the bile. It is reabsorbed rather quickly.

Jernelöv: There is quite a bit of mercury bound to the cell debris; a part of this can be extracted, a part is bound to insoluble protein and is not reabsorbed. Is this methylmercury protein, from an absorbency point of view, in a less available form?

Norseth: I would say that the methylmercury protein in bile *is* less available than the smaller molecules.

Piscator: What buffer system did you use?

Norseth: I used a phosphate buffer pH 8.0 with 1N sodium chloride. I have also used plain water with chloroform but I could not find any difference in the peaks or in the recovery.

Gage: Are you saying that you cannot account for the amount of inorganic mercury in the intestines and that you tried and failed to degrade methylmercury by incubation with gut flora? Were the conditions anaerobic? Most of the interesting reactions of the gut take place anaerobically and one could conceive that methylmercury might give methane under anaerobic conditions.

Norseth: I've tried both anaerobic and aerobic conditions but I must

---

* Altman, P.L.: *Blood and other Body Fluids.* Washington, 1961. p. 412.
† McMaster, P.D.: *J Exp Med,* 35:127, 1922.
‡ Smith, C.R. and Ivy, A.C.: *J Cell Physiol,* 10:365, 1937.
§ McMaster, P.D.: *J Exp Med,* 35:127, 1922.

admit that I reported my methods to a bacteriologist and he questioned my anaerobic methods.

Berlin: Doesn't the fact that you didn't find any difference between germ-free rats and ordinary rats speak against the idea that the shed of cells from the intestinal tract is so important for excretion?

Norseth: No, it doesn't. What it speaks against is bacterial degradation in the intestinal tract. I still think this shed of intestinal cell is the important thing for excretion of methylmercury in feces.

Berlin: To my knowledge there is a considerable difference in turnover between intestinal cells in germ-free rats and non-germ-free rats.

Norseth: I know, but I didn't test the excretion in the germ-free rats. I only tested the relative amounts of inorganic and organic mercury.

Kazantzis: Is this cellular shedding that you're referring to in the small intestine?

Norseth: Yes.

Kazantzis: Professor Booth° in London pointed out recently that small intestinal epithelium has about the highest turnover rate of any tissue in the body. This rate of turnover of cells is something absolutely fantastic. If one didn't know this one might think only very small quantities of mercury could be lost in this way.

Norseth: Calculations using established turnover rates for intestinal epithelium and my own results for cell content of mercury indicate that this may well be the important factor in the excretion of mercury.†

Berlin: This has been one of my favorite ideas. I was very disappointed when you reported that there wasn't any difference in mercury retention between germ-free rats and controls. There is a difference corresponding to a factor of two in turnover rates between germ-free rats as reported by Creamer.‡

Norseth: It could be explained by an increased reabsorption from the bile.

Berlin: Yes, I agree.

Vostal: I would like to return back to the question of the chemical form of the methylmercury excreted into the bile. A very large increase in the excretion of methylmercury into the bile occurs after the administration of BAL. Could there be other amino acids that bind methylmercury in the dithiol form and then offer a better possibility to be excreted into the bile? Perhaps the dithiol bonds are more important in mobilizing methylmercury for the biliary excretion than the monothiol linkage.

Norseth: It's possible.

Fassett: What percent of methylmercury is soluble in the feces?

° Booth, D.C.: In J.D.H. Slater (Ed.): *Sixth Symposium on Advanced Medicine*. London, Pitman Publishing Company, Ltd., 1970, p. 243.
† Norseth, T. and Clarkson, T.W.: *Arch Environ Health*, 22:568, 1971.
‡ Creamer, B.: *Br Med Bull*, 23:228, 1967.

Norseth: Thirty to 40 percent is soluble.

Fassett: How is the stool in a germ-free rat?

Norseth: They have continuous diarrhea.

Piotrowski: You have said that the excreted low molecular weight compound might be a dipeptide. We had a similar situation with urinary mercury after administration of inorganic mercury to rats.* The compound of the lowest molecular weight seemed to be mercury bound with a dipeptide. It has never been identified because of difficulties connected with great variability of the pattern of the urinary mercury. Would this compound appear in a more regular way in the bile or in the intestines so these sources could be used for the identification studies?

Norseth: Yes. The reason I suggested a dipeptide in the paper was that the compound seems to have a molecular weight between an amino acid conjugate and a glutathione conjugate. There are, of course, other possibilities.

Lucier: What kind of mercury in the liver is bound to the endoplasmic reticulum?

Norseth: The P fraction contains ten percent inorganic mercury four days after injection of methylmercury chloride.†

* Jakubowski et al.: *Toxicol Appl Pharmacology,* 16:742, 1970.
† Norseth, T. and Brendeford, M.: *Biochem Pharmacol,* 20:1101, 1971.

Chapter 16

THE *IN VIVO* KINETICS OF MERCURY BINDING OF KIDNEY SOLUBLE PROTEINS FROM RATS RECEIVING VARIOUS MERCURIALS

S.C. FANG

ABSTRACT

The kinetics of mercury-binding profiles of soluble proteins from the kidneys of rats receiving an oral dose of ethylmercuric chloride (EMC), phenylmercuric acetate (PMA), or mercury acetate (Hg^{2+}) were investigated. EMC, PMA, and Hg^{2+} treatment resulted in different ^{203}Hg binding patterns during the early stage, especially in peaks 1, 2, and 3 proteins (molecular weights > 100,000, 58,000, and 25,000, respectively) after separation on a Sephadex G-100 column. At a later stage, the binding patterns between PMA and Hg^{2+} became similar, suggesting a rapid conversion of PMA to inorganic mercury. The binding of EMC in these proteins increased continuously and did not reach its maximum until a much later stage. A very large portion of ^{203}Hg from either EMC, PMA or Hg^{2+} was bound to a soluble protein of 11,000 molecular weight. The characteristic of this protein is its capacity to bind mercury.

Two rates of removal of PMA were observed for all kidney soluble proteins. The fast rate has a half-time of 16 to 28 hours, and the slow rate was between 33 and 65 days. Only one slow rate of removal was observed with EMC and Hg^{2+} and the half-time was between 33 to 74 days for Hg^{2+} and 17 to 44 days for EMC.

REPORT

Introduction

The increased use of mercury compounds in industry, medicine and agriculture and the related risk of intoxication have stimulated interest in the metabolism of these mercury-containing compounds in mammals. A better understanding of mercurial metabolism will greatly assist in interpreting the mechanism of toxicity. Comparative studies of chronic toxicity of phenylmercuric acetate (PMA) and inorganic mercuric salt (Hg^{2+}) by Fitzhugh et al. (2) showed that as little as 0.5 ppm PMA will produce renal lesions, whereas 10 to 20 times as much mercury is required to produce similar effects with mercuric acetate. Studies on the accumulation, distribution, and rate of elimination of various mercurials in experimental animals have been conducted in an effort to explain differential toxicity (1, 8, 9, 10, 11, 12). All mercurials tested were deposited chiefly in the kidneys. The subcellular distribution of mercury in rat liver and kidneys after an oral administration of PMA and Hg^{2+} has been reported (1, 3, 6, 7, 15). The percentage distribution of mercury varied only slightly between the liver and kidney and showed no significant differences among the rats dosed with PMA or Hg^{2+}. Throughout a 72-hour post dosage period, the subcellular distribution of mercury remained quite constant even though the mercury binding in each fraction showed an increase during the early period, followed by a decrease at later periods. The binding of mercury from mercuric chloride in the soluble proteins of rat kidney was reported by Jakubowski et al (4) and Wineiwska et al (14). It was found that three different classes of compounds were present after filtration on Sephadex gels. The main fraction of mercury was bound to metallothionein.

The metabolism of mercurials and their characteristics of protein binding determine, to a great extent, the accumulation of mercury in the organ and its intracellular distribution, which may govern the sites of toxic action. Various mercurials, such as ethylmercuric chloride (EMC), PMA and Hg^{2+}, have different properties of reactivity and toxicological behavior. This paper reports a comparative study on the *in vivo* mercury binding in the soluble protein of rat kidney following a single oral dose of EMC, PMA or Hg^{2+} in an effort to obtain more information on their differential toxicity.

Materials and Methods

All animals utilized in this study were adult male rats of Wistar strain, 5 to 8 months old with an average weight 392 ± 29

g. Solutions of ^{203}Hg-labeled mercurials were prepared at a concentration of 6µmoles/ml. The mercuric acetate was dissolved in water, whereas PMA and EMC were prepared in Wesson oil. Each rat received an oral dose of 6µmoles of mercurial by intubation, and kept individually in a metabolism cage. Both food and water were available to the animals during the entire experimental period.

The rats were sacrificed at various time intervals, and the kidneys were quickly removed and weighed. After measurement of the radioactivity the kidney was homogenized in two volumes of cold 0.25 M sucrose solution with a motor driven tissue grinder with a Teflon pestle. The homogenate was subjected to differential centrifugation using a Sorvall refrigerated centrifuge. The nuclear fraction was isolated by centrifugation at $600 \times g$ for 10 minutes, and the sediment was washed twice with 2.5 ml of cold sucrose solution and recentrifuged as before. The pooled, supernatant fluids from the nuclear isolation were centrifuged at $5000 \times g$ for 20 minutes to sediment the mitochondrial particles, and the pellet was resuspended in cold sucrose solution and recentrifuged. The combined supernatant and washing from the mitochondrial isolation was centrifuged at $35,000 \times g$ for 90 minutes to sediment most of the microsomal particles. The supernatant from this step was designated as soluble fraction.

Aliquots of the soluble solutions, equivalent to 0.25 to 0.5 g of fresh kidney, were filtered through a Sephadex G-100 column (1.8×110 cm) and developed with a buffer solution containing 0.4 M NaCl and 0.1 M NaAc at pH 6.0. The flow rate was maintained at 12 ml per hour and the eluate was passed through a UV detector before it was collected automatically every 30 minutes. The entire operation was carried out in a cold room at $3°$ to $5°$. The recovery of ^{203}Hg from the soluble fraction after Sephadex G-100 separation was quantitative ($99.8 \pm 3.1\%$).

To prepare the calibration curve the elution volumes of gammaglobulin (mol wt 160,000), hemoglobin (mol wt 68,000), albumin (mol wt 67,000), chymotrysinogen A (mol wt 25,000), cytochrome C (mol wt 12,400) and Bacitracin (mol wt 1,450) were determined on the same column in order to estimate the range of molecular weight as described by Whitaker (13).

Measurement of ^{203}Hg radioactivity in all samples was carried out with a Technical Associate γ-scintillation spectrometer equipped with a 2″ NaI(Tl) well detector. The background of this instrument was between 25 to 30 cpm and the initial counting rate for all three mercurials was approximately 500 cpm per 1 nmole.

In some experiments the protein content of each subcellular fraction and each individual tube from the Sephadex G-100 separation was determined either by a micro Kjeldahl procedure or by the Lowry method using albumin as a standard.

Results

Accumulation of ^{203}Hg in the Kidney

The accumulation of mercury in the kidney at various intervals after oral dosing is shown in Figure 16–1. There is 1½ to 2-

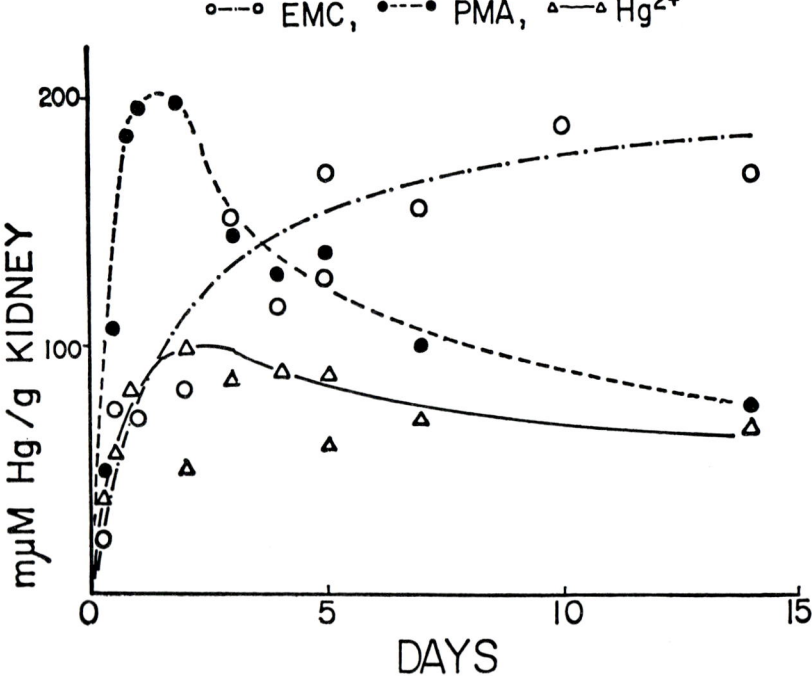

Figure 16–1. Accumulation of ^{203}Hg in the kidney after a single oral dose of 6μmoles of ^{203}Hg-labeled EMC, PMA, or Hg^{2+}.

fold greater accumulation of Hg from PMA than from inorganic mercury during the first three days. After reaching the maximum concentration of mercury in the kidney, which was two and three days for PMA and Hg^{2+}, respectively, the mercury content of the kidney from PMA dosed rats showed a greater rate of decrease at first and then continued to decline gradually which appeared to be similar to that from Hg^{2+} dosed rats. In the case of EMC, the content of mercury in the kidneys increased slowly and did not reach its peak even on the fourteenth day. At this time the mercury content of the kidney was more than two-fold greater than those of PMA or Hg^{2+} treated rats. As shown in Figure 16-2 the incorporation of mercury in the soluble fraction of the kidneys was quite similar for rats dosed with EMC, PMA, or Hg^{2+}. The average incorporation was approximately 50 percent during the initial period, and increased gradually to 60 percent by the seventh day; after which the level remained unchanged. The binding of mercury in the soluble

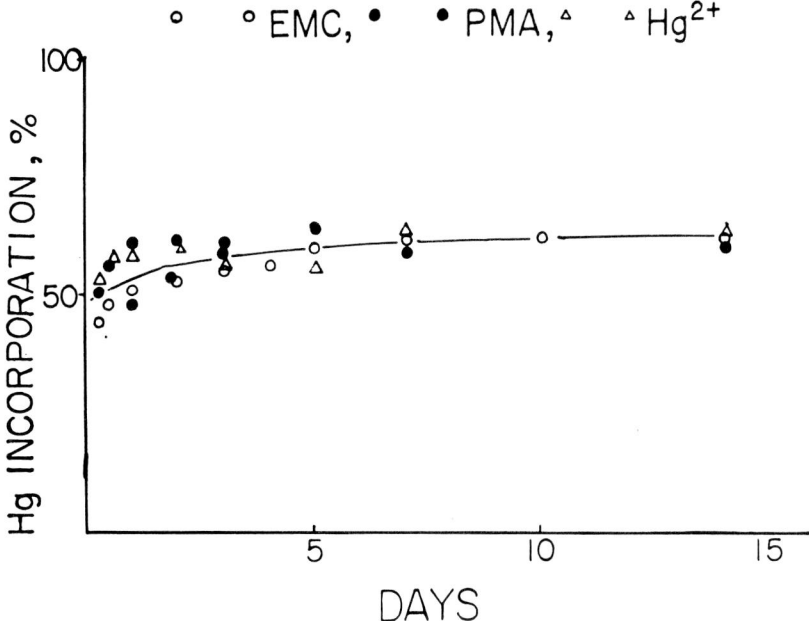

Figure 16-2. Percent incorporation of ^{203}Hg in the soluble fraction of rat kidney.

proteins, therefore, would parallel the mercury accumulation in the kidney, and varies with the mercurial used.

The $O.D._{257}$ profile and the pattern of radioactivity of kidney soluble fraction after Sephadex G-100 separation are shown in Figures 16–3 and 16–4. The ratio of elution volume, V_e, to the void volume V_o, of each radioactivity peak and of six reference markers were plotted against the logarithm of their molecular weights (Fig. 16–5). There were five distinctive optical peaks, with the major one appearing at the void volume. The relative sizes of each $O.D._{257}$ peaks from kidneys of rats with or without mercurial treatment are shown in Table 16-I. Mercurial treatment significantly increased the relative size of $O.D._{257}$ Peak 1 from 20.6 ± 2.3 percent to 31.2, 29.8 and 28.0 for EMC, PMA

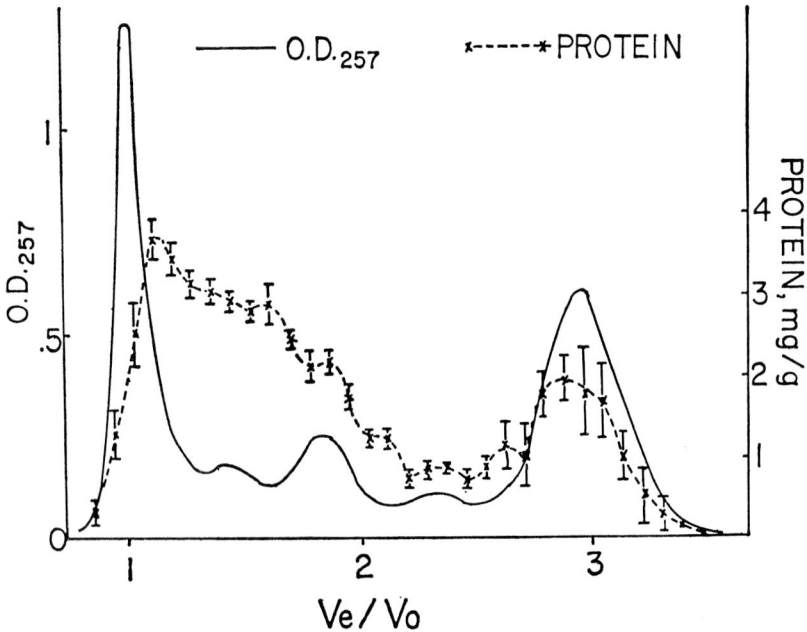

Figure 16–3. Sephadex G-100 (1.8 × 110 cm column) gel chromatography showing $O.D._{257}$ and protein profiles of kidney soluble proteins (1 g fr. weight) from rats receiving a single oral dose of 6μmoles of EMC, PMA, or Hg^{2+}. Each point in the protein profile curve was the average value of five separate runs; and the vertical bars were the standard deviation from the average. ——— $O.D._{257}$ was measured with a 5 mm light path flow cell; x – – – x protein was measured by Lowry method/6.5 ml fraction.

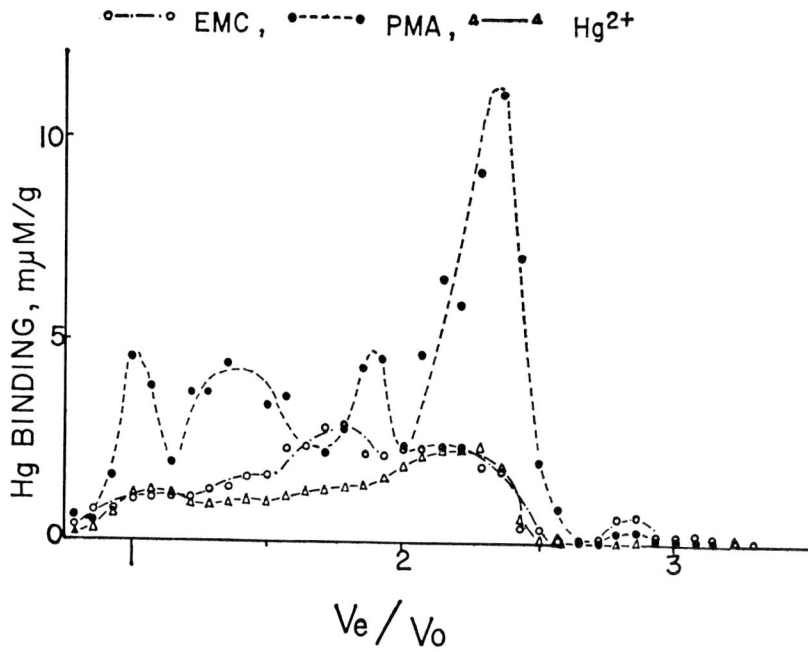

Figure 16–4. Sephadex G-100 (1.8 × 110 column) gel chromatography showing the ^{203}Hg binding patterns of kidney soluble proteins (1 g fr. weight) from rats receiving a single oral dose of 6 μmoles of EMC, PMA, or Hg^{2+}. Rats were sacrificed 24 hours after administration. o—·—·—o EMC; •——————• PMA; △ — △ Hg^{2+}. V_e = elution volume; V_o = void volume.

and Hg^{2+}, respectively. These increases were due to an actual increase of soluble nitrogen from 5.9 ± 0.7 mg/g fresh weight for kidneys of untreated rats to an average of 7.7 ± 1.8 mg for kidneys from mercurial treated rats. Part of this increase was derived from the loss of nitrogen in the nuclear and the microsomal fractions. No significant difference was observed between EMC, PMA, and Hg^{2+} treatments. As shown in Figure 16–5, the molecular weights were > 100,000, 58,000, 25,000, 11,000 and 1,590 for Peaks 1, 2, 3, 4 and 5, respectively. The last optical peak emerged with a V_e/V_o of 2.98 ± .28 and usually contained either little radioactivity for all treated with Hg^{2+} or somewhat more radioactivity for those treated with EMC or PMA. Four distinctive radioactive peaks were observed and they coincided

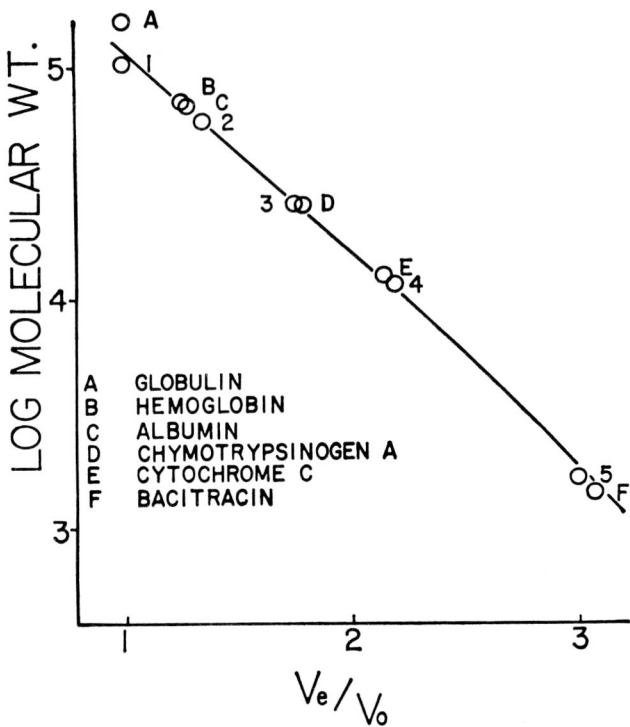

Figure 16–5. Determination of molecular weight of ^{203}Hg-labeled rat kidney soluble proteins by Sephadex G-100 gel filtration. V_o = void volume; V_e = elution volume of particular protein.

very well with the optical peaks. For comparative purposes, the mercury binding patterns were plotted on the basis of nmoles of mercury per fraction per one gram fresh tissue (Fig. 16–4). In all mercurial treated rats, the same four radioactive soluble protein peaks from kidney were observed. However, they differed greatly in the amount bound to each protein. The time course changes of the mercury binding for each peak after a single oral dose of 6μmoles of EMC, PMA, or Hg^{2+} are shown in Figures 16–6, 16–7, 16–8, and 16–9. The mercury binding from PMA in all four peaks showed an increase during the first 24 hours, reaching a maximum of 23 to 38 nmoles for Peaks 1, 2, and 3, and 75 nmoles for Peak 4. After 24 hours, Peaks 1, 2 and

TABLE 16-I

EFFECTS OF MERCURIAL TREATMENT (6μmoles) ON THE RELATIVE AMOUNT OF KIDNEY SOLUBLE PROTEINS AS MEASURED FROM THE O.D.$_{257}$ PEAKS AFTER SEPHADEX G-100 GEL SEPARATION

Treatment	Peak 1	Peak 2	Peak 3	Peak 4	Peak 5
Control (7)*	20.6 ± 2.3%†	9.3 ± 1.4%	17.1 ± 2.8%	7.9 ± 0.6%	54.1 ± 3.9%
EMC (11)	31.2 ± 5.6	6.8 ± 1.2	14.9 ± 2.2	9.3 ± 2.0	37.7 ± 4.6
PMA (7)	29.8 ± 4.4	7.5 ± 1.2	10.8 ± 1.9	9.2 ± 2.8	42.1 ± 3.6
Hg^{2+} (7)	28.0 ± 1.2	8.3 ± 1.5	14.7 ± 1.0	8.0 ± 2.2	41.5 ± 2.8

* Number of rats.
† Mean and standard deviation.

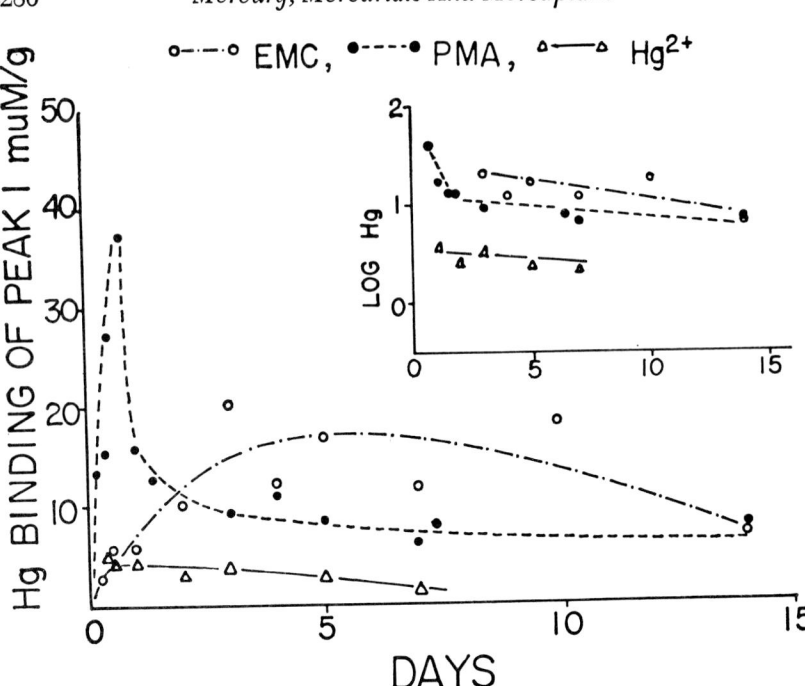

Figure 16–6. Time course study of ^{203}Hg binding of peak 1 protein (mol wt > 100,000) following a single oral dose of ^{203}Hg-labeled compounds (6μmoles).

3 showed a rapid reduction of ^{203}Hg binding, while Peak 4 still showed accumulation of Hg. In the case of inorganic mercury, the mercury binding of Peaks 1, 2, and 3 was small and never reached as high a concentration found in PMA. The mercury level in these three soluble proteins began to decline after one day even though the total mercury in the kidney still showed an increase (Fig. 16–1). The accumulation of ^{203}Hg from Hg^{2+} in Peak 4 did not reach its maximum until the fifth day and then declined slowly. At this stage the level of mercury binding of Peak 4 was comparable to that between the amounts for PMA and Hg^{2+}. In the case of EMC, the binding of mercury in all peaks increased continuously and reached its maximum on the fifth day for Peaks 1, 2 and 3; while the binding in Peak 4 did not reach its maximum until the fourteenth day. Although the maximum levels of mercury binding from EMC in the soluble

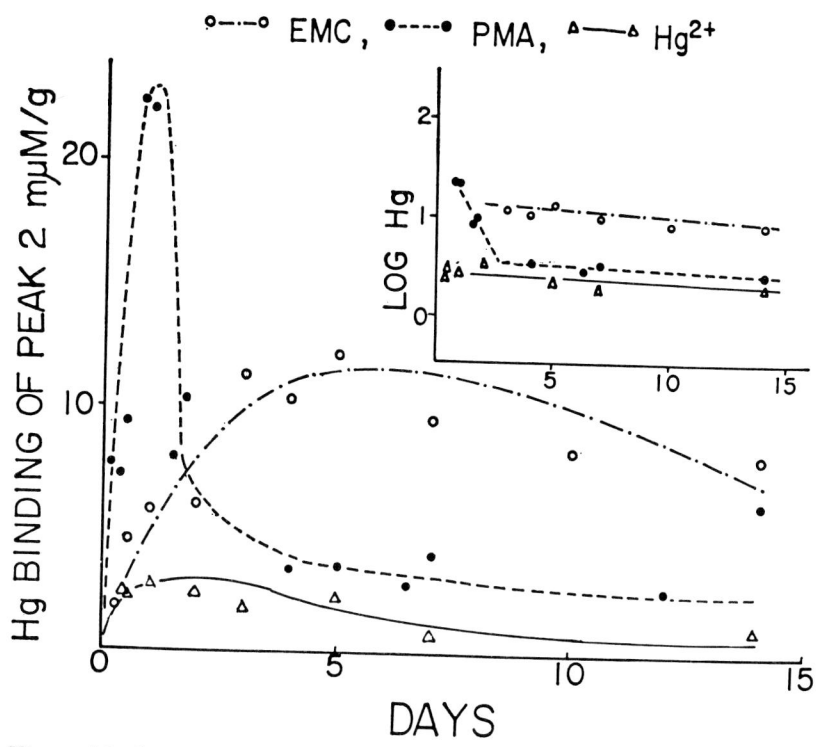

Figure 16-7. Time course study of ^{203}Hg binding of Peak 2 protein (mol wt 54,000) following a single oral dose of ^{203}Hg-labeled compounds (6μmoles).

proteins were not as high as those found with PMA, they were retained much longer.

As revealed by a semilog plot of ^{203}Hg binding against the time after oral administration, there were two rates of ^{203}Hg removal in all four soluble proteins for PMA; and there was only a single rate for EMC and Hg^{2+}. The slopes of the lines for each protein and mercurial, and their half-life are shown in Table 16-II. The half-life ($t\frac{1}{2}$) of the fast rates of removal for PMA ranged between 16 to 28 hours, while the $t\frac{1}{2}$ for the second rates was between 33 and 65 days. The $t\frac{1}{2}$ for EMC was between 17 and 44 days and for Hg^{2+} between 33 and 74 days. The experiment was of insufficient duration for estimating of the half-life of EMC in Peak 4.

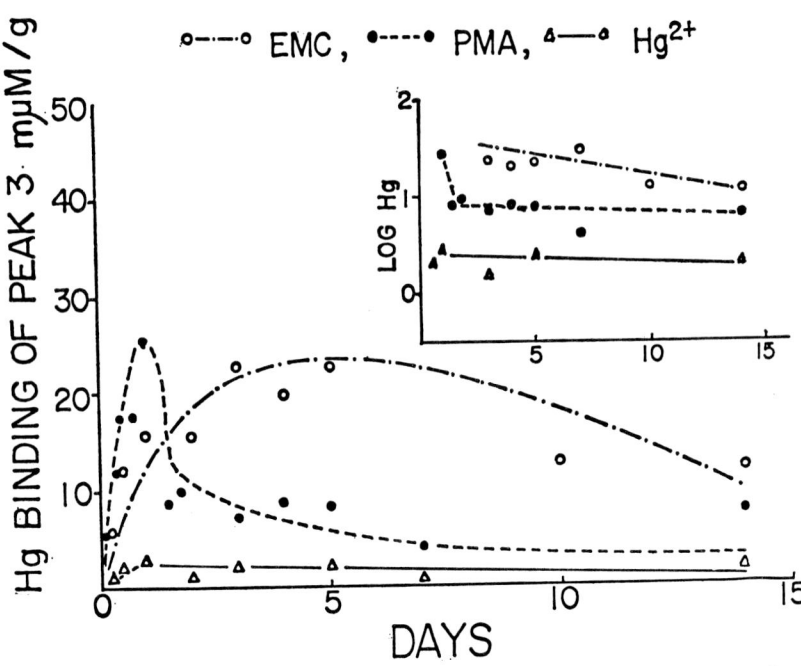

Figure 16–8. Time course study of ^{203}Hg binding of Peak 3 protein (mol wt 25,000) following a single oral dose of ^{203}Hg-labeled compounds (6μmoles).

Discussion

In vivo metabolism of ^{203}Hg from EMC, PMA or Hg^{2+} resulted in a very large portion of ^{203}Hg being bound to a soluble protein of 11,000 molecular weight (Peak 4). The most characteristic feature of this mercury protein is its capacity to bind mercury. This protein has been identified by Wisniewska *et al.* to be metallothionein (14). The rates of removal of ^{203}Hg from this protein as estimated from a 14-day experiment were approximately 0.0106 day^{-1} and 0.0094 day^{-1} for PMA and Hg^{2+}, respectively. The half-life was 65 days for PMA and 74 days for Hg^{2+}. Rothstein and Hayes (8) studied the metabolism of inorganic mercury and concluded that the clearance of mercury from the rats took place in three phases. The two slow phases of excretion represented mainly the clearance from the kidney, and had a half-life of 30 days and approximately 100 days,

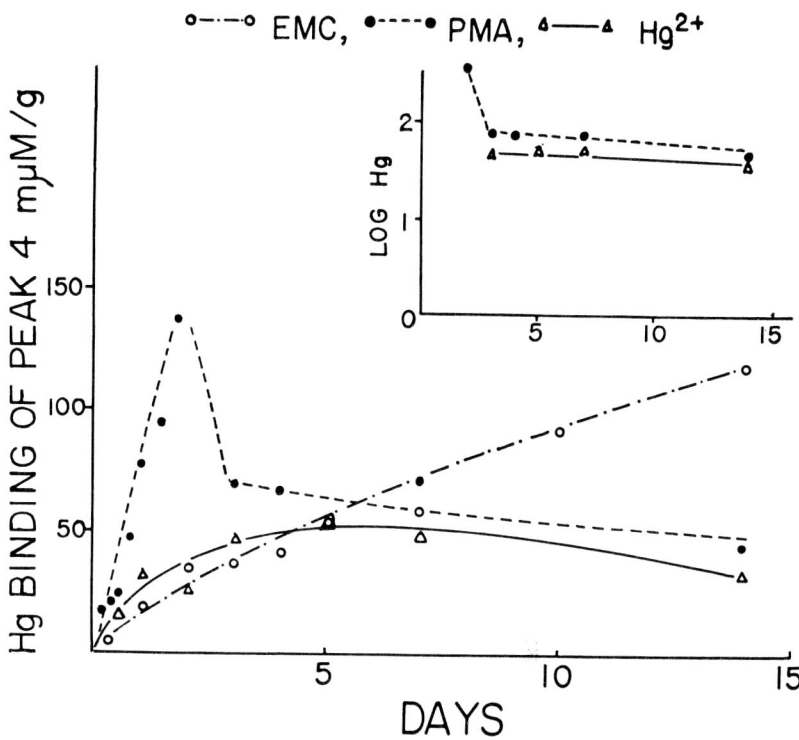

Figure 16-9. Time course study of ^{203}Hg binding of Peak 4 protein (mol wt 11,000) following a single dose of ^{203}Hg-labeled compounds (6μmoles).

respectively. This is in good agreement with our present finding which showed the half-lifes of 33 days for Peak 1 protein and 74 days for Peak 4 protein.

EMC, PMA and Hg^{2+} treatments resulted in different ^{203}Hg binding patterns, particularly during the early stages. As the post-treatment time progressed, the binding patterns of PMA and Hg^{2+} became similar, probably due to the conversion of PMA into an inorganic form (5). Takeda et al. (11) stated that in the early period after the administration of mercurials, the distribution patterns of PMA in several organs and the blood were more similar to those of alkylmercury compounds than those of mercuric chloride, but in the later period they became quite similar to those of mercuric chloride. This would suggest that the rat can convert PMA to inorganic mercury more quickly

TABLE 16-II
THE RATE OF ^{203}MERCURY REMOVABLE FROM KIDNEY SOLUBLE PROTEINS

Proteins	EMC		PMA		MERCURY^{2+}	
	Rate of Removal	Half-time	Rate of Removal, k	Half-time	Rate of Removal, k	Half-time
Peak 1 > (mol wt 100,000)	k = 0.0267 day^{-1}	26 days	k_1 = 0.043 hr^{-1} k_2 = 0.0208 day^{-1}	16 hr 33 days	k = 0.208 day^{-1}	33 days
Peak 2 (mol wt 54,000)	k = 0.0156 day^{-1}	44 days	k_1 = 0.025 hr^{-1} k_2 = 0.0125 day^{-1}	28 hr 55 days	k = 0.0125 day^{-1}	55 days
Peak 3 (mol wt 25,000)	k = 0.0416 day^{-1}	17 days	k_1 = 0.042 hr^{-1} k_2 = 0.0125 day^{-1}	16 hr 55 days	k = 0.0094 day^{-1}	74 days
Peak 4 (mol wt 11,000)			k_1 = 0.025 hr^{-1} k_2 = 0.0106 day^{-1}	28 hr 65 days	k = 0.0094 day^{-1}	74 days

than it can with EMC. A greater binding of mercury from PMA at an early stage and a moderate mercury concentration for a much longer period from EMC in these proteins may be the basis for higher toxicity. This experiment, although it did not provide a clear cut distinction between the differential toxicities of EMC, PMA and Hg^{2+}, did nonetheless indicate different behaviors of organic mercurials on protein binding as compared to that of inorganic mercury.

REFERENCES

1. Ellis, R.W. and Fang, S.C.: *Toxical Appl Pharmacol, 11:104,* 1967.
2. Fitzhugh, O.G., Nelson, A.A., Lang, E.P. and Kunze, F.M.: *Arch Ind Hyg Occup Med, 2:*433, 1950.
3. Grief, R.L., Sullivan, W.J., Jacobs, G.S. and Pitts, R.F.: *J Clin Invest, 35:*38, 1956.
4. Jakubowski, M., Piotrowski, J. and Trojanowska, B.: *Toxicol Appl Pharmacol, 16:*743, 1970.
5. Miller, V. L., Klavano, P.A. and Csonka, E.: *Toxicol Appl Pharmacol, 2:*344, 1960.
6. Norseth, T.: *Biochem Pharmacol, 16:*1645, 1967.
7. Norseth, T.: *Biochem Pharmacol, 17:*581, 1968.
8. Rothstein, A. and Hayes, A.D.: *J Pharmacol Exp Ther, 130:*166, 1960.
9. Swensson, A., Lundgren, K.D. and Lindstrom, O.: *AMA Arch Ind Health, 20:*432, 1959.
10. Swensson, A., Lundgren, K.D. and Lindstrom, O.: *AMA Arch Ind Health, 20:*467, 1959.
11. Takeda, Y., Naguchi, T. Mori, T. and Kitagawa, H.: *Toxicol Appl Pharmacol, 13:*156, 1968.
12. Vostal, J. and Heller, J.: *Environ Res, 2:*1, 1968.
13. Whitaker, J.R.: *Anal Chem, 35:*1950, 1963.
14. Wisniewska, J.M., Trojanowska, B., Piotrowski, J. and Jakubowski, M.: *Toxicol Appl Pharmacol, 16:*754, 1970.
15. Yoshino, Y., Mozai, T., and Nakao, K., J.: *Neurochem, 13:*397, 1966.

DISCUSSION

Carpenter: Can you explain the rather unusual observation that large doses cause no effect and smaller doses cause a severe one in the rat?

Fang: There's a very quick removal of phenylmercury acetate from the kidney which may play an important role; such toxic effects are usually not observed under ordinary conditions.

Carpenter: In other words, this gives some support to the idea that in clinical observations we would not expect them to be as toxic as some of the other things.

Gage: Were these single dose injections?

Fang: Yes.

Piscator: I think it should be pointed out that the material will contain many proteins. It's conceivable that we'll have a relatively high increase in fraction four of some other proteins and there we can find muramidase, ribonuclease cytochrome C, amylase and many other low molecular weight proteins coming from serum.

Kazantzis: What is the molecular weight of fraction three.

Fang: Twenty-five thousand.

Rothstein: Why doesn't your fraction four bind the metal *in vitro?*

Fang: It did.

Rothstein: But in some of your curves you got ratios of 0.3 and 0.4?

Fang: Yes, that means the amount of mercury reaching the kidney is very small.

Rothstein: But *in vitro?*

Fang: *In vitro* is very high.

Rothstein: Not on Peak 4. Some of them were less than 1 to 1.

Fang: Peak 4 is much less and my interpretation is that they contain a mixture of these proteins including metallothionein.

Vostal: What is the stability of the sulfhydryl groups in the metallothionein? It might be possible that the sulfhydryl groups of the metallothionein are oxidized and then unable to bind the mercury.

Fang: I presume that is due to the quantity, the amount of the metal binding present in the soluble fraction.

Vostal: Don't you expect that the amount of metallothionein is the same in peak four *in vivo* as *in vitro?*

Fang: No, because there is mercury-induced biosynthesis.

Piotrowski: If you add metallothionein to the homogenate, the binding capacity increases. Depending upon the kind of gel and conditions of chromatography the number and shape of the peaks of mercury complexes may vary to some degree. So I would not pay too much attention to the differences between the diagrams of this report and those of mine. There is, however, much in common in both these data. For instance, *in vitro* studies with inorganic mercury: in our experiments the bulk of mercury was also bound on the proteins of high molecular weight, whereas, *in vivo,* binding by metallothionein usually played the major role.

The only discrepancy I actually see would be connected with the behavior of phenyl acetate. From your data binding of this compound to metallothionein was high at the beginning and then had a tendency to diminish with time. Just the opposite had been found in our experiments (Piotrowski and Bolanowska; *Med Pracy,* 21:338, 1970). Our explanation was that this was due to the breakage of the carbon-mercury bond resulting in "inorganic mercury" which probably has higher affinity to methallothionein.

Gage: One has to realize that homogenization destroys the compartmentalization that occurs in the kidney and you may be associating substances that are never normally associated. What are the stability constants of these various complexes? And by that I mean the equilibrium constant between the mercury compound and the proteins. If the mercury complexed with, say, Peak 4, and brought together with other components is there any redistribution?

Fang: I don't know.

Fassett: Have you looked at fractions in the liver?

Fang: Yes.

Fassett: Do you see anything in a peak similar to Peak 4?

Fang: Their optical density patterns are rather similar to the kidney. There are only slight differences of relative amounts. Peak 5 and Peak 1 in the liver are much larger. If we treat a given animal with these labeled mercurials and then sacrifice at different times and fractionate the soluble protein in the same way, we note that the binding of mercury is quite similar to the kidney. We also find binding in the metallothionein-binding area. But the half-life of these is very short; we think about two days. If you let the rat go for longer than that, you never find it.

Fassett: What would you get if one fed the animals orally over a period and then got out the same fractions to see whether you had the same distribution between the different peaks?

Fang: This part we haven't done.

Rothstein: Have you ever spiked your kidney before you've homogenized with mercury and then looked at the distribution of mercury?

Fang: No.

Piotrowski: What about the recovery in your experiments?

Fang: The recovery of mercury through the column is about 100 percent.

Chapter 17

PATHOLOGY OF EXPERIMENTAL METHYLMERCURY INTOXICATION: SOME PROBLEMS OF EXPOSURE AND RESPONSE

C.A. GRANT

ABSTRACT

Our studies on squirrel monkeys, rats and cats are concerned with the uptake and distribution of methylmercury and the clinical and pathological manifestations of methylmercury intoxication under defined exposure conditions. Analysis of exposure time for squirrel monkeys has demonstrtaed a redistribution of mercury in the brain to correspond with sites of cerebral cortical damage. In squirrel monkeys, slow instead of rapid accumulation resulted in a clinical pattern resembling that of intoxicated humans but coupled with more widespread cortical damage. Clinically inapparent damage to the nervous system has been encountered in the squirrel monkey and rat. Cats exposed to a simple methylmercury compound acquired blood and brain Hg levels and cerebellar damage very similar to that encountered in cats fed methylmercury-containing fish from a contaminated lake.

REPORT

Introduction

The material presented here has been drawn from current projects on the experimental toxicity of methylmercury (CH_3Hg) to illustrate some of the pathological problems encountered in exploring the relation between exposure and response. These problems relate to the following:

Those collaborating in the projects reported here are L. Albanus, F. Berglund, M. Berlin, L. Frankenberg, Ulla von Haartman, J. Hellberg, A. Jernelöv, G. Nordberg, S. Skerfving and A. Sundvall representing the National Institute for Public Health, Stockholm; Department of Hygiene, University of Lund; Department of Hygiene, Karolinska Institutet, Stockholm; the Defense Research Institute, Stockholm; and the Institute for Water and Air Pollution, Stockholm.

Some Problems of Exposure and Response

1. Different intensities of exposure giving differences and similarities in clinical and pathological patterns at comparable blood and tissue Hg levels (in squirrel monkeys).

2. Clinically inapparent but morphologically evident damage to the nervous system (in squirrel monkeys and rats) and

3. Whether toxicity studies with simple CH_3Hg compounds are pathologically and otherwise relevant for assessing the toxicity (in cats) of CH_3Hg accumulated naturally in biological material such as fish.

Experimental Background

Details of the procedures applied to the **squirrel monkeys** can be found in our first report on work with this species (7) and in chapter 11. The methods described for **rats** by Berglund and Berlin (1) at the first Rochester Conference also apply to the rats described here except for the dose levels (see Table 17-II, full details to be published by Ulla von Haartman).

Background information concerning the **cat** experiments is given by Skerfving (9) but is not yet available in English. The purpose of the experiment was firstly to determine whether a fish-eating animal could be poisoned by CH_3Hg in the form present in fish captured from contaminated water and, secondly, to compare the pathological and metabolic effects of CH_3Hg accumulated naturally in fish with a simple methylmercury compound added to an uncontaminated fish diet.

Five cats were fed exclusively on fish supplemented with appropriate vitamins and minerals, particularly vitamin E, thiamine and iron. One group of cats received homogenized pike from a lake contaminated with phenyl Hg from a pulp mill. The fish homogenate contained 5.7 mg Hg/kg, now practically entirely as CH_3Hg. This group was compared with a group of five cats given homogenized pike with a Hg content of about 0.1 mg/kg from a "clean" lake. To this was added CH_3HgOH to 5.3 to 5.7 mg Hg/kg. A control group of five cats was fed on the "clean" pike diet. The fish given both Hg groups was labeled with a small amount of ^{203}Hg as radiochemically pure (99 percent) CH_3HgOH corresponding to less than one percent of the total CH_3Hg content. The total activity in the fish homogenates was about 5μ Ci/kg at the beginning of the experiment. The Hg contents listed in Table 17-III were calculated from activity measurements in comparison with appropriate standards; chemical analyses are underway and presently confirm the activity results.

The cats were sacrificed to accomplish pathological examination as soon as possible after unequivocal clinical signs developed (ataxia with loss of related reflexes, convulsions).

The methods applied for **pathological examination of the nervous system** in all three animal species necessarily involved a compromise to permit concomitant pathological examination and mercury determinations, and in some instances autoradiographical studies on the same material. To do this limited us to the use of a neutral fixative—neutral formol-calcium (loss of Hg to this fixative was negligible)—and necessitated special care during autopsy to prevent contamination with extraneous Hg from blood, hair and intestinal contents. A modified Cammermeyer technique (2) was used for the squirrel monkeys and cats for retrograde perfusion and fixation of entire anaesthetized animals from the kidneys to the cephalic structures. For rats, however, it proved impracticable to flush them without contaminating the internal organs and skin with blood; they were skinned and the brain and spinal cord fixed *in situ* by immersion. Peripheral nerves—the sciatics and the median nerves—together with the brachial plexus plus the eyes, optic nerves, and a wide selection of the other organs and tissues were also sampled.

In all species the brains were sectioned midsagittally and the left half taken for Hg determination or autoradiography or both. To ensure topographical mapping of brain damage the right half was sectioned transversely, the cats and squirrel monkeys at 2 mm intervals in a mitre box, the rats at shorter intervals by freehand razor slicing. The blocks representing an entire transverse section were marked in such a manner that the microscopical sections represented even intervals from the olfactory bulb to the medulla; corresponding sections from different brains could therefore easily be compared.

The brain and other tissues fixed in formol-calcium were prepared for microscopy by washing in running water overnight, dehydration in increasing concentrations of ethanol, clearing in methylbenzoate, rapid rinsing in xylol, and embedding in Fibrowax® under vacuum. The brains were sectioned primarily at 8μ and stained with azure A eosinate at pH 6.5 and with Nissl's cresyl violet.

Intensity of Exposure and Response

Slow accumulation of CH_3Hg by squirrel monkeys (over periods of 81 to 122 days) ultimately resulted in a clinical pattern differing from that seen in monkeys accumulating CH_3Hg more rapidly (21 to 36 days) to comparable blood and brains levels (Table 17-I, cf. ref. 2).

Under both circumstances of exposure to CH_3Hg visual disturbances dominated clinically. Rapid acumulation, however, was accompanied by abrupt onset of obvious visual impairment, usually to apparent blindness over the course of two days at the most and sometimes within a few hours. No unequivocal signs of ataxia were seen; that the monkeys moved clumsily may have been related to their inability to judge distances (cf. refs. 2, 7).

The visual impairment resulting from slow accumulation of CH_3Hg was first detected as sluggish responses in the visual discrimination test described by Berlin et al. (2). Over the course of a week or so the test pattern became consistent with constriction of visual fields. The monkeys also became clumsy in grasping the raisin rewards; this has been provisionally interpreted as impairment of tactile and coordinating functions.

Regardless of whether exposure to CH_3Hg was rapid or more prolonged and regardless of the clinical pattern, the pathological changes in the squirrel monkeys were topographically similar and limited to the cerebral cortex. In our experiments to date we have not encountered relevant lesions involving the cerebellum, the spinal cord with nerve roots, the peripheral nerves, eyes and optic nerves, or other organs of the squirrel monkeys.

Rapid exposure of monkeys Nos. 1, 2 and 3 resulted in damage to the cortical areas abutting upon the calcarine sulcus and its major ramifications (Fig. 17-1, Table 17-I). Monkey No. 4, also

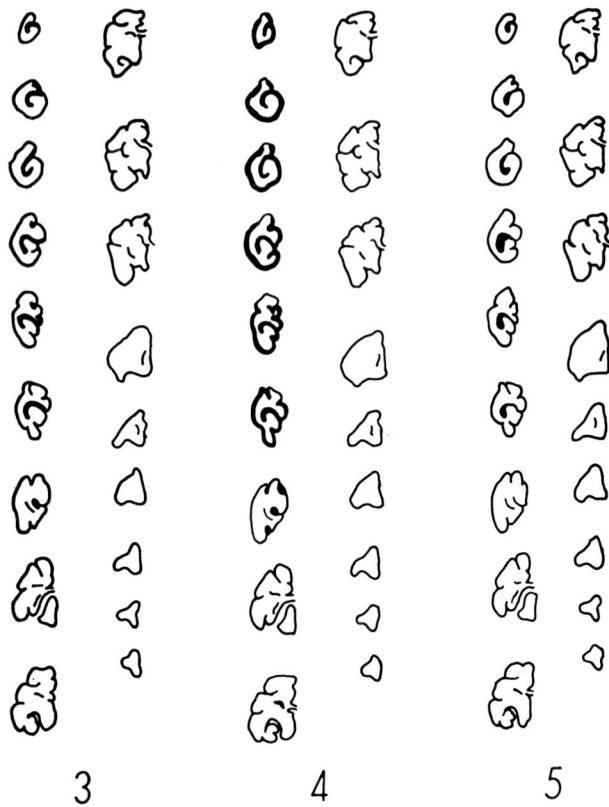

Figure 17-1. Distribution of cerebral cortical damage (heavy line) in equidistant slices through the right half of the brain of squirrel monkeys with the occipital pole at upper left, frontal pole at lower right. Monkey No. 3 represents the usual distribution of cortical damage in rapidly exposed monkeys. Monkey No. 4 had the most extensive lesions among the rapidly exposed monkeys. Monkey No. 5 illustrates the extent of clinically silent damage. Compare with Table 17-I.

belonging to this group in duration of exposure (28 days) and maximum blood Hg level (1.4µg/g) but differing in a much longer post-exposure latent period (37 days), had the most extensive cortical damage involving the entire circumference of the occipital cortex and adjacent areas of the parietal and temporal lobes at the posterior origin of the Sylvian fissure.

More prolonged exposure (81 to 122 days) to comparable maximum blood Hg levels and brain levels (Table 17-I) re-

TABLE 17-I

SOME ACCUMULATION AND DISTRIBUTION DATA FOR CH$_3$Hg IN SQUIRREL MONKEYS IN RELATION TO INTENSITY OF EXPOSURE AND RESPONSE

Monkey No.	Time to (days)	Max. Blood Hg ($\mu g/g$)	Max. Brain Hg ($\mu g/g$)	Distribution of of Brain Lesions
Intensive accumulation				
1	36	1.6	19.3	See Fig. 17-1
2	36	1.8	16.4	
3	28	1.8	14.4	
4	28	1.4	∼ 14	
Slow accumulation				
01	81	1.9	14.5	See Fig. 17-2
15	122	1.2	13.5	
Clinically silent brain damage				
5	21	1.2	∼ 12*	See Fig. 17-1
No brain damage				
6	21	0.75	∼ 7*	

* Brain Hg levels based on extrapolation from values found at conclusion of experiment and the assumption of a similar half-life for CH$_3$Hg in brain and blood.

sulted in much more extensive cortical damage as exemplified in Figure 17-2. Under these exposure conditions damage was more extensive and within contiguous fields involving the entire occipital cortex at the pole to recede gradually and to ebb out along the lateral surface of the hemisphere as the Sylvian fissure is approached.

Regardless of intensity of exposure and the extent of damage the microscopial appearance of the damaged areas of the cortex was basically similar in all monkeys and followed the pattern of concomitant neuron changes and status spongiosus, presumably representing damage to the processes of neurons and glia cells (Figs. 17-3–17-6), where the most severe damage extended through practically the whole depth of the cortex leaving only the outer lamina I and the deeper strata of lamina V and lamina VI as recognizable landmarks (Figs. 17-4, 17-5). The slightest degrees of damage became apparent among the smaller neurons of lamina II and III, leaving pyknotic and karyorhectic nuclear remnants. As the damage extended deeper the larger pyramidal neurons underwent lytic changes (Fig. 17-6).

Atrophy of the cortex, glia cell reaction and, to a lesser extent, infiltration of overlying leptomeninges with lymphocytes and eosinophils appear to be related to vertical depth and horizontal extent of cortical damage. Cortical atrophy, when present (as

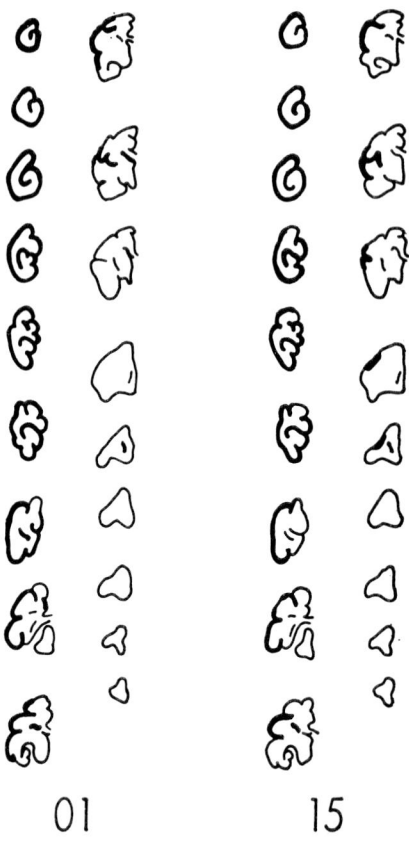

Figure 17-2. As in Figure 17-1 monkeys 01 and 15 were exposed more slowly but attained comparable blood and brain Hg levels.

in No. 4, rapid accumulation; Nos. 01 and 15, slow accumulation), was most obvious around the posterior portions of the calcarine sulci and subsided further forward. The degree of glia proliferation was more or less parallel to the degree of atrophy and the type of glia reaction also changed from predominantly astrocytosis at the occipital pole to predominantly microgliosis in the most forward damaged areas. There is, accordingly, a clear influence of the overall extent of cortical damage upon these aspects. It is less certain, however, whether cortical atrophy and glia reaction are directly associated with the age of the cortical lesions (see below, monkey No. 5).

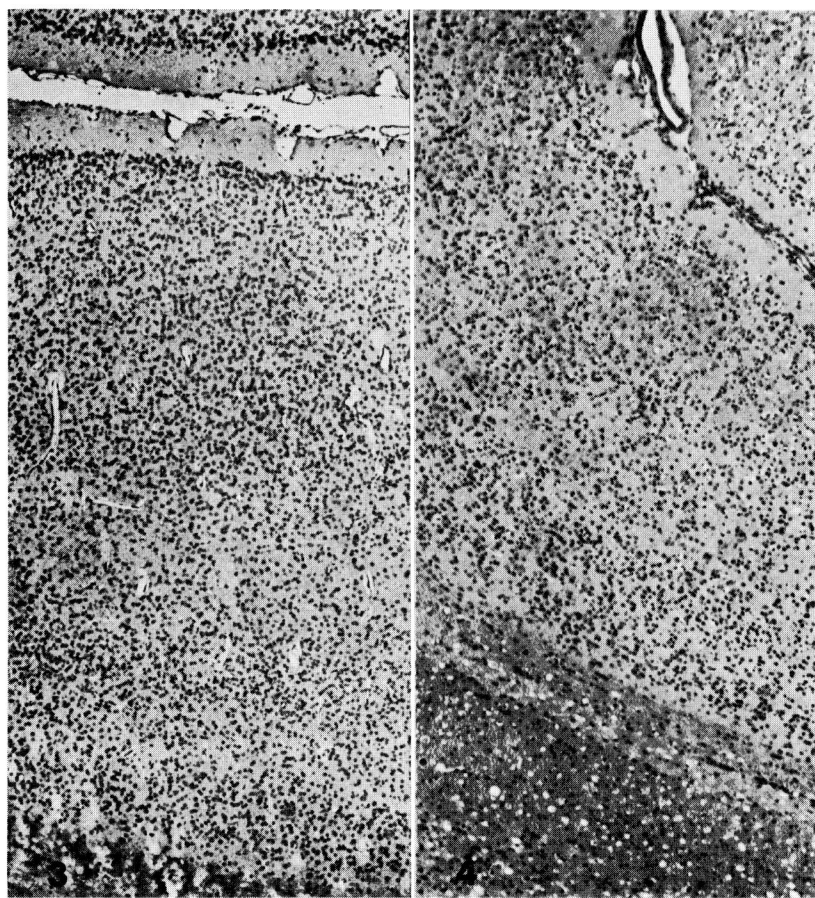

Figure 17-3. Normal calcarine cortex in squirrel monkey for comparison with Figures 17-4 and 17-5 (azure A eosinate, × 40).

Figure 17-4. Calcarine cortex from squirrel monkey No. 2 (Table 17-I), rapidly exposed to methylmercury, to demonstrate disorganization of laminar pattern (azure A eosinate, × 40).

Clinically Inapparent Damage to the Nervous System

By itself it is not unique that a toxic substance can induce damage which escapes clinical detection. But since this has not been demonstrated before for CH_3Hg and since there are clear implications for risk evaluation, our observations on squirrel

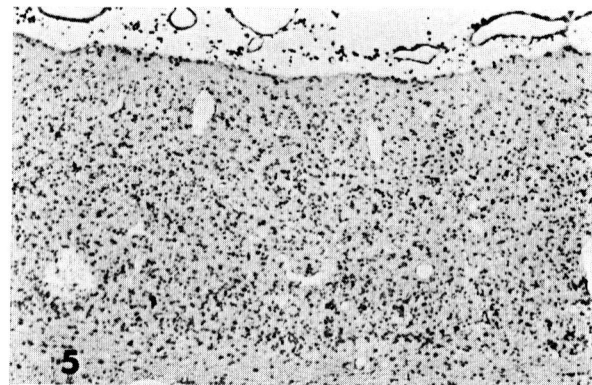

Figure 17-5. Atrophic calcarine cortex from squirrel monkey No. 4 (Table 17-1), same magnification as in Figures 17-1 and 17-2 (azure A eosinate, × 40).

monkeys and rats are pertinent to the purposes of this conference.

An incidental finding among the rapidly exposed **squirrel monkeys** described above was the presence of cortical damage in monkey 5 (Table 17-I, Fig. 17-1). This monkey had not displayed any detectable clinical signs after an exposure period of 21 days, a maximum blood Hg level of 1.2µg/g, and a post-exposure observation period of 85 days. Although this particular monkey was not subjected to visual discrimination testing it is

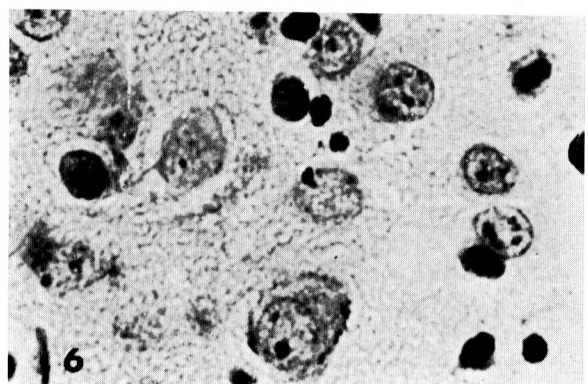

Figure 17-6. Chromatolysis of neurons deep in the calcarine cortex of squirrel monkey No. 4 (Table 17-I) (Nissl's cresyl violet, × 630).

unlikely that the tests would have detected an abnormality not apparent to an observer (cf. ref. 2). The visual discrimination results for the rapidly exposed monkeys which were tested were normal up to the point when obvious visual impairment abruptly developed.

The cortical damage in this monkey (No. 5) was relatively slight but distinct in the calcarine cortex (Fig. 17-1). There was neither atrophy nor discernible glia reaction. The absence of these changes in this monkey, examined 85 days post-exposure, illustrates the need for caution when trying to estimate the age of Hg/-induced cortical damage.

Its mate in the rapid exposure group, No. 6, attained a maximum blood Hg level of $0.75\mu g/g$ after exposure for 21 days but had no detectable cortical damage when sacrificed for examination after an observation period of 85 days. It would appear from this limited series that a threshold for cortical damage in squirrel monkeys lies somewhere between maximum Hg blood levels of 0.75 and $1.2\mu g/g$ under conditions of rapid exposure.

The rats listed in Table 17-II offer a further example of clinically-inapparent but morphologically evident nervous tissue damage induced by CH_3Hg. In this instance damage involved the peripheral nerves and their dorsal roots.

Peripheral nerve damage in rats was described in the classical paper by Hunter *et al.* (3). Since then, however, this part of the pathological spectrum of CH_3Hg toxicity has tended to be neglected in both human beings and animals (cf. ref. 5) although recently taken up again by Miyakawa *et al.* (6) and Takeuchi (10).

As a clinical test for CH_3Hg damage in mice Saito *et al.* (8) used the "dolphin kick" (or frog kick) movement of the hind legs when affected animals were picked up by the tail. Saito *et al.*, however, did not describe the morphological basis for this sign.

The rats in Table 17-II were regularly tested for this phenomenon during the course of the feeding experiment. Most of the group receiving a uniform dose of CH_3Hg (11.8 mg Hg/kg feed) developed this sign after about 130 days of exposure. The sign persisted throughout the experiment. Although carefully exam-

TABLE 17-II

SOME ACCUMULATION AND DISTRIBUTION DATA FOR CH_3Hg IN RATS IN RELATION TO EXPOSURE AND RESPONSE

Rat No.	Exposure Period (days)	Max. Blood Hg ($\mu g/g$)	Brain Hg ($\mu g/g$)	Onset of Clinical Sign "dolphin kick" (days)	Pathological Observations
11.8 mg Hg as CH_3Hg per kg feed					
351	139	87.4	10.6	~130	
352	139	111.2	14.0	~130	
353	139	104.2	17.8	~130	Degeneration
358	139	111.5	18.9	~130	of peripheral
354	224	97.8	17.4	~130	nerves and
355	224	103.9	17.4	~130	dorsal roots
356	224	102.6	17.5	~130	
357	224	99.6	17.4	~130	
7.9 mg Hg as CH_3Hg per kg feed					
251	169	60.6	9.5		
252	169	64.8	10.0		
253	169	59.6	8.3		Degeneration
258	169	56.5	8.2	No	of peripheral
254	226	68.3	10.2	signs	nerves and
255	226	70.3	9.7		dorsal roots
256	226	72.0	11.0		
257	226	64.3	10.8		
3.8 mg Hg as CH_3Hg per kg feed					
151	214	30.9	4.9		
152	214	38.5	4.8	No	No relevant
153	215	30.0	4.9	signs	lesions
154	214	34.2	4.4		

Activity measurements, data from U. von Haartman.

ined in this respect, none of the other exposed rats developed unequivocal signs of this reaction.

When the rats were examined pathologically, all those demonstrating the dolphin-kick sign in the 11.8 mg/kg group including those sacrificed after the shorter exposure period had evidence of dorsal root and peripheral nerve damage. Furthermore, two of the rats in the clinically normal 7.9 mg/kg group (Nos. 255, 256–Table 17-II) had similar distinct damage on a par with the damage seen in the 11.8 mg/kg group and the others had somewhat slighter signs of peripheral nerve damage.

As was the case for the rats described by Hunter et al. (3), damage was uniformly more severe in the sciatic nerve than in the median nerve or brachial plexus. The changes represented both recent damage—axon swelling and fragmentation, myelin sheath disintegration—and older damage with empty neurolemmal sheaths and Schwann cell proliferation (Fig. 17-7). The

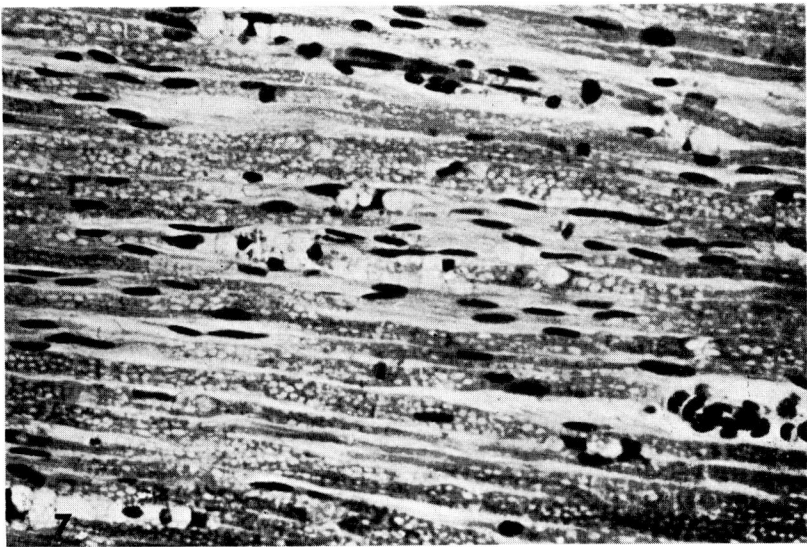

Figure 17-7. Sciatic nerve from rat No. 255 (Table 17-II) with no detectable clinical signs. Nerve fibres with swelling and fragmentation of the myelin sheath are interspersed with patchy Schwann cell proliferation and intact fibres (azure A eosinate × 250).

older changes dominated quantitatively to give the impression of a low-grade, protracted, smouldering process.

From pilot studies on rats given large amounts of CH_3Hg intraperitoneally we know that this species can develop the full spectrum of cerebral cortical, cerebellar and peripheral nerve damage. In the rats reported here, however, damage was limited to the peripheral nerves and their dorsal roots.

Relevancy of Experimental Studies to "Natural" Intoxication

The cat studies were undertaken to demonstrate the toxic potential of CH_3Hg-containing fish from a contaminated lake and to compare the effects of CH_3Hg in this form with those of a simple compound (CH_3HgOH) added to a similar diet.

Details of exposure period to the development of clinical signs, total Hg intake, maximum blood Hg levels and brain Hg levels are listed in Table 17-III. The cats were killed for pathological

TABLE 17-III
SOME ACCUMULATION AND DISTRIBUTION DATA FOR CATS FED ON FISH CONTAINING CH_3Hg (5.7 mg Hg/kg) OR ON FISH WITH CH_3HgOH ADDED TO 5.3 TO 5.7 mg Hg/kg

Cat No.	Exposure mg Hg/kg Body Weight* and day	Max. Blood Hg (µg/g)	Brain Hg (µg/g)	Onset of Clinical Signs (day)
CH_3Hg in fish				
11	0.47	30.3	15.9	68–69
12	0.55	33.9	18.6	59–60
13	0.34	28.4	15.6	59–83
14	0.44	33.8	16.9	63
15	0.44	30.9	14.8	62
CH_3HgOH added to fish				
21	0.43	28.8	15.7	72–75
22	0.45	29.5	15.6	65–70
23	0.46	36.0	18.1	69
24	0.48	36.2	16.6	58–69
25	0.54	39.5	17.8	55–73

Activity measurements, data from S. Skerfving.
* Mean body weight throughout experiment.

examination as soon as possible after distinct clinical signs appeared (ataxia, convulsions).

The relevant pathological changes in the cats of both CH_3Hg groups were limited to the central and peripheral nervous system, specifically the cerebral cortex, the granular layer of parts of the cerebellum, and the peripheral nerves with their dorsal roots.

In the peripheral nerves and dorsal roots degenerative changes involved segments of nerve fibres scattered among intact fibres. While quantitatively relatively slight, the pattern of degeneration was the usual one of swelling and fragmentation of the axon and myelin sheath. At this low quantitative level there were no obvious group differences.

In the cerebellum, damage was not only severe but also remarkably uniform both within and between groups. The topographical distribution was highly selective with preferential involvement of the midline and basal regions—the lingula, nodulus and uvula together with the basal medial areas of the anterior, medial and ansiform lobes (Fig. 17–8). Degeneration within the granular layer of the cortex followed the pattern established by Hunter et al. (3) and Hunter and Russell (4) for rat and human cerebellum. In our cats, the damage, where present, was so

Some Problems of Exposure and Response

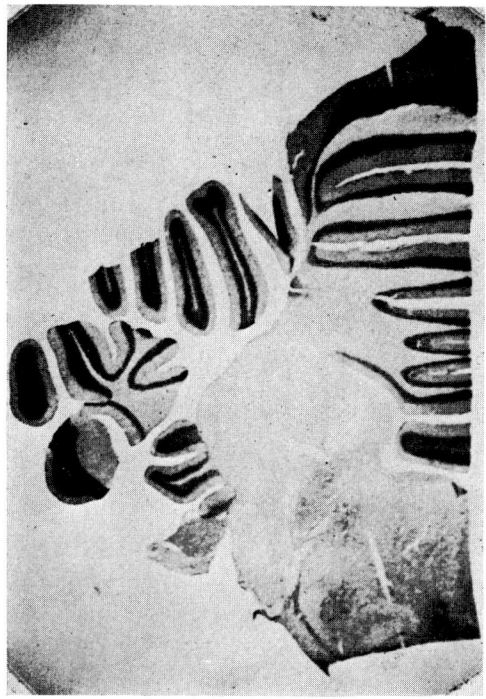

Figure 17-8. Section through right half of cerebellum of control cat for comparison with Figure 17-9 (PAS × 5).

severe that the entire depth of the granular layer was involved, often with loss of the adjacent Purkinje cells.

The cerebral cortical damage, present in all CH_3Hg cats, presents us with definite problems of interpretation. Unlike the effects in squirrel monkeys and rats (large doses IP, see above), i.e. contiguous areas of cortical damage, the cerebral cortical damage in the cats consisted of scattered small foci in practically all regions of the neocortex. The focal cortical damage involved particularly the deeper strata of lamina II and lamina III. Neuron changes were predominantly of Nissl's "ischaemic" type with shrinkage, loss of cytoplasmic detail with increased affinity for stains, and nuclear pyknosis. Accompanying these changes was a locally intense microgliosis and perivascular lymphocyte cuffing. The overall impression is of wider individual differences than group differences in respect to the cerebral cortical damage.

The problem of interpretation is the possibility that the cortical damage is not directly related to CH_3Hg but is the result of hypoxydosis during convulsions.

The cerebellar and peripheral nerve damage, at least, can be confidently related to CH_3Hg intoxication. In this respect there are no discernible differences between the cats fed CH_3Hg-containing fish and the cats fed fish with CH_3HgOH added. The blood and brain Hg levels are also remarkably consistent.

As far as our results can be interpreted, at present, it appears that they are relevant for assessing the toxicity of CH_3Hg accumulated naturally in biological material.

Comment

The information now available about the pathology of CH_3Hg intoxication in human beings and animals has amply confirmed the basic pattern postulated by Hunter *et al* (3) and Hunter and Russell (4), on the basis of only four clinical cases and a subsequent autopsy in human beings and limited experimental material. Within the general framework of involvement of the cerebral cortex, particularly the visual cortex, the granular layer of the cerebellum, and sensory nerves, together or in various combinations, there are apparent vagaries regarding the distribution of lesions under particular circumstances (for review see ref. 5). Differences in exposure conditions, survival periods, and parameters for both human beings and animals have hitherto precluded detailed analysis of CH_3Hg intoxication.

It is against this background that we try to extract as much information as possible that can be correlated with the uptake and distribution of CH_3Hg and the clinical and pathological manifestations in mammals under defined conditions of exposure.

The time factor, for example, has been approached from different angles. Redistribution of CH_3Hg from initially a diffuse cortical distribution to a subcortical and particularly occipital pattern has been followed autoradiographically in squirrel monkeys and has been found to correspond to the distribution of the cerebral cortical damage (Refs. 7; cf. Fig. 17–1). Comparison of rapid and slow exposure of squirrel monkeys to comparable blood

Some Problems of Exposure and Response

and brain Hg levels resulted in the detection of a clinical pattern more closely resembling that seen in human beings and associated with more widespread cortical damage (Ref. 7, Table 17-I, Fig. 17–2).

The demonstration of clinically silent CH_3Hg damage to the nervous system in squirrel monkeys and rats (Tables 17-I and 17-II, Figs. 17–I and 17–7) undoubtedly reflects to some extent the limited accessibility of sensory functions to experimental analysis. There remain, however, clear implications when making risk evaluation for human beings, particularly when based upon the fragmentary data available from the Minimata disaster. That the artificial conditions of experimental studies can be relevant when assessing the risks to people living in a contaminated environment seems to be a fair conclusion from the cat experiment (Table 17-III, Fig. 17–9).

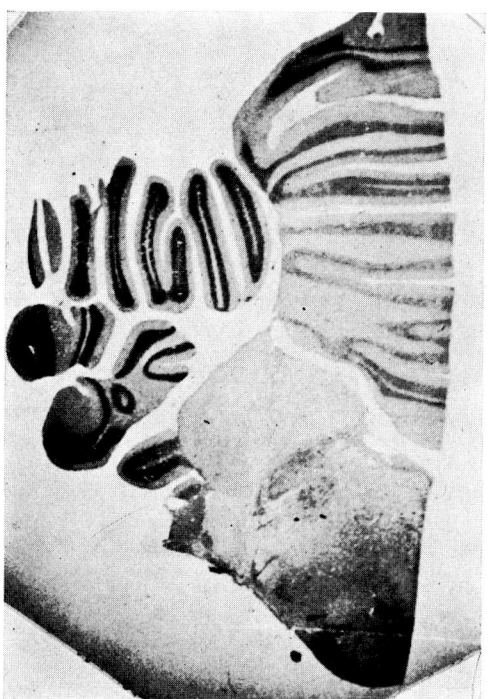

Figure 17–9. Corresponding section from cat 13 (methylmercury-containing fish, Table 17-III) to demonstrate loss of granular layer staining in the anterior and simplex lobes (PAS × 5).

The problem of pathogenesis of CH_3Hg neurotoxicity remains largely untouched. Our current projects are particularly aimed at defining the target and the means of transport of CH_3Hg to its target.

REFERENCES

1. Berglund, F. and Berlin, M.: In M.W. Miller and G.G. Berg (Eds): *Chemical Fallout*. Springfield, Thomas, 1969, chap. 22.
2. Cammermeyer, J., J.: *Neuropathology 19*:141, 1960.
3. Hunter, D., Bomford, R.R. and Russell, D.: *Q J Med, 33*:193, 1940.
4. Hunter, D. and Russell, D., J.: *Neurol Neurosurg Psychiat, 17*:235, 1954.
5. Methyl mercury in fish. Toxicological and epidemiological evaluation. Pathology of methyl mercury poisoning. *Nordisk Hygienisk Tidskrift*, (suppl 4) 1971.
6. Miyakawa, T., Deshimaru, M., Simiyoshi, S., Teraoka, A., Udo, N., Hattori, E. and Tatetsu, S.: *Acta Neuropath (Berl), 15*:45, 1970.
7. Nordberg, G., Berlin, M. and Grant, C.A.: Proc. 16th Int. Congr. Occup. Health, Tokyo, 1970, in press.
8. Saito, M., Osono, T., Watanabe, J., Yamamoto, T., Takeuchi, M., Ohyagi, Y. and Katsunuma, H.: *Jap J Exp Med, 31*:277, 1961.
9. Skerfving, S.: Svensk-finskt kvicksilversymposium, Helsinki, 1969.
10. Takeuchi, T.: Proc. Int. Conf. Environ. Mercury Contam., Ann Arbor, 1970.

DISCUSSION

Suzuki: The species' difference between animals is very clearly shown here at the minimum toxic dose regarding early damage in the nervous system. What about the mercury content in various parts of the nervous system, including peripheral nerves in relation to the symptoms?

Grant: I have to point out that our studies are still underway and that we do not yet have a final answer for all the species. In the monkey and rat experiments summarized in Tables 17–I and 17–II we were fortunate in bracketing the methylmercury levels at which clinical signs and pathological changes became manifest. The main difference between these two species is the blood/brain ratio rather than the brain levels. The cat experiments were intended to compare the effects of methylmercury from different sources; we made no attempt to bracket the toxic-nontoxic level. Cats also have characteristic blood/brain ratio judging from the results shown in Table 17–III.

Goldwater: Were there any abnormal changes in the fish brain or kidney?

Grant: We didn't look for them.

Jernelöv: We found in some of them with the high mercury contents that the spleen seemed to be larger than expected.

Miettinen: We have seen pike which have been submitted to different levels of methylmercury. At the levels of 1ppm and 5 ppm there are noticeable changes in the fish; e.g. slight inflammation of the gills and broken blood vessels in the ventral skin. Inflammation of pseudo bronchiae is observable already with Hg- levels below 1 mg/kg body weight.

Kazantzis: Were the brain sections of the mercury-dosed monkeys read blind together with similar sections from control animals?

Grant: The silent ones happened to be the only ones which were.

Joselow: In your table showing the relationship between blood concentrations of methylmercury and symptoms, you noted some morphological changes in the brain at blood concentrations of $1.2\mu g$ Hg/g, and no effect at levels of $0.75\mu g$ Hg/g. In the Niigata episode, the lowest level at which toxic symptoms in humans was observed was $0.2\mu g$ Hg/g. Does this imply that humans are more sensitive to the effects of mercury than monkeys?

Grant: There are obviously interspecies differences but the committee end of it I'll leave to the committee.

Joselow: Were these figures known to you when you set this point of $0.2\mu g$?

Grant: Yes. We considered all of the facts we knew at that time.

Joselow: In standard toxicological procedures, in determining an acceptable level for one species from data obtained for another species, a factor of 100 is usually applied. If we assume that a blood concentration of $0.75\ \mu g$ Hg/g is close to a "no effect" level in monkeys, we might then arrive at 0.007 or $0.008\mu g$ Hg/g as a maximum acceptable blood concentration for humans, which is quite different from the figure of $0.02\mu g$ Hg/g that the Swedish expert group arrived at.[*]

Berlin: Then you assume a safety factor of ten?

Jernelöv: When you lack human data you sometimes use animal data.

Berlin: We adopted the factor of ten on the 0.2. But do you want to say we should not have a factor of ten since we start from animal experiments?

Joselow: Yes; if we didn't have human data you would be caught with animal data.

Berlin: I agree. First of all, this was the monkey with no effect and the next monkey with effect was 1.2. But it's obvious that the data from the monkeys we now have fit rather well with the clinical data. These monkeys do get affected at about the same level as man.

Joselow: No, we should use both human and animal data. However, the human data used, I believe, suffers from a serious flaw. $0.2\mu g$ Hg/g in blood was calculated as being the lowest level found at which symptoms of mercury poisoning appeared, and a factor of ten was applied to this. It would have been more in accord with accepted procedures, and certainly less controversial, to determine the lowest "no effect" level in humans

---

[*] *Nord Hyg Tidsk,* (suppl 4), 1971, p. 277.

and scale down from this; or if such data were lacking, to use the animal toxicity data you developed and apply a factor of 100.

Jernelöv: I feel that there is nothing like a "no effect" level with that number of animals.

Vostal: Did you look also for some possible changes in the retina or the optical nerve?

Grant: Yes.

Hammer: I would like to congratulate Dr. Grant on his work. It points out the crudeness of using clinical illness as an index or effect of environmental pollution. We understandably don't do brain biopsies on intermediate mercury-exposed people who just have an elevated body burden as reflected by hair or blood mercury. Yet now we must be much more cautious before we say we observe no effect of mercury ingestion at these intermediate exposure levels.

Kazantzis: We have to be very careful how we use the term silent damage because we don't really know what an ill monkey is. You have been testing the monkey only by its ability to obtain access to, and to pick up a raisin, which is a crude test of function of as complex an organ as the central nervous system.

Grant: If I might make a comment. Damage to the neurons, which is what we base a diagnosis on, is quite probably a secondary effect and not a primary effect. We've got to remember that the mercury accumulates in the subcortical white substance, a very important area for the nutrition of neurons and also for the transmission of impulses in and out of the cortex.

SESSION V

BIOTRANSFORMATION OF MERCURY AND ORGANOMERCURIALS

Robert J.M. Horton, *Chairman*

Chapter 18

A NEW BIOCHEMICAL PATHWAY FOR THE METHYLATION OF MERCURY AND SOME ECOLOGICAL IMPLICATIONS

ARNE JERNELÖV

ABSTRACT

The first described biochemical pathways for the biological methylation of mercury involved methyl cobalamin and were studied in anaerobic species of Methanobacterium. Based on the work of Dr. Landner, another biochemical pathway is reported from *in vivo* studies with *Neurospora crassa*. The ability to synthesize methylmercury is found to be correlated with resistance towards high concentrations of mercury in the substrate. If this is the case among microorganisms in general the ecological consequence of a temporary release of mercury by man might be a higher recirculation rate for mercury from organogenic sediment and maintained higher concentrations of methylmercury in water-living organisms.

REPORT

A New Biochemical Pathway for the Methylation of Mercury and Some Ecological Implications

Since the first reports on biological methylation of mercury (3,4,5), studies on the biochemical pathways for the process have been made, using methanogenic bacteria (15). These bacteria are strictly anaerobic, and the emphasis on this group for biochemical experiments has sometimes lead to the idea that anaerobic conditions in mercury-contaminated water and sediment should be required for biological methylation of mercury. In natural waters and sediments, anaerobic conditions almost always mean that hydrogen sulphide is present. Mercury thus will be bound as mercuric sulphide and, even under aerobic conditions, the availability for methylation is very low (1). Under anaerobic conditions, the probability for oxidation of the sulphide to sulphate and consequent release of divalent mercury available for methylation is very low. Thus, in aquatic ecosystems, anaerobic conditions, with hydrogen sulphide present, effectively inhibit methylation of mercury.

The discussion on whether biological methylation of mercury is an anaerobic or an aerobic process is further confused by the fact that the terms have one limnological and one microbiological meaning. In the limnological sense, an organic sediment often has an aerobic surface layer and an anaerobic deeper layer (Fig. 18–1). Aerobic means that the interstitial water contains oxygen. However, the sediment consists of particles. The surface of these might be oxidized while the conditions in the inside of the particle might be reducing (Fig. 18–2). The limnologically "aerobic" sediment thus, from a more strict microbiological point of view, is a mixture of anaerobic and aerobic conditions.

To this discussion can be added some observations from Sweden: A general positive correlation exists between methylation rate and microbiological activity (6), and very high methylation rates in land deployed for drying dredged mercury-rich sediments (8) have been reported.

From *in vivo* experiments with *Neurospora crassa*, Dr. Lars Landner from the Swedish Institute for Water and Air Pollution Research has described another biological pathway for the methylation of mercury (9):

> In this work, studies were performed on the relationship between the resistance of *Neurospora* towards inorganic mercury and its ability to produce methylmercury. The finding that loci determining the resistance towards Hg^{2+} and the synthesis of methionine are closely associated in one complex gene in Staphylococcus (11) further suggested a possible relationship between methylation of mercury and methionine biosynthesis.
>
> The possibility that the methylation of mercury *in vivo* is achieved by means of a direct transmethylation involving methionine or S-adenosyl-methionine can be considered less probable from preliminary experiments with *Neurospora* (Jernelöv and Landner, unpublished; Kitamura, personal communication). In no case an increased yield of methylmercury was found when L-methionine was added to the $HgCl_2$-containing medium. The addition of L-methionine to the medium even rendered the $HgCl_2$ more toxic to the

A New Biochemical Pathway 317

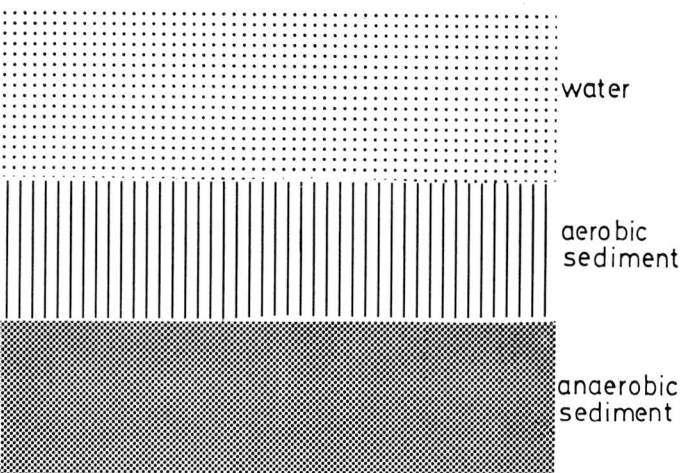

Figure 18–1. Sediment with an aerobic surface layer and an anaerobic deeper layer.

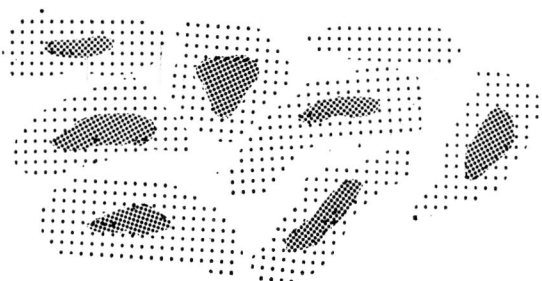

Figure 18–2. Sediments that limnologically are aerobic but microbiologically partly anaerobic.

fungus. Many strains do not tolerate 10 ppm of Hg^{2+} in the presence of other methyl donors like choline or betaine.

In order to investigate the connection of resistance towards Hg^{2+} with its methylation, a series of selection experiments were carried out, starting from a homocaryotic isolate of the wild type strain 74–OR8–1 a, with a low tolerance to Hg^{2+}. The procedure was done in two steps by means of mutation induction by x-rays (20,000 R) and plating of irradiated conidia on Hg^{2+}-containing media in Petri dishes. Mutants with a tolerance level of 200 ppm Hg^{2+} were collected.

A preliminary screening test of the mercury-methylating capacity of the isolates from the dishes revealed that the isolates from the 200 ppm dishes yielded a higher amount of methylmercury than the original strain ($0,05 > P > 0,02$). In a second series of experiments, the tolerance level was increased to 225 ppm Hg^{2+}. The highly tolerant isolates as well as nonselected strains were then cultivated in Erlenmeyer flasks containing liquid Fries' minimal medium (2) with Hg^{2+} and a thiol added at different concentrations. (In preliminary experiments, addition of thiols was found to enhance the yield of methylmercury in tolerant as well as nontolerant strains.) The mycelial pads were harvested after growth during four weeks at 22°C, washed in distilled water, dried at room temperature and analyzed for methylmercury content. These analyses were routinely made by a modification of a gas chromatography technique with E.C., according to Westöö (12).

The results are seen in Table 18-I. A definite increase in yield of methylmercury is apparent with the isolate tolerant towards 225 ppm Hg^{2+} as compared to the starting material with a lower tolerance. In Table 18-II, it can also be seen that the highly tolerant isolate (IV-1) has a much higher methylating efficiency than other strains. These results suggest that the methylation of mercury in *Neurospora* indeed implies a detoxification of Hg^{2+}.

The relationship between yield of methylmercury and presence of thiols in the medium, as revealed in Tables 18-I and 18-II, was further studied, using less tolerant isolates.

TABLE 18–I
CONCENTRATION OF METHYLMERCURY (ng Hg/g CELLS ± S.E.) IN MYCELIA AFTER FOUR WEEKS OF GROWTH

	DL-homocysteine Added		*L-cysteine Added*	
	40 ppm Hg^{2+}	*80 ppm Hg^{2+}*	*40 ppm Hg^{2+}*	*80 ppm Hg^{2+}*
Original strain	269 ± 55	106 ± 79	79 ± 19	694 ± 214
Strain tolerant to 225 ppm Hg^{2+}	3675 ± 862	98 ± 65	680 ± 288	5125 ± 1209

From Landner, 1971.

TABLE 18-II
CONCENTRATION OF METHYLMERCURY (ng Hg/g CELLS ± S.E.) IN MYCELIA AFTER FOUR WEEKS OF GROWTH

Strain	DL-homocysteine Added		L-cysteine Added	
	40 ppm Hg^{2+}	80 ppm Hg^{2+}	40 ppm Hg^{2+}	80 ppm Hg^{2+}
IV-1 Selected	3675 ± 862	98 ± 65	680 ± 288	5125 ± 1209
V-2 Selected	423 ± 24	43 ± 11	155 ± 43	743 ± 213
740R8-1a Unselected	176 ± 66	18 ± 4	109 ± 22	933 ± 264
Fiji Unselected	30 ± 12	15 ± 8	28 ± 5	230 ± 26
Costa Rica Unselected	59 ± 4	31 ± 5	54 ± 12	205 ± 19

From Landner, 1971.

From Table 18-III, it can be seen that a certain surplus of DL-homocysteine over the concentration of Hg^{2+} (on a molar basis) gave the most efficient synthesis of methylmercury. A large surplus of DL-homocysteine reduced the efficiency and a surplus of Hg^{2+} prevented growth. No similar relationship was obtained when L-cysteine replaced DL-homocysteine, even if there was a definite stimulation of the mercury-methylation also by L-cysteine. Other thiols tested, like mercaptoacetic acid, dimercaptopropanol and glutathione, did not cause an increase in the methylmercury concentration in the cells. Further, when a large surplus of DL-homoserine

TABLE 18-III
CONCENTRATION (UPPER NUMBER) IN ng Hg/g CELLS AND TOTAL AMOUNT (LOWER FIGURE) IN ng OF METHYLMERCURY IN MYCELIA AFTER FOUR WEEKS

Hg^{2+} mM	DL-homocysteine 0,06	0,32	0,64	3,2	mM
0,05	58	11	7	11	
	33,4	4,9	3,9	5,6	
0,10	5	29	24	12	
	3,9	18,0	11,3	5,7	
0,20	—	13	380	132	
		8,7	232	67	
0,40	—	—	37	710	
			19,3	214	
0,60	—	—	—	455	
				121	
1,0	—	—	—	345	
				101	

From Landner, 1971.
Each number is the mean of 2 replicates; a dash (—) means no growth.

was added together with approximately equimolar amounts of DL-homocysteine and Hg^{2+}, a doubling of the methylmercury yield was observed. No similar increases in yield were noted when DL-homocysteine occurred in large surplus over Hg^{2+} (other factors unchanged), or when DL-homocysteine was substituted by L-cysteine.

From the reported results, the following conclusions can be drawn:

1. The detoxificating methylation of mercury involves one or several steps of the biosynthesis of methionine.

2. The finding that out of five different thiols tested, only cysteine and homocysteine increase the amount of methylmercury per cell weight, rules out the explanation that thiols stimulate methylation of mercury only by facilitating the uptake of Hg^{2+} to the cells.

3. The relationship between yield of methylmercury and concentration of homocysteine and homoserine in the medium indicates that a negative control of the methylating enzyme [e.g., a transmethylase (13,14)] is affected presumably by methionine, (in analogy with 10). The methionine can probably be formed in sufficient amounts only when a large surplus of homocysteine (over Hg^{2+}) is available.

Consequently, a tentative model for one type of biological methylation of mercury can be proposed: The methyl group, whether synthesised *de novo* or not, is transferred to the mercury atom, this latter being complexed to homocysteine. The methylation of mercury might then be regarded as an 'incorrect' synthesis of methionine. As the methyl-mercury-homocysteine complex presumably cannot execute the feed-back control of the methylating enzyme(s), the synthesis goes on until unloaded homocysteine molecules in sufficient amount have been methylated to methionine. This latter can apparently inactivate the enzyme or repress the synthesis of it. This is indicated to happen when methionine is added to the medium, leading to a break of the detoxification of mercury by methylation. In this connection, it is quite plausible to consider the mutants resistant towards Hg^{2+} as constitutive mutants. The control of one of the last enzymes in the methionine biosynthesis would then be impaired, giving rise to a continuous methylation of Hg^{2+}.

The correlation between high methylation rate and resistance towards high concentrations of mercury in the substrate might lead—if it is supposed to be valid for microorganisms in general— to far-reaching ecological consequences (7).

The principal transport of mercury in a closed aquatic ecosystem can be illustrated in the following way (Fig. 18–3)

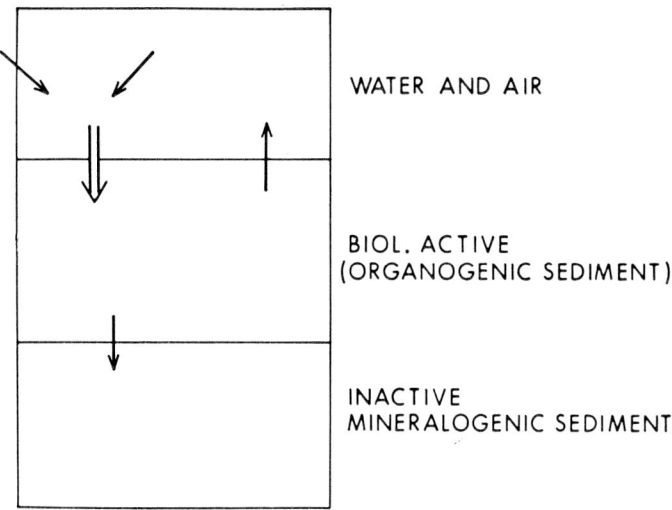

Figure 18–3. Schematic picture of mercury transport in an aquatic ecosystem. Mercury entering the ecosystem will end up in the biological active sediment (or on suspended particles in the water). Most of the mercury will finally be bound in an inorganic form (e.g. to sulphide or iron complexes) and be transported to the mineralogenic layers. Part of the mercury will be recirculated through biological conversion to mono- or dimethylmercury. In due time this mercury will be returned to the sediment.

Mercury entering the ecosystem will end up in the biological active sediment (or on suspended particles in the water). Most of this mercury will finally be bound in an inorganic form (e.g. to sulphides or iron complexes) and transported to the mineralogenic layers. Part of the mercury will be recirculated through biological conversion to mono- or dimethylmercury. In due time, this mercury will be returned to the sediment.

The relative rates of these processes will determine the mercury concentration in the organogenic sediment. If the concentration of mercury is increased in the organogenic sediment through more or less temporary human activities, the organisms resistant to higher mercury concentrations will be favored, compared to others. Within species—as well as between species—competition will result in a selection pressure more in favor of the resistant individuals, which means that these will increase in relative numbers. Under the assumption of a correlation between good ability to methylate mercury and resistance towards higher mercury concentrations in the substrate, the methylating organisms will be favored and a temporarily increased mercury concentration in the sediments will result in a higher degree of recirculation. As a consequence, the percentage of mercury fixed to the mineralogenic sediment is decreased. The resulting principal transport of mercury in the aquatic ecosystem can thus be illustrated as in Figure 18-4.

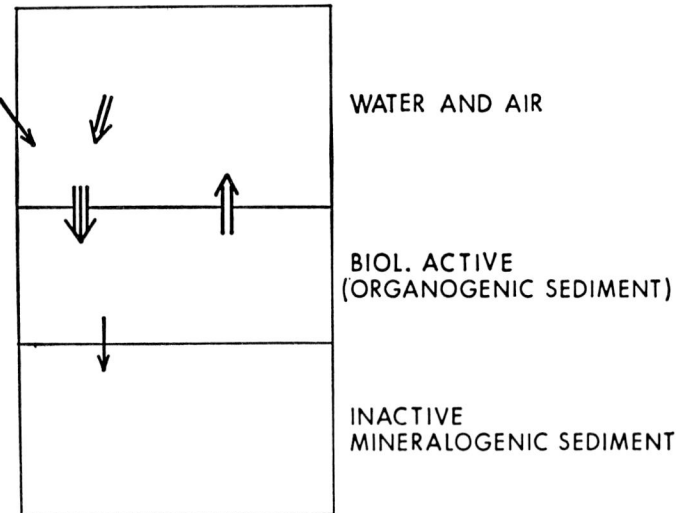

Figure 18-4. Schematic picture of mercury transport in an aquatic ecosystem when methylating microorganisms are favored. The recirculation rate is now higher and the degree of fixation of mercury in the mineralogenic sediment lower. The larger amounts of mercury returned to the sediment per unit of time maintain the higher concentrations of mercury in the organogenic layer.

The lower degree of fixation, the higher degree of recirculation and the larger amount of mercury that is transported to the sediment per unit of time, will result in maintaining higher concentrations of mercury in the organogenic sediment. Thus, the selective advantage of high methylation ability on the part of the microorganisms will remain. The process can be described as a "negative feedback." Of course, this model is very much simplified, and the argument can be made much more sophisticated by the addition of questions like: "How does dilution affect the process?" "What does it mean that the mercury returning to the sediment will largely be in the form of methylmercury bound to dead organisms, while the mercury entering the ecosystem presumably was in an inorganic form?"

Time does not permit me, here and now, to try to go into these complications. The principal, however, will be the same even after quantitative modifications: A temporary increase in mercury concentrations in the sediments might cause (have caused) a change in the selective pressure on the microbiological community leading to a remaining higher recirculation rate for mercury and thus maintaining higher methylmercury concentrations in water-living organisms.

REFERENCES

1. Fagerstrom, T. and Jernelöv, A.: *Water Research*, 1971, in press.
2. Fries, N.: Sumbolae bot. *Upsaliensis*, 3:188, 1938.
3. Jensen, S. and Jernelöv, A.: Biocidinformation. *14*, Nordforsk, May, 1967.
4. Jensen, S. and Jernelöv, A.: Biocidinformation. *14*, Nordforsk, February, 1968.
5. Jensen, S. and Jernelöv, A.: *Nature*, 223:753, 1969.
6. Jernelöv, A.: *Vatten*, 25:304, 1969.
7. Jernelöv, A.: *Zoologisk Revy*, 1971, in press.
8. Jernelöv, A., Lann, H. and Lord, M.: *Vatten*, 1971, in press.
9. Landner, L.: *Nature*, 1971, in press.
10. Marzuluf, G.A. and Metzenberg, R.L.: *J Mol Biol*, 33:423, 1968.
11. Miller, M.A. and Harmon, S.A.: *Nature*, 215:531, 1967.
12. Westöö, G.: *Acta Chem Scand*, 20:2131, 1966.
13. Wiebers, J.L. and Garner, H.R.: *J Bacteriology*, 80:51, 1960.
14. Wiebers, J.L. and Garner, H.R.: *Biochim Biophys Acta*, 117:403, 1966.
15. Wood, J.M., Kennedy, F.S. and Rosen, C.G.: *Nature*, 220:173, 1968.

DISCUSSION

Clarkson: Do microorganisms producing elementary mercury occur in sediments and will they affect this sort of positive feedback mechanism you're speaking of?

Jernelöv: Elementary mercury is also volatilized so it really would not change the situation because they also contribute to the recirculation. If we could find that biological process that tends to fix mercury to the inorganic layer then it would work against itself.

Barber: Is there any evidence that dimethylmercury is released from the surface of the water?

Jernelöv: It depends very much on the conditions. On the one hand, if it's formed together with methane gas or something like that it could easily go with the bubbles. If we look at the gas coming from fibrous sediments where they use phenylmercury we find varying but still comparatively large amounts of dimethylmercury. On the other hand, if dimethylmercury is formed by itself without any gas carrying it through the water one might expect it to accumulate in organisms in the water.

The question then is how stable will it be there? From experiments with birds and mammals we know that dimethylmercury is not very stable within the organisms. Of course it's very hard to go from there to fish. When you go to the next step in the nutrition chain, when someone eats it, it's very likely that dimethylmercury would be degraded.

Norseth: You mentioned an increase of mercury as a factor in keeping amounts of mercury circulating. How would you judge the connection between an increased general pollution with organic waste?

Jernelöv: This is a very typical ecological question. If you change one thing, anything, this will literally effect all the others; the sub-result is very, very hard to predict. For example, let's take a trout and make a physiological experiment with it. If we increase the temperature and the amount of food, we find that our trout will grow very much faster and multiply very much more effectively in warmer water and with a good supply of food.

This would lead us to the conclusion of course that if we increased the amount of nutrients and the temperature in the water we would get more trout. But, of course, we don't. This is where the ecology comes in. Pike would take over and they would be even more effected by this and they would win in the competition with trout.

We have a somewhat similar situation here. If we do a simple experiment in our test tubes (increase the nutrient standards) we'd increase methylation rate considerably. On the other hand, if we look at the fish, it gets the mercury from the methylation either by the direct uptake or through the nutrition change but that's just another step on the way.

The methylation rate here will be very important for the uptake. Then the fish has two ways to get rid of or to decrease it's mercury concentration. It can excrete it, which is a fairly slow process at least in the Scandinavian country temperatures or it can grow and dilute the mercury by growth. The first question here evidently is, which process, the biological methylation or the growth rate of the fish, will be most effected by a change in nutrition level?

Chapter 19

SOME ASPECTS OF MERCURY UPTAKE BY PLANT, ALGAL AND BACTERIAL SYSTEMS IN RELATION TO ITS BIOTRANSFORMATION AND VOLATILIZATION

J. BARBER, W. BEAUFORD AND Y. J. SHIEH

ABSTRACT

Mentha spicata absorbed mercury into its roots from solutions of mercuric chloride or acetate. Small quantities of this element were translocated to the leaves but there was no evidence that it entered the transpiration effluent. On the other hand *Pseudomonas fluorescens* and another bacterium isolated from mercury-rich soil were shown to facilitate the volatilization of mercury. As a result it is argued that the anomalous atmospheric mercury levels detected over heavily vegetated areas associated with mercury-rich soils could be partly due to bacterial action on mercury-containing organic debris.

Addition of $HgCl_2$ at concentrations of $10^{-4}M$ and above to suspensions of *Chlorella pyrenoidosa* resulted in a breakdown in the permeability of the cell membrane as indicated by a rapid net efflux of internal K^+. Between $10^{-5}M$ and $5 \times 10^{-5}M$ $HgCl_2$ there was a stimulation of the $K^+ - K^+$ exchange across the cell surface without any change in the cytoplasmic level of this cation. Maximum stimulation occurred at about $3.0 \times 10^{-5}M$ $HgCl_2$. Enhancement of the rate of K^+ turnover was associated with a fast component of the mercury uptake which could be removed by washing with cysteine. The mercury-stimulated $K^+ - K^+$ exchange was inhibited by low temperature and by the uncoupler CCCP at $5 \times 10^{-5}M$. Concentrations below $6 \times 10^{-6}M$ $HgCl_2$ had no effect on the K^+ transport system.

The authors are indebted to Barringer Research Ltd. and to the Scientific Research Council for financial support. The work was carried out while one of us (Y.J. Shieh) held an International Atomic Energy Agency Fellowship.

REPORT

Introduction

Although low concentrations of mercury can be extremely toxic to biological cells, for example by altering the properties of the cell membrane, there seems to be increasing evidence that certain organisms can bring about a transformation of many of its compounds into other forms (7,9,11), some of which are volatile (13,15,19). To date it has generally been found that microorganisms are responsible for this biotransformation and in some cases mercury-resistant strains of certain species have been identified with this process (1,9,14,19).

The first half of this paper is concerned with the possibility that whole plants, as well as microorganisms, may be able to induce a biotransformation of absorbed mercury into a volatile form which could lead to its transpiration from the leaf surface. The remaining part of this paper deals with a separate, but related, topic involving studies of the action of mercury on the permeability properties and ionic transport mechanisms of the alga *Chlorella pyrenoidosa*.

Absorption, Translocation and Transpiration of Mercury by Plants

Many plants give off volatile organic compounds from their leaves. Went (21) estimated that a total of 10^8 tons of volatile organic matter is released into the atmosphere by the world's vegetation per annum. It is conceivable that some of these organic volatiles could act as carriers for heavy metals such as mercury. To test this possibility, *Mentha spicata* has been used as experimental material. This plant is a member of the mint family and transpires several organic compounds including the terpenes, 1-menthanol and carvone.

Material and Methods

Experimental plants of *Mentha spicata* were obtained by taking cuttings from stock plants and rooting them in 125 ml growth tubes containing 25% Long Ashton Nutrient. They were grown in a greenhouse at about 25°C and were well rooted after 12 days or so.

For transpiration experiments the plants were placed in perspex plant chambers as shown in Figure 19-1. The volume of the upper foliage chamber was about 550 cm³ while the root chamber was about 100 cm³. A split diaphragm held together with Vaseline made it possible to seal off the foliage from the solution bathing the roots. Detection of mercury transpiration was ac-

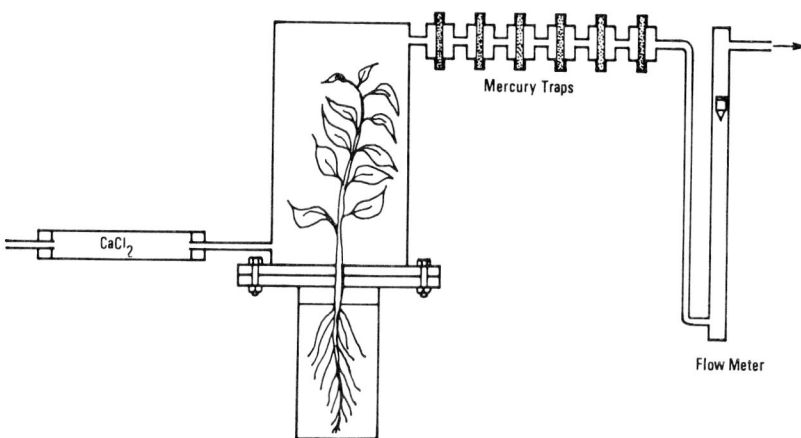

Figure 19–1. Apparatus for measuring mercury transpiration.

complished by using either the radioisotope ^{203}Hg purchased as mercuric chloride and mercuric acetate from the Radio-chemical Centre, Amersham, England, or by using cold mercury. In the former case six glass fibre pads impregnated with cadmium sulphide were used as mercury traps on the outlet side of the upper chamber. These pads were prepared as outlined by Christie et al. (8) and held in 'Delrin' In-line filter holders (Gelman-Hawksley). Radioactivity on the pads was assayed using a Nuclear Chicago gas flow counter. When cold mercury was used, the incoming air was scrubbed free of trace atmospheric mercury by passing through an 8-inch column of iodized activated carbon granules (Sutcliffe Speakman, Leigh, Lancs. U.K.), while the outlet had attached to it six mercury traps in series consisting of 2.5 cm diameter silver gauzes 30 to 60 mesh (B.D.H. Prod. No. 11021) held in the above filter holders. Mercury absorbed on the silver gauzes was released by R.F. heating and measured on a double-beam atomic absorption instrument with a detection limit of 0.1 ng/Hg (10). Transpiration studies with microorganisms used similar flow and trapping systems.

Whole plant autoradiographs were obtained by drying plants which had been previously treated with ^{203}Hg and exposing the flattened material to Kodak Crystallex X-ray Film (CR-54).

Studies with microorganisms involved cultures of *Pseudomonas fluorescens* obtained from the National Collection of Industrial Bacteria, Torry Research Institute, Aberdeen and other bacterial species isolated from high-mercury-containing soil obtained near Keel in Eire. For experiments the organisms were grown in shake cultures on a standard nutrient broth (Oxoid CM1).

Results

Absorption and Translocation

Autoradiographical studies clearly indicated that ^{203}Hg fed to the roots either as mercuric chloride or acetate could enter the plant and be translocated to the leaves. In Figure 19-2 the approximate leaf levels are given in ppm on a fresh weight basis for a plant whose roots had been incubated in 2×10^{-6}M HgCl$_2$ (0.4 ppm Hg) for 48 hours. In general it was found from the autoradiographs that, with reference to the ^{203}Hg levels in the leaves, a slight accumulation occurred in the vascular system and at the nodes. In contrast to the levels usually found in the foliage the roots contained 100 to 500 times more mercury; the concentration being in the region of 100 ppm (fresh weight basis) after a 48-hour incubation in 0.4 ppm Hg fed as mercuric chloride. In fact, although mercury was translocated to the upper parts of the plant the levels in the xylem sap were considerably lower than that of the external medium as shown in Table 19-I.

The roots apparently act as a barrier to the free movement of mercury into the vascular system and extraction of root material previously treated with ^{203}Hg has shown that the majority of the radiomercury occurs in a water-insoluble complex.

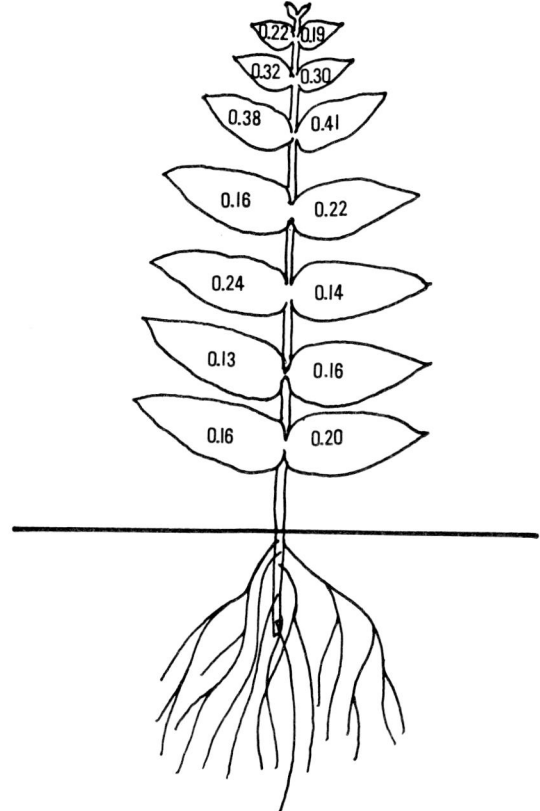

Figure 19–2. Levels of mercury in leaves given in ppm on a fresh weight basis. Plant roots incubated 48 hours in 2×10^{-6}M HgCl$_2$ (0.4 ppm Hg).

TABLE 19-I
MERCURY LEVELS IN XYLEM EXUDATE

External concentration	Exudate concentration $\pm$ S.E.
2.0×10^{-6} moles Hg/liter	$2.8 \pm 0.4 \times 10^{-8}$ moles Hg/liter
400 ppb Hg	6 ppb Hg

The roots of eight plants were bathed in 2.0×10^{-6}M HgCl$_2$ for 24 hours before decapitation. The xylem exudates were then collected over a period of 12 hours from glass capillaries attached to the stems.

Possible Transpiration

Growth experiments indicated that levels greater than 1 ppm Hg added as mercuric chloride or acetate to the liquid culture medium inhibited growth and caused a darkening of the root system. For this and other reasons we choose to use an external concentration of 0.4 ppm Hg to test the possibility of transpiration of this element from plants. As shown in Table 19-II very small amounts of mercury were trapped but, unlike water transpiration, this was independent of whether or not the plants had any leaves.

TABLE 19-II
TEST FOR MERCURY TRANSPIRATION FROM *MENTHA SPICATA*

	With leaves	Without leaves
Mercury trapped (ng)	4.6 ± 2.2	5.0 ± 1.8
Rate of trapping (ng/h)	0.09 ± 0.04	0.10 ± 0.04
Conc. in air (pg/m^3)	7.7 ± 3.8	8.4 ± 4.2
Water transpired (ml)	9.5 ± 0.7	3.1 ± 0.2

Results are the mean ± S.E. of six experiments. The plants were illuminated and the experiments conducted at 25°C with the roots bathed in culture medium containing 2×10^{-6}M HgCl$_2$. Air flow through the chamber was 12.5 m^3/hour and the mercury collected for 48 hours on cadmium sulphide pads.

Very similar results were also obtained with mercuric acetate using both the cadmium sulphide and silver gauze trapping systems. When the same type of experiment was conducted with no plant present, so that air from the lower chamber could diffuse through the 3 mm diameter hole in the diaphragm into the flow system, higher mercury levels were detected on the traps as compared with the plant experiments. The rates of collection on the pads in this case were in the region of 10^{-8} g/hour and the level in air corresponded to 0.9 ng/m^3.

Transpiration from Microorganisms

From the above experiments there was no clear indication that *Mentha spicata* could induce a release of mercury via its transpiration effluent. However, it was considered possible that microorganisms could act as intermediaries between plant and feed solution. In pursuing this idea we looked at the ability of cultures of microorganisms to induce volatilization of mercury.

Initially, culture of *Pseudomonus fluorescens* were used. Suspensions of this bacterium which had been pretreated with mercuric chloride were placed in two identical chambers incorporated in a flow system similar to that used for the plant experiments. One of these served as the experimental suspension and the other as a control. The control cells were killed by treatment with 1 percent toluene or 2 percent of a commercial disinfectant (Dettol) and measurements of respiration with an oxygen electrode indicated that complete death occurred with these concentrations. The experiments were carried out in this manner since the organism was found to bind mercury to levels of about 80 ppm on a fresh weight basis and as a result to reduce the original external mercury level from 0.4 ppm to a value at least 30 to 50 percent lower. Thus at the beginning of the transpiration experiment the binding was about the same in both the control and experimental suspensions.

Table 19-III shows some selected results from a series of experiments using both cadmium sulphide pads and silver gauzes as traps. The amounts of mercury trapped varied from experiment to experiment reflecting differences in experimental conditions but there was always a stimulation of mercury release with the living bacterial suspensions, ranging from four to 30 times greater than that released from the control. This stimulation was

TABLE 19-III
MERCURY VOLATILIZATION FROM BACTERIAL CULTURES

Exp.	Details		Mercury Trapped		Mercury in Airstream pg/m^3
			Total ng	Rate ng/hr	
A	Pseudomonas f. ^{203}Hg/CdS/Toluene	Culture	7.4	0.31	26.0
		Control	0.8	0.03	2.5
B	as A	Culture	10.0	0.42	35.0
		Control	0.4	0.02	1.4
C	as A	Culture	16.0	0.67	56.0
		Control	3.9	0.16	14.0
D	Pseudomonas f. Hg/Ag/Toluene	Culture	28.0	1.17	98.0
		Control	3.0	0.13	10.5
E	as D	Culture	39.0	1.62	137.0
		Control	7.0	0.29	25.0
F	Pseudomonas f. ^{203}Hg/CdS/Dettol	Culture	5.0	0.21	17.0
		Control	0.5	0.02	1.7
G	Keel bacterium ^{203}Hg/CdS/Toluene	Culture	12.0	0.50	42.0
		Control	0.7	0.03	2.4

The bacteria were preincubated for 24 hours in a culture medium containing 2×10^{-6}M HgCl$_2$ before testing for mercury volatilization. The trapping time was 24 hours and the total volume of air passed over the surface of the culture during that time was 285 m^3. ^{203}Hg/CdS means that radiomercury and cadmium sulphide traps were used while the use of cold mercury and silver gauze traps is indicated by Hg/Ag. The agent used to kill the control suspension after the pretreatment time is also shown.

also detected with cultures of a bacterium isolated from a high mercury containing soil (10 ppm) obtained near Keel in Eire. This organism had the general characteristics of the genus *Pseudomonas* and showed a resistance to mercury being able to tolerate at least 10 ppm Hg as mercuric chloride when cultured on agar.

Although cultures of microorganisms seem to induce a biotransformation of mercuric chloride into some more volatile form, it is unlikely that they can act as an intermediary for inducing volatilization from the leaf surface. Experiments on *Mentha spicata* designed with this possibility in mind did not show any significant mercury transpiration.

Discussion

Although mercury fed either as mercuric chloride or as acetate to plants can enter and accumulate to appreciable amounts in the roots, relatively small quantities are translocated into the leaves. At this stage there is no evidence, at least with *Mentha spicata*, that the mercury which is translocated to the leaves can be transformed into volatile forms and transpired from the leaf surface.

The above studies were initiated in order to gain some understanding of the influence of the biosphere on the volatilization of mercury from soils rich in this element. Such soils are often found in the region of hydrothermal orebodies (10); with the development of sensitive aircraft-mounted spectrometers it has been possible to locate some minerialized areas by analyzing atmospheric mercury levels (6). This type of surveying has indicated that anomalous mercury volatilization occurs from heavily vegetated areas growing on mercury-rich soils (Barringer, personal communication). From the experiments reported in this paper it would seem unlikely that this is due to the complexing of mercury with organic volatiles normally released from the surface vegetation but could result from microbial volatilization of mercury during the bacterial breakdown of mercury-containing organic debris shed by the plants and trees growing in these areas.

Support for this comes from field studies near Keel in Eire, where soils with mercury concentrations as high as 10 ppm are found in the region of sulphide mineralization (D. Evans, personal communication). Evans has identified the anomalous mercury levels with the topsoil, humus-containing horizons. He has also found that soil-air mercury concentrations in these soils are very high, being in the region of 20 to 200 ng/m^3 as compared with background levels less than 5 ng/m^3. As yet there is no direct evidence that bacterial action is responsible for these soil-air mercury levels but the bacterial profile through the soil closely follows that of mercury and we do have the indirect evidence from laboratory experiments that bacteria isolated from this soil can induce mercury volatilization. At present we have no idea in what form mercury occurs in the soil-air since the levels have been determined using silver gauze collectors which unfortunately do not distinguish between inorganic, organic or metallic forms. Studies on the relationship of soil microorganisms with mercury distribution in the environment and also on the nature of the volatilization products are continuing.

Uptake of Mercury by *Chlorella* and its Effect on the Cation Transport System

During our investigations into the possibility of volatilization of mercury by plants we were interested in obtaining some idea of how and at what levels mercury affects normal plant cell processes. It has been known for some time that mercury, as well as other elements, can change the cell membrane properties of nonphotosynthetic systems (17) and for this reason we have investigated the effect of mercury on the permeability and transport properties of *Chlorella pyrenoidosa*. This is a nonvacuolated unicellular alga which like many other microorganisms can accumulate K^+ to a high intracellular concentration (2). In fact, comparison of the electrical and concentration gradients across its surface (2) together with tracer studies (3,4,5) suggested that this accumulation resulted from an active transport mechanism probably located in the plasma membrane.

Recently it has been found that under nonsteady-state conditions a net K^+ influx is balanced by an extrusion of internal Na^+

and H^+. However, under conditions of no net movement a closely coupled exchange was found to control the passage of K^+ in and out of the cell. Specific exchange systems of this type seem to be characteristic of microorganisms and are usually associated with relatively impermeable membranes. In the case of *Chlorella pyrenoidosa*, the rate of K^+ leak, even into solutions containing other ions including Na^+, is extremely slow (5). Estimation of the permeability coefficient (P_k) for passive K^+ movement across the *Chlorella pyrenoidosa* cell membrane based on the Goldman theory gave a value of about 2.0×10^{-8} cm sec^{-1}.

The experiments presented below show that mercury affects both the ability of *Chlorella pyrenoidosa* to maintain a K^+ gradient and the mechanism by which K^+ is transported across the cell membrane.

Materials and Methods

The alga, *Chlorella pyrenoidosa*, was an Emerson strain obtained from the Indiana University Algae Collection and the culturing and harvesting procedures have previously been given (2). Tracer fluxes using ^{42}K and ^{203}Hg were measured by taking 1 ml aliquots from suspensions of 1 percent packed cell volume, filtering and assaying internal radioactivity; see Reference 4. Net fluxes were also measured on 1 percent suspensions; 2 ml aliquots were taken, centrifuged and the resulting cell pellets washed and digested in HNO_3. The extracts were analysed for K^+ and Na^+ by flame photometry (18). Except where stated the experiments were conducted at 25°C.

Results

Mercury Uptake

The addition of $3 \times 10^{-5}M$ $HgCl_2$ (6 ppm Hg) to a suspension of *Chlorella pyrenoidosa* cells results in a rapid uptake of mercury by this alga (Fig. 19-3). At 25°C the time course saturated at about 15 minutes and the final concentration bound to the cell was in the region of 1.4µmoles/ml packed cells which corresponded to 488 ppm Hg on a fresh weight basis. The up-

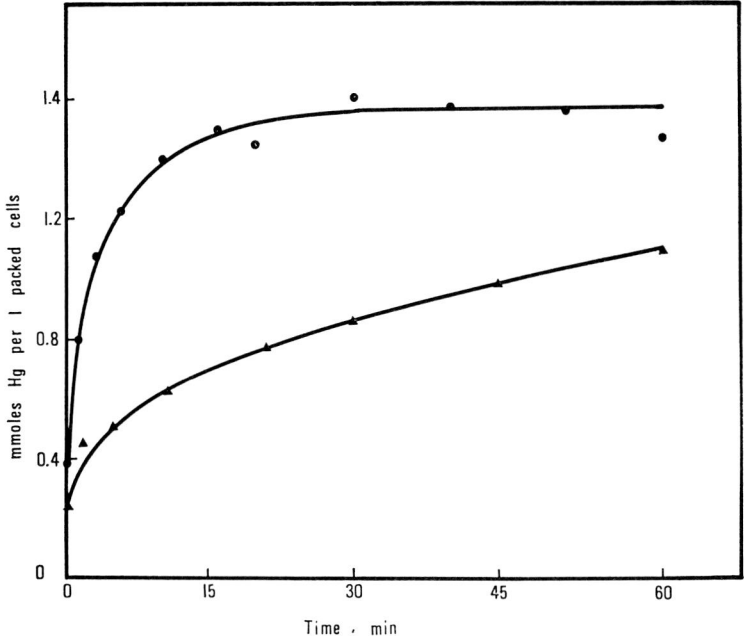

Figure 19–3. Time course of mercury uptake by illuminated *Chlorella* cells measured at 25°C (closed circles) and 0°C (closed triangles). Cells were suspended in culture medium and 3×10^{-5}M HgCl$_2$ labeled with ^{203}Hg added at zero time.

take was not affected by illumination but was reduced by lowering the temperature. At 0°C the influx seems to consist of two components, a rapid phase corresponding to about 0.4μmoles Hg/ml packed cells and a slower uptake which did not saturate within 60 minutes. The presence of two components of the mercury uptake was also shown by resuspending cells preloaded with ^{203}Hg in nonradioactive solutions. There was little or no efflux into normal, mercury-free culture medium, while additions of 3×10^{-5}M HgCl$_2$ or 2×10^{-4}M cysteine to the suspending medium induced a rapid efflux (Fig. 19–4). This removable fraction also corresponded to a reduction of the bound mercury by about 0.4μmoles Hg/ml packed cells. The remaining internal radioactivity representing 1.1μmole Hg/ml packed cells was not readily removed.

Unlike the findings with yeast (16), levels of mercury in

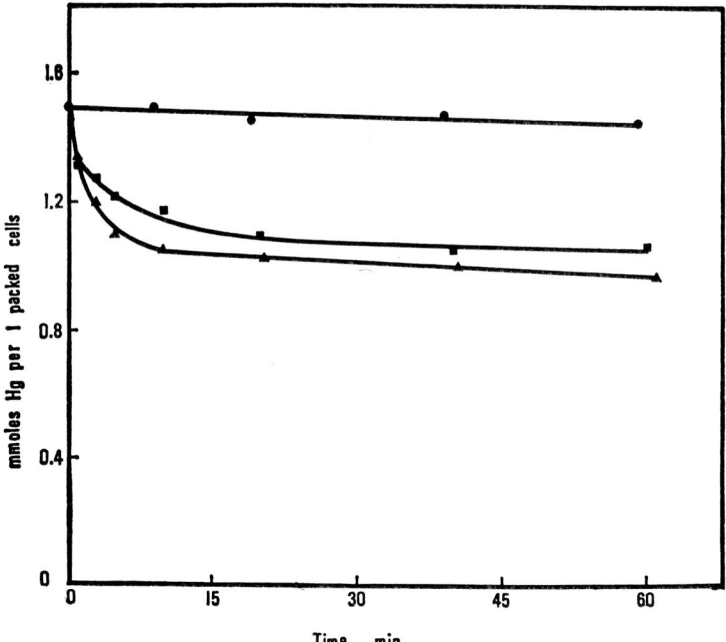

Figure 19-4. Mercury efflux from illuminated *Chlorella* cells preloaded with ^{203}Hg by incubating in 3×10^{-5}M $HgCl_2$ for 30 minutes. The labeled cells were washed and then suspended in either culture medium (closed circles), culture medium plus 3×10^{-5}M $HgCl_2$ (closed triangles) or culture medium plus 2×10^{-4}M cysteine (closed squares).

Chlorella measured after an equilibrium period of 60 minutes in solution ranging from 10^{-8}M to 10^{-3}M $HgCl_2$ showed a linear relationship, with respect to the external concentration on a log-log plot with saturation finally occurring at initial $HgCl_2$ levels above 10^{-3}M.

Effect of Mercury on Internal K^+ Levels

Under normal conditions the internal K^+ concentration in *Chlorella* is approximately 120 mM while the level of this cation in the culture medium is 4 mM. However, as Table 19-IV shows, on adding $HgCl_2$ at concentrations of 7×10^{-5}M (14 ppm Hg) and above to suspensions of *Chlorella* containing 10μl packed cells/ml there is a breakdown in the permeability barrier and a leak of internal K^+. With 5×10^{-4}M $HgCl_2$ (100 ppm Hg) the

TABLE 19-IV
THE EFFECT OF VARIOUS CONCENTRATIONS OF $HgCl_2$ ON THE INTRACELLULAR LEVELS OF K^+ IN *CHLORELLA SPICATA*

$HgCl_2$ Treatment (M)	Time (min)			
	5	30	60	90
5×10^{-6}	100	100	100	100
1×10^{-6}	100	100	100	100
3×10^{-5}	100	100	100	100
5×10^{-5}	100	100	100	100
7×10^{-5}	100	100	100	74
1×10^{-4}	100	86	59	12
5×10^{-4}	48	20	5	3

The data are the intracellular K^+ levels given as a percentage of the control at various times after the addition of $HgCl_2$ to the cell suspension.

net K^+ efflux had a $t_{1/2}$ of about 5 min but with $HgCl_2$ additions in the region of $7 \times 10^{-5}M$ there was a lag period before observing a drop in the internal level of K^+.

Effect of Mercury on the K^+/K^+ Exchange

Using ^{42}K we were surprised to find that concentrations of $HgCl_2$ ranging from $10^{-5}M$ to $5 \times 10^{-5}M$, which did not induce a net K^+ loss, altered the steady-state K^+- K^+ exchange. In fact, as Figure 19-5 shows, the ^{42}K influx was considerably stimulated in the presence of $3 \times 10^{-5}M$ $HgCl_2$ when compared with normal rates of uptake of this cation into illuminated cells. It was found that this stimulation of the influx was balanced by a similar stimulation of the efflux and as such resulted in no net change of the internal K^+ level. Levels of $HgCl_2$ of $2.5 \times 10^{-5}M$ to $3 \times 10^{-5}M$ gave the maximum stimulation with the cell density used ($10\mu l$ packed cells/ml). Figure 19-5 also shows the effect of injecting $5 \times 10^{-4}M$ $HgCl_2$ into a suspension of cells. In this case there was a drop in the internal radioactivity indicating a net loss of K^+.

Recovery of K^+/K^+ Exchange After Mercury Treatment

Figure 19-6 shows that cells pretreated with $3 \times 10^{-5}M$ $HgCl_2$ for 30 minutes retained their ability to show a stimulated K^+- K^+ exchange even after washing with distilled water followed by resuspension in culture medium. However, if the distilled water wash was replaced by a wash with $2 \times 10^{-4}M$ cys-

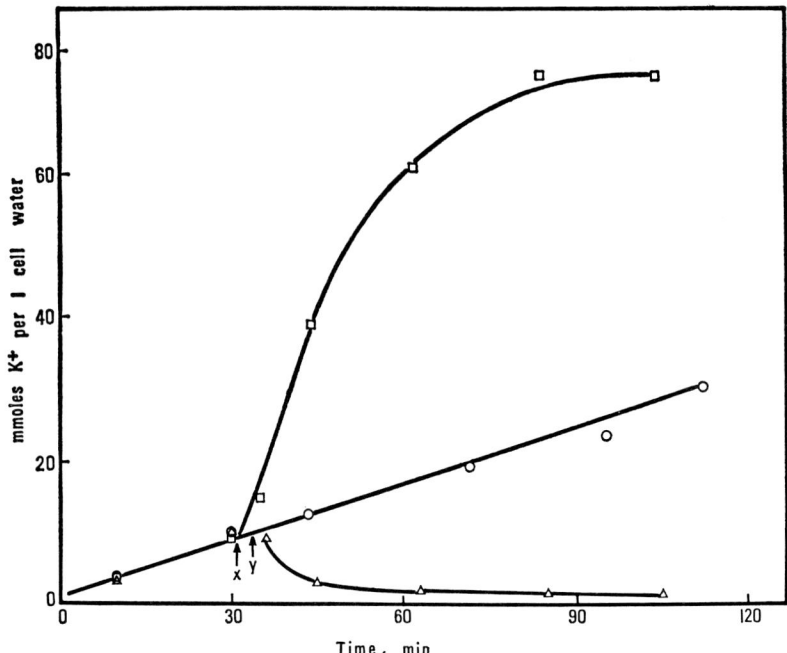

Figure 19-5. The effect of low (3×10^{-5}M) and high (5×10^{-4}M) concentrations of $HgCl_2$ on ^{42}K influx into illuminated *Chlorella* cells. The isotope was added at time zero to three identical suspensions. 3×10^{-5}M $HgCl_2$ was injected into one suspension as shown (open squares) and 5×10^{-4}M $HgCl_2$ into another (open triangles). The remaining suspension acted as a control (open circles).

teine then the ^{42}K influx returned to a value similar to that found in cells not pretreated with $HgCl_2$. Similarly the addition of $HgCl_2$ to a suspension of cells containing 2×10^{-4}M cysteine did not stimulate the K^+ - K^+ exchange as it did in cells which had no cysteine present.

Effect of Carbonyl Cyanide m-Chlorophenylhydrazone (CCCP) and Low Temperature on the Mercury Stimulated K^+ - K^+ Exchange

Surprisingly, the K^+ - K^+ stimulation by 3×10^{-5}M $HgCl_2$ was inhibited by CCCP, an uncoupler of both oxidative and photosynthetic phosphorylation and also reduced by low temperature; see Table 19-V.

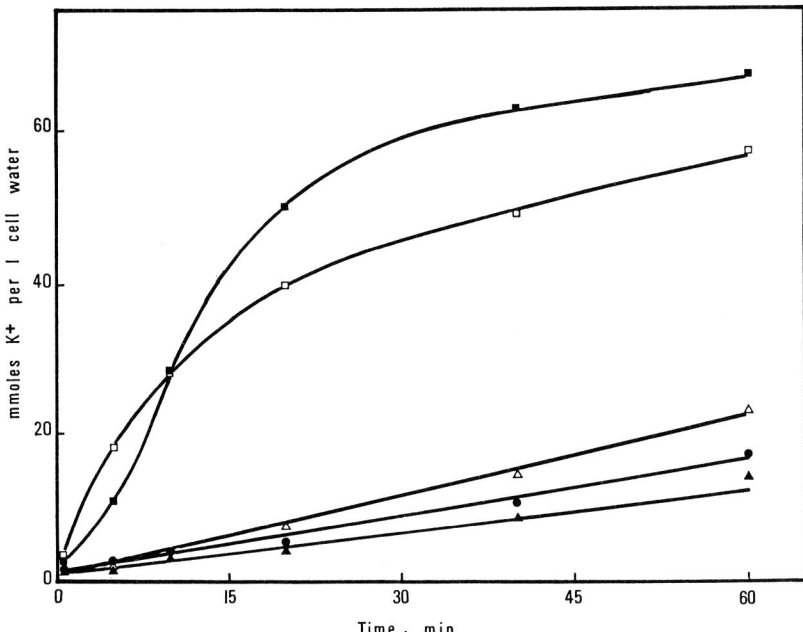

Figure 19–6. Effect of $HgCl_2$ pretreatment and washing on the ^{42}K influx into illuminated *Chlorella* cells. Some cells were incubated for 30 minutes with $3 \times 10^{-5}M$ $HgCl_2$. The Hg-treated suspension was divided in two; one-half washed with distilled water (open squares) and the other half washed with $2 \times 10^{-4}M$ cysteine (open triangles). Finally both were washed with distilled water, suspended in culture medium and the ^{42}K influx followed. Another suspension of cells were treated in the same way as above but without the $HgCl_2$ present. This suspension was divided into three for measuring ^{42}K influx; cells in, culture medium (closed circles), culture medium plus $3 \times 10^{-5}M$ $HgCl_2$ (closed squares) and culture medium plus $3 \times 10^{-5}M$ $HgCl_2$ and $2 \times 10^{-4}M$ cysteine (closed triangles). The $HgCl_2$ and cysteine were added to the suspension five minutes before injecting the ^{42}K.

TABLE 19-V

THE EFFECT OF LOW TEMPERATURE AND $5 \times 10^{-5}M$ CCCP ON THE $K^+ - K^+$ EXCHANGE STIMULATED BY $3 \times 10^{-5}M$ $HgCl_2$

Treatment	Initial Rate ($pmoles\ K^+\ cm^{-2}\ sec^{-1}$)	Total K^+ Exchange in 60 min ($mmoles/liter\ cell\ water$)
Control	1.0	19.0
$HgCl_2$	4.6	91.0
$HgCl_2$ at 0°C	0.2	4.2
$HgCl_2$ + CCCP	1.3	20.0

Efflux of ^{42}K

Normally the efflux of ^{42}K from *Chlorella* cells preloaded with this radioisotope requires an exchange with external K^+. This is clearly shown in Figure 19–7 where it can be seen that there was little or no leakage of ^{42}K into the suspending medium which contained Na^+ but no K^+. Only with the addition of external K^+ did the ^{42}K leave the cells. As Figure 19–6 shows, this efflux was considerably stimulated by injecting $3 \times 10^{-5}M$ $HgCl_2$ while a similar addition of $HgCl_2$ to cells suspended in the K^+–free medium resulted in a much slower leak.

Na^+/K^+ Exchange

If K^+ is added to a suspension of *Chlorella* cells which have a high internal Na^+ concentration there is a rapid efflux of Na^+

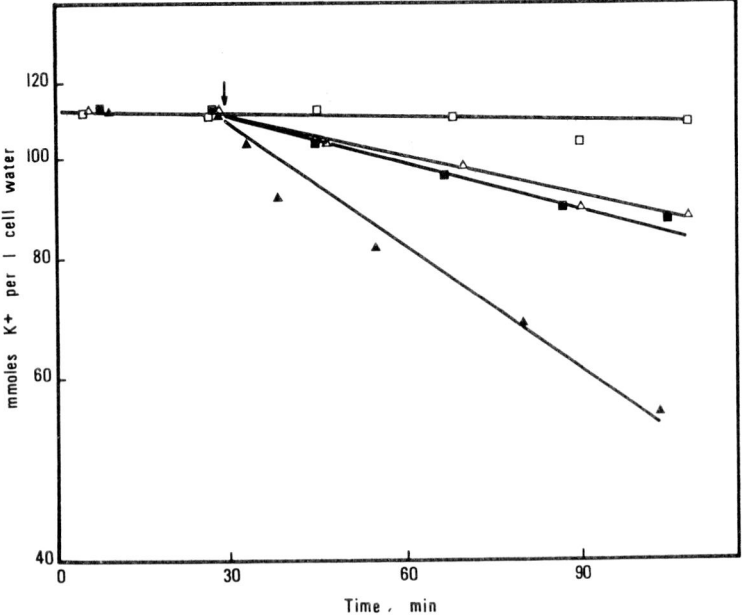

Figure 19–7. The effect of $HgCl_2$ on ^{42}K efflux from *Chlorella* cells. Cells preloaded with ^{42}K were suspended in modified culture medium which had all its K^+ salts replaced by the corresponding Na^+ salts and divided into four. The arrow indicates the addition of 3mM NaCl (open squares), $3 \times 10^{-5}M$ $HgCl_2$ plus 3 mM NaCl (closed squares), 3 mM KCl (open triangles) and $3 \times 10^{-5}M$ $HgCl_2$ plus 3 mM KCl (closed triangles) to separate suspensions.

and concomitant uptake of K^+ (18). The Na^+-rich cells are obtained by growing *Chlorella* in a K^+-free, but Na^+-containing, culture medium. It has been found that $HgCl_2$ concentrations which stimulate the K^+-K^+ transport system inhibit the net K^+-Na^+—Na^+ exchange.

Discussion

The above results show that $HgCl_2$ not only induces a breakdown in the permeability of the cell membrane as already reported for yeast and other cells (17), but also acts directly on the cation transport systems. In the case of the $K^+ - K^+$ exchange it seems quite clear that the fast component of the uptake, which can be readily removed by washing with cysteine, is responsible for the stimulation. Webb (20) has questioned if the effect of mercury on K^+ levels in yeast and other cells is a membrane phenomenon or due to an inhibition of some intracellular process. Since with *Chlorella* the removal of the fast fraction of the mercury uptake which almost certainly corresponds to a surface adsorption, results in a recovery in the normal rate of $K^+- K^+$ turnover, we conclude that the mercury is acting on a membrane-located transport site. The slower mercury uptake seen at low temperature probably corresponds to its entry into the cytoplasm. The temperature sensitivity of this second component suggests that it is controlled by a diffusion barrier which is likely to be the cell membrane.

The mechanism of the stimulation of the $K^+- K^+$ exchange is not clear. To gain some understanding why the stimulation is reduced by low temperature and the uncoupler CCCP may require acceptance of a hypothesis previously put forward (5) that the $K^+ - K^+$ exchange is regulated by carriers which require the presence of ATP but not its hydrolysis. In fact the $K^+- K^+$ exchange in *Chlorella* is normally sensitive to light, uncouplers and temperature which means that this organism can maintain a constant internal K^+ level under a variety of conditions. Presumably the action of mercury at about $3 \times 10^{-5}M$ $HgCl_2$ is to relieve some rate-determining process which normally controls the turnover of the $K^+- K^+$ carrier. It is well known that heavy metals like mercury can stimulate an enzymic reaction before inhibiting it at higher concentrations (17). Stimulation of the active K^+ influx into red blood cells has been observed after mercury and lead treatment (12).

One common feature of the above and earlier work on other systems is that high concentrations of mercury act on the outer cytoplasmic membrane such as to induce a leak of intracellular solutes. For *Chlorella*, $10^{-4}M$ $HgCl_2$ (which corresponded to a binding concentration of 15μmoles Hg/ml packed cells) and above was found to induce relatively rapid net effluxes of K^+ which resulted in cell death. Whether bacterial and fungal membranes, especially those associated with mercury-resistant strains, are also as sensitive to mercury as the cell membrane of *Chlorella* has yet to be investigated.

REFERENCES

1. Ashworth, L.J. and Amin, J.V.: *Phytopathology*, 54:1459, 1964.
2. Barber, J.: *Biochim Biophys Acta*, 150:618, 1968.
3. Barber, J.: *Biochim Biophys Acta*, 150:730, 1968.
4. Barber, J.: *Biochim Biophys Acta*, 163:143, 1968.
5. Barber, J.: *Biochim Biophys Acta*, 163:531, 1968.
6. Barringer, A.R.: Paper 20, Ninth Comm. Min. and Metall. Cong., 1969.
7. Booer, J.R.: *Ann Appl Biol*, 31:340, 1944.
8. Christie, A.A., Dunsdon, A.J., Marshall, B.S.: *Analyst*, 92:185, 1967.
9. Greenaway, W.C.: Mercury resistance in *Pyrenophora avenae* Ph.D thesis, University of London, King's College, 1970.
10. James, C.H., Webb, J.S.: *Trans Inst Min Met*, 73:633, 1964.
11. Jensen, S., Jernelov, A.: *Nature*, 223:753, 1969.
12. Joyce, C., Moore, R.B., Weatherall, M.: *J Pharmacol*, 9:463, 1954.
13. Kimura, Y., Miller, V.L.: *Agri Food Chem*, 12:253, 1964.
14. Landner, L.: *Nature*, 230:452, 1971.
15. Magos, L., Tuffery, A.A., Clarkson, T.W.: *Br J Ind Med*, 21:294, 1964.
16. Passow, H., Rothstein, A.: *J Gen Physiol*, 43:621, 1960.
17. Passow, H., Rothstein, A. and Clarkson, T.W.: *Pharmacol Rev*, 13:185, 1961.
18. Shieh, Y.J. and Barber, J.: *Biochim Biophys Acta*, in press.
19. Tonomura, K. *et al.*: *Nature*, 217:644, 1968.
20. Webb, J.L.: *Enzyme and Metabolic Inhibitors* vol. 2, London and New York, Academic Press, 1966, Chap. 7.
21. Went, F.W.: *Proc Nat Acad Sci*, 46:212, 1960.

DISCUSSION

Rothstein: Did you try some sodium loaded cells to see whether the sodium efflux against the passing influx was stimulated by mercury?

Barber: Yes, as I have mentioned in the paper, we have found that concentrations of mercury which stimulated the steady-state K^+ influx inhibited the Na^+-K^+ exchange. This may mean that the K^+-K^+ exchange does not operate through the carrier responsible for the net fluxes.

Rothstein: Then what good is it?

Barber: Well, an exchange diffusion process of this type could function so as to maintain a constant cellular potassium level under a wide variety of environmental conditions. Of course, our results with mercury cannot be taken as conclusive evidence that the K^+-K^+ exchange cannot convert to an active K^+-Na^+ mechanism during net cation transport. There is always the possibility that mercury has an additional inhibitory effect on the energy requiring Na^+-K^+ pump.

Gage: Has the mercury content of the leaves of vegetation at high soil levels been measured? Any analyst who has tried to measure mercury knows that mercury will volatilize from solutions of its salts. So I imagine that in your mint plants there must have been some mercury vapor above the leaves even though it be only a few molecules.

When you do the calculation based on the determination of mercury vapor in the atmosphere above your vegetation and relate that to the amount of the vegetation in the area, would you expect a measurable amount of mercury to emanate from one mint plant?

Barber: Levels of mercury in leaves of trees and plants growing on high mercury containing soils are relatively low but it depends on the sort of plant you pick and the age of the tissue you analyze. I think it is a case of growth versus a slow accumulation. On a dry weight analyses it is usually found that leaf and stem levels are in the region of 1 ppm where soil levels range from 10 to 40 ppm.

With regard to the second part of your question we have simply investigated the possibility whether plants are capable of inducing significant volatilization of mercury from their leaves. As you can see from the data presented the method we used was extremely sensitive giving us a means of detecting mercury levels in the region of pg/m^3 of air or less. With this sensitivity we have not detected any volatilization of mercury from our plants. I am confident if there had been any we would have detected it. For this reason we cannot explain the anomolous mercury levels, which are in the region of ng/m^3 of air, detected over heavily vegetated areas by a plant transpiration mechanism.

Gage: My question was, would you detect it had it been there? Had the vegetation been contributing to the measure of mercury in the atmosphere could you have measured it in your experiments? Are your methods sufficiently sensitive?

Barber: It's a very sensitive method of detecting mercury and the amount that we catch is very small over a long period of time. I think that we should if there had been any volatile forms coming from our plants. Whether or not they would account for the increased level of mercury that one detects over vegetated areas I'm not certain.

Gage: You're still left with the problem of why the mercury is present.

Nechay: Do you think that animals with bacterial flora for digestion can pick up enough mercury from leaves to play a role in the system of mercury circulation in the biological systems? And, is it known whether urbanant flora and other intestinal bacterial flora or micro flora contribute to methylation of mercury?

Barber: As I have already pointed out, as far as I am aware, levels of mercury in the leaves of higher plants are generally very low. However, I would suspect that some mercury is introduced into the food chain this way. I am not sure whether a methylation of mercury can occur in the gut. Yesterday Dr. Norseth reported that he found no difference in the ability of rats with or without intestinal flora to induce a biotransformation of mercury.

Miller: Were the stomata on the mint leaves open?

Barber: We passed dry air over our plants so as to encourage stomatal opening and induce a maximum rate of water transpiration.

Miller: Did you analyze for materials that are normally transpired by the mint leaf and show that it is occurring?

Barber: We did not directly analyze the volatile organics released from our plants but we could smell them.

Fassett: I'd like to get back to Dr. Gage's question since your argument that there was no release of volatile mercury would depend on the efficiency of this trapping system. Has anyone investigated the ability of the cadmium-rich plants to trap dimethylmercury or some other volatile mercurys?

Barber: Yes, this has been done by Christie *et al*, and I have given the reference in the paper. They have found cadmium sulphide pads to be reasonably efficient for trapping a range of organomercurials. I may also point out that we used six to 12 of these pads in series and found them satisfactory for detecting bacterial induced mercury volatilization.

Rothstein: Radioactive mercury labeling studies can probably sort this out.

Dr. Clarkson and I performed an experiment* very similar to yours except we did it with a rat which exhales a certain amount of volatile mercury. I wonder if airplanes equipped with detectors and flying over population centers could detect an increase of volatile mercury?

Barber: Yes, you certainly would detect higher mercury levels over populated areas, but not for the reason you suggest. I think it more likely that the background level of mercury is increased by general pollution

---

* Clarkson, T. and Rothstein, A.: *Health Physics,* 10:115, 1964.

of the atmosphere, particularly from industrial and domestic burning of fossil fuels.

McDuffie: In your experiment the ratio of the mercury in the leaves to the mercury in the solution was roughly 1 : 0.1 ppm. Do you know what the ratio is between mercury in the leaves and mercury in the soil where the plant may be growing? Is it in the ratio of 0.20 or of 1 to 100?

Barber: In our experiments the roots were bathed in a solution of 0.4 ppm mercury. In the soil the mercury in free solution will vary considerably on conditions and may be very much lower than that indicated by total soil analyses. In general the levels in the leaves are lower than levels in the soil, but it depends on the units you compare. For example, in the mercury rich soils which we have been investigating at Keele, Northern Ireland, it has been found that the total soil level of mercury in about 10 ppm and that levels of 1 ppm on dry weight basis occur in the grass growing on this soil. However, soil water levels could easily be as low as 0.4 ppm.

McDuffie: Does someone know what the forms of mercury are, or the principle form, in sewage sludge from municipal treatment plants?

Jernelöv: Very little of the sludge is methylmercury, almost always below one percent. On the other hand, you have the formation of dimethylmercury which goes with the gases.

Chapter 20

THE METABOLISM OF METHOXYETHYLMERCURY AND PHENYLMERCURY IN THE RAT

J.C. GAGE

ABSTRACT

The metabolism of methoxyethylmercury chloride (MEMC) has been studied in the rat, using a sample labeled with ^{14}C in the ethyl group. After a subcutaneous dose, 60 percent of the radioactivity appears in the exhaled air as ethylene, with a little as carbon dioxide. The urine contained 12 percent as a neutral mercury-free metabolite. It is probable that MEMC is widely distributed in the tissues where it is broken down by a nonenzymic process to release ethylene, and mercury which migrates to the kidneys.

When phenylmercury acetate (PMA), uniformly labeled with ^{14}C in the benzene nucleus, is administered subcutaneously to rats, most of the radioactivity appears in the urine as phenol conjugates. A metabolite is found in bile and urine which may be a derivative of o- or p- hydroxyl phenylmercury. A metabolite with these properties together with inorganic mercury is produced when PMA is incubated with a microsomal preparation. It is likely that PMA is hydroxylated in the liver, and that this is followed by a nonenzymic breakdown to phenol and inorganic mercury.

REPORT

The interest of our laboratories in the metabolism of organic mercurials was aroused about ten years ago, when methylmercury dicyandiamide (MMD) was introduced as a seed dressing. It soon became apparent, from reports of cases of human poisoning, that MMD was considerably more neurotoxic than the hitherto used phenylmercury acetate (PMA), although animal experiments at that time had not clearly demonstrated this difference. We succeeded in establishing this (6) by showing that MMD could produce paralysis of the limbs in mice after repeated subcutaneous administration of doses that produced no ill-effects with PMA.

We then considered whether this greater toxicity of MMD was due to its ease of penetration into the nervous system or to a greater reactivity with the receptor site, or whether it could be attributed to a greater resistance to metabolic attack. It had already been shown by investigators using a radioactive mercury isotope that the distribution of radioactivity in the body after the administration of PMA resembled that observed after the administration of inorganic mercury, and that similar experiments with MMD led to a quite different distribution. In these experiments no attempt had been made to distinguish between organic and inorganic mercury in the tissues and excreta, so an extraction technique was developed capable of separately determining organic mercury, and which permitted a measure of the extent of breakdown in the body (4). By using this technique, it was possible to demonstrate (5) that PMA after subcutaneous injection into rats was rapidly broken down in the tissues to inorganic mercury. Only a small proportion of the dose appeared in the urine as organic mercury, and on repeated dosing there was no accumulation in the tissues after the first week, and the brain concentration was very low. With MMD the pattern was quite different; there was little breakdown to inorganic mercury and over a six-week-period of repeated injections there was a progressive accumulation in the tissues, including the brain. The major reservoir in the body was the red cells, and there was also a high concentration in hair. As a result of this investigation it was concluded that it was not necessary to postulate that MMD had a greater affinity for the nervous system than PMA, but that the absence of the neurotoxic effect with PMA could be attributed to the low plasma concentration arising from its rapid metabolism in the body.

Metabolism of Methoxyethylmercury Chloride

We subsequently turned our attention to methoxyethylmercury salts, when it was proposed to use these in seed dressings.

Methoxyethylmercury chloride (MEMC) had been used for some time without any reports of a neurotoxic action in man, but this appeared never to have been investigated in animals. Unpublished work from our laboratory demonstrated that MEMC, like PMA, did not produce paralysis in mice as did MMD. MEMC labelled with ^{203}Mg had been studied by Ulfvarson (9), who showed that the distribution of radioactivity in rats resembled that attained after the administration of labelled mercuric chloride, although he came to the rather surprising conclusion that it was not readily broken down in the body. For our investigation (3) we used MEMC labeled with ^{14}C on the ethyl group, and a more sensitive method of measuring mercury based on the procedure of Magos and Cernik (8) in which elemental mercury is released by the addition of stannous chloride and measured with a mercury vapor meter. The method was modified for the separate determination of organic and inorganic mercury in tissues and excreta (7) by making use of the lability of MEMC after addition of an acid cysteine reagent (10).

Our results showed that after subcutaneous injection of 1.4 mg/kg MEMC, most of the radioactivity appeared in the exhaled air (Fig. 20–1), the half-time for this excretion being about one day. A small proportion of this radioactivity was present as carbon dioxide, but by scrubbing the exhaled air with a mercury perchlorate solution followed by gas-liquid chromatography, we demonstrated that the major metabolite was ethylene in an amount equivalent to about one-half the dose administered.

The urinary excretion of radioactivity, total and inorganic mercury is shown in Figure 20–2. There was a considerable excess of radioactivity over organic mercury throughout the four days of the experiment, indicating the presence in urine of metabolites not containing mercury. We have not succeeded in identifying this fraction, which corresponds to about 12 percent of the dose, but it is likely to contain neutral metabolites of ethylene. During the first 24 hours after dosing the mercury in urine was mainly organic, identified as MEMC; thereafter inorganic mercury predominated.

A much larger proportion of organic mercury was found in bile. About 20 percent of the dose was excreted in this form dur-

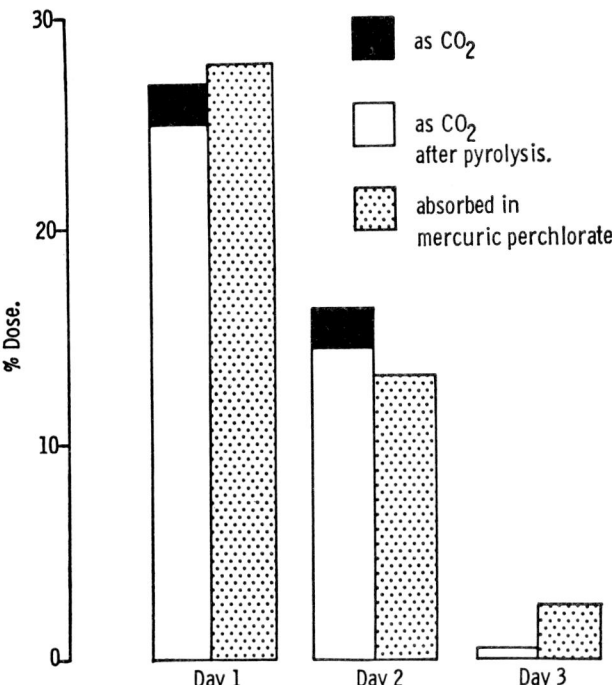

Figure 20–1. Excretion of radioactivity in exhaled air after administration of MEMC.

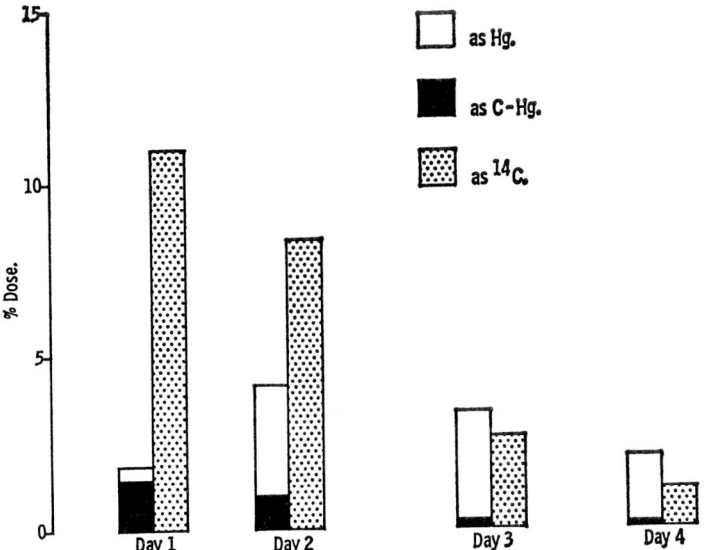

Figure 20–2. Excretion of radioactivity and mercury in urine after administration of MEMC.

ing the first 24 hours; this tailed off to a trace by the fourth day. The excretion of inorganic mercury by this route remained low. No radioactivity was detected in feces, so that MEMC must be degraded in the gut, and there is some evidence of reabsorption. There is also evidence for the presence of a small amount of a mercury-free metabolite in bile.

Analyses of liver and kidney tissue at two, six and 24 hours after dosing showed that after two hours the liver contained about 1 percent of the dose as inorganic mercury, and 1 percent as organic mercury. The kidneys contained about four times these amounts. After 24 hours the total mercury in liver was about 1 percent with more than half inorganic, whereas the kidney contained 14 percent of the dose as inorganic and none as organic.

These results suggest that MEMC, after a subcutaneous dose, is distributed in the tissues. Part is excreted in bile and a little in urine, the remainder is broken down in the tissues to inorganic mercury, which migrates to the kidneys where it is stored then excreted in urine. It is possible that this tissue degradation is not due to an enzyme mechanism, but involves the reverse of the reaction used in synthesizing MEMC, which Weiner et al. (10) have shown to occur in the presence of acid and cysteine:

$$CH_2 = CH_2 + CH_3OH + HgCl_2 \rightleftharpoons CH_3O\text{-}CH_2\text{-}CH_2\text{-}HgCl + HCl$$

The rate of breakdown of MEMC *in vitro* in the presence of cysteine at pH 6 is about 1 percent/hour which is about the same as the *in vivo* breakdown of MEMC. Although there is some uncertainty concerning intracellular pH values, Carter et al. (1) have indicated that the pH of skeletal muscle may be as low as six.

Metabolism of Phenylmercury Acetate

Our success with MEMC encouraged us to reinvestigate the fate of PMA, as this compound is unaffected by cysteine at pH 6, and must presumably involve some metabolic transformation. We administered to rats 2.5 mg/kg of a sample uniformly labelled with ^{14}C in the benzene ring. Only 5 percent of the radioactivity appeared in the exhaled air; about 85 percent was excreted in urine. Most of the mercury (50 to 60 percent) was excreted in feces, with about 12 percent in urine. The daily urinary excretion of radioactivity and mercury is shown in Figure 20–3. Only a small proportion of the mercury appeared in the first 24-hour urine, and about one-third of this was organic, thereafter all was inorganic.

The considerable excess of radioactivity over organic mercury indicates the presence of mercury-free metabolites, and the major

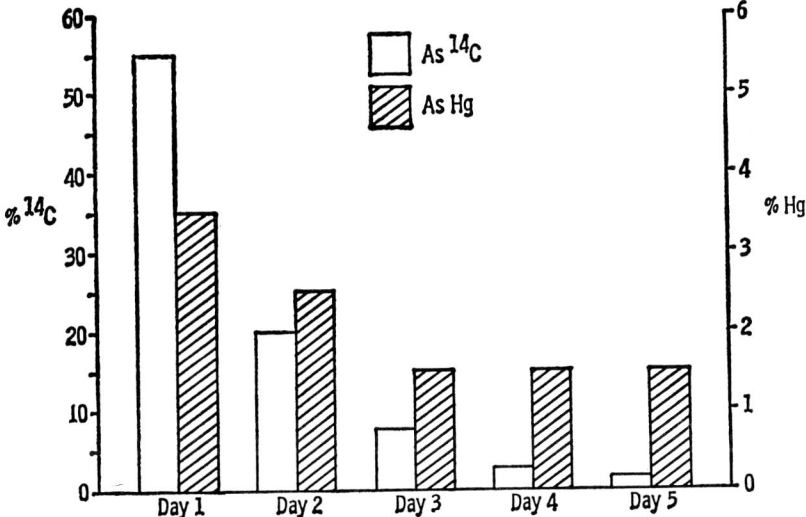

Figure 20–3. Excretion of radioactivity and mercury in urine after administration of PMA.

components have been identified as the sulphate and glucuronide of phenol. Weiner *et al.* (10) have stated that *p*-hydroxyphenyl-mercury acetate is labile in the presence of acid and cysteine, and this has led us to the tentative hypothesis (2) that PMA is first hydroxylated and then undergoes nonenzymic degradation in the tissues in the same way as MEMC. Such a hypothesis is attractive as it is difficult to conceive an enzymic system which will attack a C-Hg bond. Investigations to confirm this hypothesis are at present in progress. We have found in urine, bile and tissues a component which is reduced to elemental mercury by a strong solution of stannous chloride in sodium hydroxide. PMA does not release mercury by this reagent, but o- and p- hydroxyphenylmercury acetate do so. A metabolite with this property, together with inorganic mercury, can also be obtained in good yield from a rat liver microsomal preparation in the presence of PMA. However, unlike hydroxy-PMA, the metabolite in bile cannot be extracted by a solution of dithizone in an organic solvent. It seems probably that the metabolite is a derivative of o- or p-phenylmercury.

It is surprising that PMA, which is regarded as an effective enzyme inhibitor, readily undergoes this transformation by the microsomal enzyme system. It is possible that the molecule is orientated with the aryl ring in the microsomal lipid membrane, and that the greater stability of the alkylmercury salts is due to their inability to be so orientated.

REFERENCES

1. Carter, N.W., Rector, E.C. and Seldin, D.G.: *J Clin Invest, 46*:920, 1967.
2. Daniel, J.W. and Gage, J.C.: *Biochem J, 122*:24p, 1971.
3. Daniel, J.W., Gage, J.C.: and Lefevre, P.A.: *Biochem J 121*:411, 1971.
4. Gage, J.C.:*Analyst, 86*:457, 1961.
5. Gage, J.C.: *Br J Ind Med, 21*:197, 1964.
6. Gage, J.C. and Swan, A.A.B.: *Biochem Pharmacol, 8*:77, 1961.
7. Gage, J.C. and Warren, J.M.: *Ann Occup Hyg, 13*:115, 1970.
8. Magos, L. and Cernik, A.A.: *Br J Ind Med, 26*:144, 1969.
9. Ulfvarson, V.: *Arch Gerwerbepath Gewerbehyg, 19*:412, 1962.
10. Weiner, I.M., Levy, R.I. and Mudge, C.H.: *J Pharm Exp Ther, 138*:96, 1962.

DISCUSSION

Gage: We did some studies with humans who were exposed primarily to phenylmercurials in measuring the relative amount of mercury in cells versus serum; we found that there was relatively little in the cells.

Clarkson: Have you looked at the effect of stimulators of drug metabolism?

Gage: No; they may have an effect, although the rate of breakdown of PMA *in vivo* is fairly rapid. PMA is quite effectively metabolized by a rat-liver microsomal preparation. The rate of conversion is very surprising.

Norseth: I have tested the intracellular distribution in the liver after exposure to inorganic mercury, methylmercury and to methoxyethylmercury. Even if the mercury is fairly rapidly broken down the distribution within the liver cells, even after a very long period of time, isn't exactly the same for methoxyethylmercury and inorganic mercury. So there may be that some of the organic mercurials enter part of the cell and stay there, and thus, even if it breaks down, it may have a different distribution.

Fang: We did C-14 distribution studies with ^{14}C-phenylmercuric acetate and were not able to find even any trace amounts of the ^{14}C in respiratory CO_2. Would you like to make a comment on that?

Gage: It may be a matter of technique, or the rats behave differently. In order to trap CO_2 efficiently it is necessary to pass the air up a column packed with rings down which trickles sodium hydroxide. Have you used a technique of that nature?

Fang: Yes.

Gage: It seems probable that most organic compounds will be metabolized to yield some carbon dioxide.

Suzuki: You mentioned that the majority of the mercury which was excreted in the urine was in the inorganic form and also that there was a small amount of organic mercury excreted the first day. What was the form of this organic mercury?

Gage: The amount in urine is too small to permit identification. There is a larger amount of a PMA metabolite in bile, and we hope that we may be able to identify this to provide evidence for the metabolic route.

Suzuki: I think there is some possibility of comparing the metabolism of phenylmercury with benzene.

Gage: One hypothesis for this breakdown is that the molecule is first split to give benzene and which is then hydroxylated. This would postulate an enzyme capable of clearing the C-Hg bond, which seems a little improbable.

Kazantzis: From your current knowledge of the metabolism of PMA are you in position to speculate on the nature of the acute exudative reaction of PMA which gives rise to blistering following skin contact?

Gage: It would seem very likely to be a direct action with certain enzymes in the skin affecting changes in cell permeability.

Piotrowski: I would like to emphasize that with phenylmercury any mechanism of biotransformation would probably lead to the excretion of phenol. Either way it comes via a breakdown by the microsomes.

Gage: I am inclined to reject the idea of an enzyme attacking a carbon-mercury bond. With the hydroxylation of the intact molecule it is not necessary to presuppose this; the molecule is possibly orientated in the lipid membrane resulting in the formation of p-hydroxy-PMA. There is fairly good evidence that this is the route of metabolism.

Vostal: Has anybody tried to isolate the metabolites produced by a microsomal preparation?

Gage: We have shown in experiments, as yet unpublished, that rat liver microsomes can degrade PMA to a metabolite, probably hydroxy-

PMA, and also to a considerable amount of inorganic mercury. If we administer hydroxyl-PMA to rats we get a lot of inorganic mercury in the bile.

Vostal: It is possible to confirm the importance of the oxidative step by preventing microsomal aromatic hydroxylation *in vivo*, for example, with diethylaminoethyl diphenylpropylacetate (SKF 525A)? If your theory is correct the breakdown to inorganic mercury must be blocked by inhibition of the hydroxylation of phenylmercury molecule.

Gage: This might be possible, and also as Dr. Clarkson has suggested, the effect might be confirmed by metabolic stimulators.

Chapter 21

BIOLOGICAL OXIDATION OF ELEMENTAL MERCURY

F. Nielsen Kudsk

ABSTRACT

The resistance of elemental mercury vapor to oxidation in air has been previously demonstrated. The importance of the fairly high lipoid solubility of the metal for its pulmonary absorption and subsequent distribution in the body is emphasized. The processes involved in the absorption of mercury vapor from the lungs in man are briefly reviewed. The uptake of mercury vapor in human blood *in vivo* and *in vitro*, as well as the inhibitory effect of ethyl alcohol upon the uptake, are discussed on the basis of previous investigations. The possible mechanisms of oxidation of mercury in the blood are discussed. It is demonstrated by the results of *in vitro* experiments that increasing concentrations of crystalline beef liver catalase cause a very pronounced acceleration in the rate of uptake of mercury vapor in 3 mM glutathione solutions which were equilibrated with diffusing vapors from metallic mercury and a 1 percent W/W solution of hydrogen peroxide. The catalase-stimulated rate of mercury uptake is markedly inhibited by 0.2 percent W/V ethyl alcohol and by increasing concentrations of the catalase inhibitor, 3-amino-1,2,4-triazole. No correlation was found between the rate of oxidation of glutathione caused by hydrogen peroxide vapor and the rate of mercury uptake. These investigations indicate that the primary hydrogen peroxide catalase complex is of major importance in the biological oxidation of mercury. The consequence of a more general catalase induced oxidation of mercury related to the distribution and desposition of the metal in the body is briefly discussed. The need for further investigations to settle which (if any) other enzymes might influence the oxidation of elemental mercury in biological systems is emphasized.

REPORT

Introduction

Mercury is unique as a metal in that at room temperature it has an appreciable vapor pressure which may easily cause toxic concentrations of mercury in the air. The metal vapor is very resistant to oxidation in the air. It has previously been shown by the author (26) that the mercury vapor clearance of workrooms, defined as the volume of air containing the amount of mercury vapor removed from the room per minute, corresponds to the ventilation of the rooms (Fig. 21–1). Increased concentrations of a trace substance such as ozone do not cause any appreciable oxidation of mercury vapor in the air.

The solubility of mercury in the body lipoids is, according to Hughes (17), about 1.5 mg/liter at 40°C. Approximately 60µg of mercury dissolves in one litre of air at the same temperature; hence, the partition coefficient for mercury between lipoids and air is of the magnitude of 25 : 1. The partition coefficient between lipoids and water is about three times higher. These properties favor the absorption of metallic mercury across the alveolar membranes by diffusion, its dissolution in the blood lipoids and its subsequent passage across body membranes, especially the blood-brain barrier.

It has been shown in man that the uptake of mercury vapor from inhaled air amounts to 75 to 85 percent at mercury concentrations ranging from 50µg to 350µg/m^3 (33,24). The size of the respiratory dead space in man for mercury vapor uptake has been found to correspond to the physiological dead space and can be correlated to the size of the tidal volume. It has further been demonstrated that mercury retention in the nasal and oral cavities is insignificant and that expired alveolar air is free of mercury in humans respiring mercury-contaminated air. Thus, these investigations show that mercury hardly reacts with the respiratory mucosae, but probably is completely absorbed in elemental form from the alveoli of the lungs (24,29). Experiments with animals have confirmed that mercury vapor rapidly passes the alveolar membranes by diffusion (2,20).

Oxidation of Mercury in Biological Liquids

Oxidation of elemental mercury may cause the formation of two types of ions, the mercurous ion, Hg_2^{++} and the mercuric

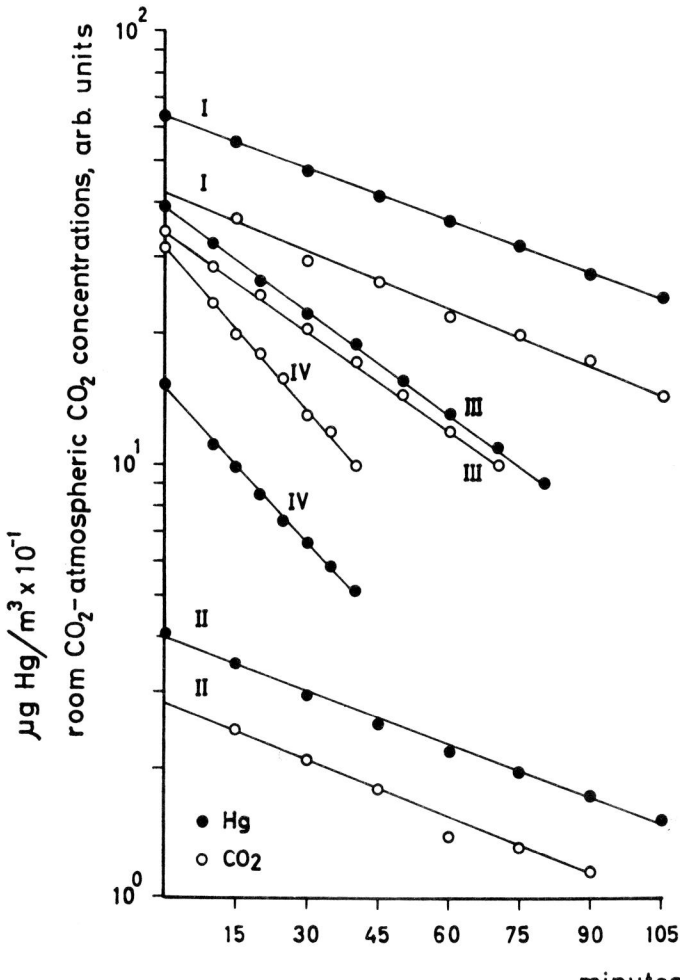

Figure 21-1. Relation between the decreasing concentrations of mercury vapor and carbon dioxide related to time in the air of a 40 m³ working room. Lines I and II are based on measurements taken during naturally occurring ventilation, while lines III and IV demonstrate the effect of artificially increased ventilation rates. The curves representing mercury concentrations express the following equation: $\ln c = -ut/V + \ln c_o$, where c is the mercury vapor concentration at time t (minutes), c_o the concentration at zero time, V the volume of the room in m³ and u the mercury vapor clearance of the room (m³/min). Mercury vapor clearance may be expressed as: $u = 0.69\ V/H$, where H is the half-life for mercury vapor in the room. The ventilation of the room, expressed in a similar way as its carbon dioxide clearance, does not deviate significantly from the mercury vapor clearance.

ion, Hg^{++}. The mercurous ion, however, is unstable and dissociates according to the following equation:

$$Hg_2^{++} \rightleftharpoons Hg + Hg^{++}$$

At equilibrium with excess elemental mercury the ratio of $(Hg_2^{++}) : (Hg^{++})$ is 116 : 1. The dissociation of mercurous ions in biologically aqueous solutions is favoured by the poor solubility of elemental mercury and by the presence of chloride ions and organic ligands such as thiol groups, which are capable of forming weakly dissociated salts or bio-complexes with mercuric ions. Thus, mercury in biological liquids is eventually oxidized to mercuric ions.

Stock (32) found that mercury in water is slowly oxidized to mercuric oxide and that animal blood was capable of absorbing mercury vapor from the air. Horwitz (16) showed that mercury vapor passed through plant and bacterial membranes unchanged and that it subsequently was oxidized in the cells or bacteria.

Uptake of Mercury Vapor in Human Blood

Clarkson et al. (7,8) demonstrated a fairly rapid uptake and oxidation of mercury vapor in human blood samples *in vitro* at physiological oxygen tensions. The rate of uptake increased as the oxygen tension was raised. Dialysis studies showed that oxidized mercury, probably as mercuric ions, was strongly bound chiefly to the sulfhydryl groups of hemoglobin and plasma albumin.

However, recent investigations by Magos (19) have shown that 0.5 to 1.3 percent of the mercury absorbed by blood *in vitro* during exposure to mercury vapor is still in the metallic form 15 minutes after discontinuance of the exposure. This finding probably explains the high rate of mercury uptake by the brain during pulmonary absorption of mercury vapor or after intravenous injection of metallic mercury (1,18,20).

The author (29) has previously shown that the *in vitro* uptake of mercury vapor in blood, which was equilibrated with mercury at 37°C in an atmospheric air phase, occured at a high rate during the first hour ($0.4\mu g/ml/hour$). This was followed by a slower rate of uptake which showed a minor, but distinct individual variation (0.13 to 0.20µg/ml/hour). A high rate of uptake in plasma during the first hour (0.20µg/ml/hour) was probably responsible for the initial high rate of uptake in whole blood. The initial high rate of mercury uptake in whole blood may be related to a rapid dissolution of elemental mercury in the plasma lipoids.

It has been found that increasing concentrations of ethyl alcohol inhibit the uptake of mercury vapor in blood both *in vivo* and *in vitro* (25,26,29). A blood alcohol concentration of about 0.04 percent W/V produces a maximum decrease of 30 percent in the pulmonary absorption of mercury *in vivo*, while the inhibition in blood *in vitro* reaches a maximum of 60 percent at a blood alcohol concentration of 0.2 percent W/V.

Possible Mechanisms of Oxidation of Mercury in Blood

Horwitz (16) postulated that the oxidation of elemental mercury in biological tissue was catalysed by thiol-containing compounds, and the results of Clarkson *et al.* (7) indicated that oxyhemoglobin was a possible oxidant which contained dissociable oxygen and a high number of free thiol-groups.

The finding that methyl and ethyl alcohol inhibit mercury vapor uptake in blood (26,29), raised the question whether the primary hydrogen peroxide catalase complex could be involved in the oxidation of mercury in the erythrocytes, as this complex has a high affinity for these alcohols (34,10). Aminotriazole (3-amino-1,2,4-triazole), which is a strong inhibitor of the primary hydrogen peroxide catalase complex, does not influence the *in vitro* uptake of mercury in human erythrocytes, either when used alone or in conjunction with glucose and methylene blue as a probable hydrogen peroxide generating system (29). Thus, erythrocyte catalase seemed to be without importance in the oxidation of mercury in erythrocytes.

Methylene blue, especially when it is combined with glucose, causes a pronounced acceleration of the rate of uptake of mercury in human erythrocytes and whole blood. Menadione has a similar but even more pronounced stimulating effect upon the rate of uptake (28). It has been demonstrated that menadione causes a generation of hydrogen peroxide (28) which rapidly is destroyed, predominantly in a reaction with glutathione catalysed by glutathione peroxidase (21,9). Both methylene blue and menadione strongly stimulate the pentose shunt activity (4,5). Methylene blue accepts electrons and hydrogen from NADPH generated by the pentose shunt, probably by way of a diaphorase and presumably the NADPH-dependent methemoglobinreductase. By this means leucomethylene blue is formed which then is autooxidized with the formation of hydrogen peroxide (15).

The methylene blue and menadione stimulation of the rate of mercury uptake in blood suggested that hydrogen peroxide is an important factor in the naturally occurring oxidation of mercury in the erythrocytes. A normal generation of hydrogen peroxide in erythrocytes has never been directly demonstrated. But erythrocytes contain several compounds (hydrogen donators) which by autooxidation or in a coupled reaction with the oxygen of oxyhemoglobin could produce hydrogen peroxide. Thus, ascorbic acid probably causes a constant generation of small amounts of hydrogen peroxide in a coupled reaction with oxyhemoglobin (13,21,22). Addition of ascorbic acid to human blood *in vitro* causes an increase in the rate of uptake of mercury vapor. Glutathione added to blood samples also produces a pronounced stimulation of the mercury uptake, but it is not definitely known if this is caused by a generation of hydrogen peroxide, due to the oxidation of glutathione in plasma (28).

It has been demonstrated that the highly increased rate of mercury uptake in menadione stimulated blood samples can be very strongly inhibited by ethyl alcohol (28). Investigations have shown that the stimulatory effect of methylene blue and ascorbic acid upon the mercury uptake also is strongly counteracted by ethyl alcohol. It has further been found by using ^{14}C labeled glucose that ethyl alcohol does not inhibit the stimulatory effect of menadione and methylene blue upon the pentose shunt activity. This seems to indicate that ethyl alcohol does not inhibit the NADPH dependent diaforase, methemoglobin reductase, in the erythrocytes (27).

Thus, the questions arose whether the glutathione peroxidase-glutathione system in the erythrocytes and other body tissues and the extra-erythrocytic catalase in conjunction with hydrogen peroxide could be involved in the oxidation of elemental mercury.

Liver Catalase as a Factor in the Oxidation of Elemental Mercury *in vitro*

In view of previous experiments which showed that the catalase inhibitor, 3-amino-1,2,4-triazole, was unable to inhibit a methylene blue-stimulated uptake of mercury in erythrocytes (29), experiments have been carried out in order to investigate a possible influence *in vitro* of liver catalase and hydrogen peroxide upon the uptake of mercury vapor in glutathione solutions. The effect of ethyl alcohol on such a system was also investigated.

Material and Methods

The uptake of mercury vapor in 3mM glutathione solutions with and without the addition of other compounds was studied in simple equilibration experiments in which the solutions were exposed to diffusing vapors from mercury and hydrogen peroxide. The glutathione solutions were buffered with phosphate to pH 7.0. To the solutions were added increasing concentrations of a crystalline beef liver catalase preparation ("Fluka," Switzerland, catalogue number 60630), 0.2 percent W/V ethyl alcohol or combinations of both.

The equilibrations took place in 15 ml Warburg vessels containing 0.1 ml double-distilled mercury in the side arm and 0.1 ml 1 to 30 percent W/W hydrogen peroxide in the center well. Samples of 3 ml of the glutathione solutions were introduced into the Warburg vessels which were placed in a constant-temperature water bath at 37°C, and equilibrated for three hours in an atmospheric air phase at normal pressure. The vessels were shaken during the period of equilibration.

The mercury uptake in the glutathione solutions during the equilibration period was determined by chemical dithizone analyses on 2 ml samples (29). Glutathione concentrations before and after the equilibration were determined according to the method of Beutler *et al.* (3).

Results

Table 21-I and Figure 21-2 show the results of a study of the effect of catalase on the *in vitro* uptake of mercury in glutathione solutions exposed to hydrogen peroxide vapor diffusing from a 1 percent solution in a closed vessel.

The uptake in the glutathione solutions without any addition of catalase is about 0.08μg Hg/ml/hour, which is approximately the same as previously found in glutathione solutions not exposed to hydrogen peroxide (28). The glutathione concentration decreased about 34 percent during the equilibration period.

Increasing concentrations of catalase to about 15μg/ml caused a pronounced acceleration in the rate of mercury uptake. At higher catalase concentrations there was a slower increase in the rate of uptake which asymptotically approached a maximum of about 1.11μg Hg/ml/hour. The glutathione concentrations decreased about 18 percent during the equilibration period, obviously independent of the catalase concentration.

In the presence of 0.2 percent W/V ethyl alcohol an initial, rather steep, increase in the rate of mercury vapor uptake is

TABLE 21-I

RESULTS OF EXPERIMENTS DEMONSTRATING THE UPTAKE OF MERCURY VAPOR IN 2 ml/3 mM GLUTATHIONE SOLUTIONS IN PHOSPHATE BUFFER (pH 7.0) WITH AND WITHOUT THE ADDITION OF CRYSTALLINE BEEF LIVER CATALASE (CAT) OR ETHYL ALCOHOL (ETOH) OR BOTH COMPOUNDS IN COMBINATION.

	Hg Uptake μg	GSH, mg% Before	GSH, mg% After	% Change
No addition	0.45 ± 0.05	92.0 ± 2.4	60.9 ± 6.6	−33.9
ETOH 0.2% W/V	0.33 ± 0.04	92.3 ± 1.9	65.4 ± 4.2	−29.2
CAT 5μg/ml	2.70 ± 0.25	91.6 ± 2.0	75.0 ± 1.7	−18.3
10μg/ml	4.52 ± 0.22	92.3 ± 1.3	73.5 ± 2.4	−20.2
20μg/ml	6.24 ± 0.30	90.8 ± 2.4	76.4 ± 2.6	−15.9
100μg/ml	6.54 ± 0.41	93.2 ± 1.3	76.9 ± 3.0	−17.5
200μg/ml	6.68 ± 0.20	92.0 ± 1.1	73.3 ± 1.4	−20.3
ETOH 0.2% W/V				
CAT 5μg/ml	2.16 ± 0.12	91.8 ± 3.2	87.2 ± 0.8	− 5.0
10μg/ml	3.03 ± 0.10	92.3 ± 1.4	84.8 ± 1.1	− 8.1
20μg/ml	2.87 ± 0.10	90.5 ± 2.7	83.0 ± 1.0	− 8.3
100μg/ml	2.65 ± 0.24	92.3 ± 0.7	85.8 ± 2.2	− 7.1
200μg/ml	2.53 ± 0.10	92.3 ± 1.1	85.8 ± 1.2	− 7.1

The solutions were equilibrated in closed Warburg vessels for three hours with diffusing vapors from mercury and a 1% solution of hydrogen peroxide at 37°C in an atmospheric air phase at normal pressure. Glutathione (GSH) concentrations before and after the equilibrations are also shown. Standard deviations are given next to the individual results.

Biological Oxidation of Elemental Mercury

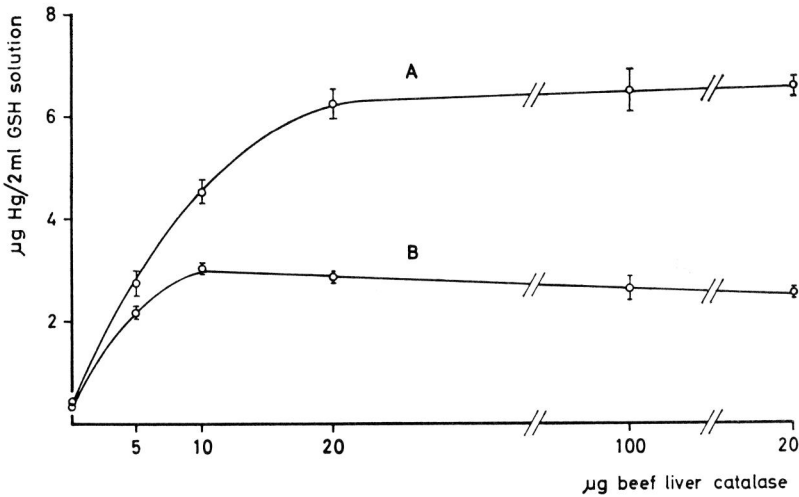

Figure 21-2. Curve A indicates the influence of increasing concentrations of crystalline beef liver catalase on the rate of mercury vapor uptake (μg Hg/2 ml/3 hours) at 37°C in 3 mM glutathione (GSH) solutions exposed to hydrogen peroxide vapor diffusing from a 1 percent solution. Curve B shows the Hg uptake with further addition of 0.2 percent W/V ethyl alcohol to the GSH solution. The curves are plotted from numbers given in Table 21-I. Standard deviations are shown as vertical bars.

recognized with increasing catalase concentrations to about 7μg/ml, while a smaller, but definite decrease occurs at higher catalase concentrations. Thus, the maximum rate of uptake is about 0.5μg Hg/ml/hour at a catalase concentration of 10μg/ml, while the maximum inhibition caused by ethyl alcohol occurs at a 20-fold higher catalase concentration. The glutathione concentration decreases only about 7 percent, or less than half the decrease seen without the addition of ethyl alcohol.

The results of similar equilibration experiments with glutathione solutions containing catalase at a fixed concentration, but exposed to solutions of hydrogen peroxide with increasing concentrations are shown in Table 21-II and Figure 21-3.

The rate of uptake of mercury vapor increases only slowly, but rectilinearly with increasing exposure to hydrogen peroxide vapor. The inhibition of the uptake caused by the presence of ethyl alcohol in a concentration of 0.2 percent W/V decreases

TABLE 21-II

EXPERIMENTAL RESULTS SHOWING THE AVERAGE RATE OF UPTAKE OF MERCURY VAPOR (μg Hg/2 ml/3 HOURS) IN 3 mM GLUTATHIONE SOLUTIONS IN PHOSPHATE BUFFER (pH 7.0) WITH AND WITHOUT THE ADDITION OF 200μg/ml BEEF LIVER CATALASE (CAT) OR 0.2 PERCENT W/V ETHYL ALCOHOL (ETOH) OR BOTH COMPOUNDS.

	Hg Uptake μg	GSH, mg%		
		Before	After	% Change
H_2O_2 solution 2.5% W/W				
No addition	0.58	90.0	58.6	− 34.9
ETOH 0.2% W/V	0.39	90.4	59.8	− 33.9
CAT 200μg/ml	6.78	93.3	81.6	− 12.6
CAT and ETOH	2.80	93.1	85.2	− 8.5
H_2O_2 solution 10% W/W				
No addition	0.60	88.9	5.2	− 94.3
ETOH 0.2% W/V	0.38	89.1	9.3	− 89.6
CAT 200μg/ml	6.93	90.0	78.8	− 12.5
CAT and ETOH	4.36	91.1	86.1	− 5.5
H_2O_2 solution 30% W/W				
No addition	0.63	92.7	0	−100.0
ETOH 0.2% W/V	0.34	91.6	0	−100.0
CAT 200μg/ml	7.83	92.1	83.2	− 9.7
CAT and ETOH	6.04	92.8	83.4	− 10.1

The solutions were equilibrated for three hours with diffusing vapors from mercury and solutions of hydrogen peroxide in increasing concentrations at 37°C and normal oxygen pressure. Average concentrations of glutathione (GSH) measured before and after the equilibrations are also shown.

with increasing hydrogen peroxide exposure. The decrease takes place with two rates, an initial faster rate up to a hydrogen peroxide concentration of 10 percent in the equilibration liquid, and a slower rate at higher concentrations.

A complete oxidation of glutathione during the equilibration period occurs at the highest hydrogen peroxide concentrations used without addition of catalase, while the oxidation at lower concentrations is only partial. There is obviously no correlation between the degree of glutathione oxidation and the mercury vapor uptake. Addition of catalase efficiently protects glutathione from being oxidized during the exposure to hydrogen peroxide.

The mercury taken up in the glutathione solutions in the above described experiments was in a nondiffusible form as ascertained by the fact that no loss in the mercury content in the solutions occurred during prolonged exposure to mercury-free air. Thus, the mercury in the solutions undoubtedly exists in an oxidized form, and presumably as mercuric ions bound to glutathione.

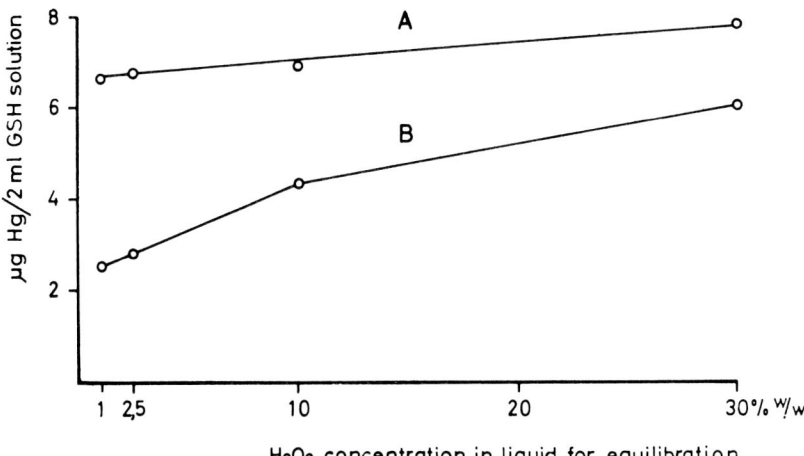

Figure 21-3. The rate of uptake of mercury vapor (μg Hg/2 ml/3 hours) in 3 mM glutathione solutions in phosphate buffer (pH 7.0) which were exposed to hydrogen peroxide vapor diffusing from solutions with increasing hydrogen peroxide concentration and to which were added 200μg/ml of crystalline beef liver catalase (A) or 200μg/ml catalase and 0.2 percent W/V ethyl alcohol (B). The curves are plotted from numbers given in Table 21-I and Table 21-II.

Equilibration experiments of three hours duration with glutathione solutions to which had been added catalase (100μg/ml) and 3-amino-1,2,4-triazole (AT), have shown that AT in the presence of diffusing hydrogen peroxide (1 percent solution) produces a decrease in the mercury vapor uptake of about 66 percent at a 60 mM concentration and 81 percent at a 120 mM concentration of AT. Amino-triazole thus distinctly inhibits the stimulatory influence of liver catalase upon *in vitro* mercury vapor uptake in glutathione solutions at the experimental conditions.

Discussion

The experimental results given in Table 21-I and Figure 21-2 clearly indicate that liver catalase strongly stimulates the oxidation of mercury in the presence of hydrogen peroxide and glutathione in aqueous solutions. The inhibitory effect of a low concentration of ethyl alcohol and increasing concentrations of the catalase poison 3-amino-1,2,4-triazole strongly suggest that the primary hydrogen peroxide catalase complex is involved, especially since these inhibitors have high affinities for this complex (34, 10).

No correlation has been found between the rate of oxidation of glutathione caused by exposure to hydrogen peroxide vapor alone and the rate of mercury vapor uptake in the glutathione solutions. The decrease of about 18 percent in the concentration of glutathione in the samples to which had been added catalase, and which were exposed to the lowest concentration of hydrogen peroxide, may partly be related to the oxidation of mercury and binding of mercuric ions and partly to an incomplete protection offered by catalase against oxidation of glutathione. The decrease in the glutathione concentration seems to be independent of the concentration of catalase, but ethyl alcohol considerably diminishes the fall in the glutathione concentration. Further investigations are needed to settle the question about the possible participation of a free radical of glutathione in the oxidation of mercury.

The results presented in Table 21-II and Figure 21-3 suggest a competition for catalase between ethyl alcohol and hydrogen peroxide, changing in favor of the latter with increasing hydrogen peroxide concentrations.

It is still unknown why it is not possible to inhibit the uptake of mercury vapor in blood, erythrocyte suspensions, or erythrocyte lysates by means of 3-amino-1,2,4-triazole in the presence of a hydrogen peroxide generating system or hydrogen peroxide itself (29, 27). The reason may be that erythrocyte catalase differs somewhat from liver catalase. It is possible that glutathione peroxidase, which is considered most important in the erythrocytic destruction of hydrogen peroxide (9) also plays a major role in the oxidation of mercury. Because of difficulties in the purification of this enzyme the author has not yet succeeded in solving this problem.

In view of the finding that a small fraction of mercury exists in elemental form in the blood for a considerable time (19), one of the consequences of a more general catalase-induced oxidation of the metal may be that the initial oxidation and deposition of a certain amount of mercury takes place in close relation to the location of this enzyme. Catalase is a constituent of many cells in the body, and is found in high concentrations in the peroxisomes (6, 12, 23), which in the liver (and probably elsewhere) also contains several oxidases which may generate hydrogen peroxide (11). The deposition of mercury in the liver, kidneys and brain after exposure to mercury vapor (14) may in part be related to a peroxisomal oxidation of elemental mercury. A hepatic deposition of inorganic mercury in the peroxisomes and lysosomes has been demonstrated by Norseth (30).

The well-known deposition of mercury in the eye-lens capsule may possibly be related to an oxidation of mercury by means of glutathione peroxidase. This enzyme has been found in the lens together with hydrogen peroxide which probably is generated from an oxidation of ascorbic acid (31).

Further investigations of the ability of peroxidizing and oxidizing enzymes other than catalase to oxidize elemental mercury will presumably contribute to our knowledge of the initial oxidation and deposition at the cellular level and probably also to a better understanding of the general toxicity of metallic mercury vapor.

REFERENCES

1. Berlin, M., Jerksell, L.G. and Ubisch, H. von: Uptake and retention of mercury in the mouse brain. *Arch Environ Health*, *12*:33, 1966.
2. Berlin, M., Nordberg, G.F. and Serenius, F.: On the site and mechanism of mercury vapor resorption in the lung. *Arch Environ Health*, *18*:42, 1969.
3. Beutler E., Duron, O. and Kelly, B.M.: Improved method for the determination of blood glutathione. *J Lab Clin Med*, *61*:882, 1963.
4. Brin, M. and Yonemoto, R.H.: Stimulation of the glucose oxidative pathway in human erythrocytes by methylene blue. *J Biol Chem*, *230*:307, 1958.
5. Carver, M.J. and Ryan, W.L.: Stimulation of erythrocyte metabolism by menadione. *Proc Soc Exp Biol Med*, *104*:710, 1960.
6. Chang, C-H., Schiller, B. and Goldfischer, S.: Small cytoplasmatic bodies in the loop of Henle and distal convoluted tubule that resemble peroxisomes. *J Histochem Cytochem*, *19*:56, 1971.
7. Clarkson, T.W., Gatzy, J. and Dalton, C.: Studies on the equilibration

of mercury vapor with blood. Univ. Rochester AEC Report No. 582, 1961.
8. Clarkson, T.W.: Toxicological aspects. *Ann Occup Hyg*, 8:73, 1965.
9. Cohen, G. and Hochstein, P.: Glutathione peroxidase: The primary agent for the elimination of hydrogen peroxide in erythrocytes. *Biochemistry*, 2:1420, 1963.
10. Cohen, G. and Hochstein, P.: Generation of hydrogen peroxide in erythrocytes by hemolytic agents. *Biochemistry*, 3:895, 1964.
11. Essner, E.: Localization of peroxidase activity in microbodies of fetal mouse liver. *J Histochem Cytochem*, 17:454, 1969.
12. Fahimi, H.D.: Cytochemical localization of peroxidase activity in the rat hepatic microbodies (peroxisomes). *J Histochem Cytochem*, 16:547, 1968.
13. Foulkes, E.C. and Lemberg, R.: The formation of choleglobin and the role of catalase in the erythrocyte. *Proc Roy Soc Biol*, 136:435, 1949–1950.
14. Gage, J.C.: The distribution and excretion of inhaled mercury vapor, *Br J Ind Med*, 18:287, 1961.
15. Gibson, Q.H.: Inhibitors of gas transport. In R.M. Hochster and J.H. Quastel (Eds): *Metabolic Inhibitors*, vol. II. Academic Press, London, 1963, pp. 539–558.
16. Horwitz, L.: Observations on the effect of metallic mercury upon some microorganisms. *J Cell Comp Physiol*, 49:437, 1957.
17. Hughes, W.L.: A physiochemical rationale for the biological activity of mercury and its compounds. *Ann NY Acad Sci*, 65:454, 1957.
18. Magos, L.: Mercury-blood interaction and mercury uptake by the brain after vapor exposure. *Environ Res*, 1:323, 1967.
19. Magos, L.: Transport of elemental mercury by blood. *Arch Pharmacol Exp Pathol*, 259:183, 1968.
20. Magos, L.: Uptake of mercury by the brain. *Br J Ind Med*, 25:315, 1968.
21. Mills, G.C.: Hemoglobin catabolism I. Glutathione peroxidase, an erythrocyte enzyme which protects hemoglobin from oxidative breakdown. *J Biol Chem*, 229:189, 1957.
22. Mills, G.C.: Glutathione peroxidase and destruction of hydrogen peroxide in animal tissues. *Arch Biochem Biophys*, 86:1, 1960.
23. Morikawa, S. and Harada, T.: Immunohistochemical localization of catalase in mammalian tissues. *J Histochem Cytochem*, 17:30, 1968.
24. Nielsen Kudsk, F.: Absorption of mercury vapor from the respiratory tract in man. *Acta Pharmacol et Toxicol*, 23:250, 1965.
25. Nielsen Kudsk, F.: The influence of ethyl alcohol on the absorption of mercury vapour from the lungs in man. *Acta Pharmacol et Toxicol*, 23:263, 1965.
26. Nielsen Kudsk, F.: Kviksølvforurenet luft (mercury-contaminated Air), Thesis, Aarhus, 1966.

27. Nielsen Kudsk, F.: Unpublished investigations, 1968.
28. Nielsen Kudsk, F.: Factors influencing the in vitro uptake of mercury vapour in blood. Acta Pharmacol et Toxicol, 27:161, 1969.
29. Nielsen Kudsk, F.: Uptake of mercury vapor in blood in vivo and in vitro from Hg-containing air. Acta Pharmacol et Toxicol, 27:149, 1969.
30. Norseth, T.: The intracellular distribution of mercury in rat liver after a single injection of mercuric chloride. Biochem Pharmacol, 17:581, 1968.
31. Pirie, A.: Glutathione peroxidase in lens and a source of hydrogen peroxide in aqueous humour. Biochem J, 96:244, 1965.
32. Stock, A.: Über Verdampfung, Löslichkeit und Oxidation des metallischen Quecksilbers. Z Anorg Chem, 217:241, 1934.
33. Teisinger, J. and Fizerova-Bergerova, V.: Pulmonary retention and excretion of mercury vapors in man. Ind Med Surg, 34:580, 1965.
34. Tephly, T.R., Mannering, G.J. and Parks, R.E. Jr.: Studies on the mechanism of inhibition of liver and erythrocyte catalase activity by 3-amino-1,2,4-triazole (AT). J Pharmacol Exp Ther, 134:77, 1961.

DISCUSSION

Foulkes: I wonder whether the inhibition of catalase with sodium azide would not be much stronger than the one you can get with DAT?

Kudsk: I have tried this and I can tell you that sodium azide in the erythrocytes causes a very highly increased uptake of mercury vapor (unpublished observation). I think it's because it probably reacts with the oxyhemoglobin in the erythrocytes causing hydrogen peroxide formation. Jacob and Jandl* found that azide had an oxidative effect in erythrocytes and caused a marked stimulation of the pentose shunt activity in addition to its catalase inhibiting properties.

I may add that it has been shown that toluene acts on oxyhemoglobin and accelerates its reaction with ascorbic acid and hence the generation of hydrogen peroxide.† This possibly explains the highly increased uptake of mercury vapor in toluene extracted erythrocyte hemolysates found by Clarkson et al.‡ The toluene stimulated uptake is strongly inhibited by ethyl alcohol.§

Foulkes: How high is the catalase activity in renal parenchyma?
Kudsk: I don't know.
Foulkes: But there is definite evidence for the activity of the enzyme?
Kudsk: Yes, it has been shown histologically by means of immunochemical and color reactions both in the proximal and distal tubular cells.

* Biol Chemistry, 241:4243, 1966.
† Proc Roy Soc Biol, 136:435, 1949–50.
‡ Univ. Rochester AEC Report No. 582, 1961.
§ Acta Pharmacol et Toxicol, 27:149, 1969.

Magos: Inhibitors like azide or cyanide inhibit not only catalase activity but to a certain degree promote through peroxidation the oxidation of other compounds.

Kudsk: Yes, but cyanide distinctly inhibits the uptake of mercury in the erythrocytes and it has been shown by other workers that cyanide and azide do not inhibit the glutathione-peroxidase.

Magos: When you use blood, the presence of different systems besides catalase certainly must influence the oxidation of metallic mercury. It has been reported* that methemoglobin can act as peroxidase. How can you reconcile this finding with your theory?

Kudsk: I can only say that I have shown that sodium nitrate induced formation of methemoglobin does not increase the uptake of mercury. Only redox systems with a very high redox potential are able to oxidize metallic mercury directly.

I think that it possibly is free hydroxyl radicals which are capable of oxidizing metallic mercury in the erythrocytes and that this is brought about by means of catalase and possibly glutathione-peroxidase.

Berlin: Is there any indication that there is a reverse of this reaction that could be a release of mercury vapor from the systems you have studied?

Kudsk: No. I have tried to equilibrate for a similar time with mercury-free air and there's no significant loss of mercury from the system.

Berlin: I had a favorite theory some years ago that the erythrocytes transport to and release mercury in the brain, a function which should be inhibited by ethanol. However, we were never successful in showing a difference in uptake in the brain between alcohol treated mice and controls. We injected the animals (mice) with alcohol and some of them were really affected by the dose given but we never found any difference in brain uptake of mercury.

Miettinen: In which form is mercury in the brain? After 10 to 15 years people who have been submitted to metallic mercury still have 8 ppm in the brain and yet they do not show any clinical symptoms. Is it stored in the brain in a very curious way because it is not excreted at all and therefore it does not cause any clinical symptoms?

Berlin: We now expose animals to vapor with radioactive vapor and take out the brain and study the distribution after different exposure times. We are thus looking into this problem now and we hope to be able to produce radiograms showing distribution after several months of exposure and this may throw some light on the subject.

Kudsk: In your experiments with ethanol and mice what was the influence of ethanol upon the total brain burden of mercury? Was it lower, higher or about the same as in the control group?

Berlin: We didn't see any difference between the two groups.

Kudsk: And no difference in the distribution of mercury either?

* *Nature,* 173:720, 1954.

Berlin: We haven't looked into that so far.

Fassett: It might be appropriate at this point to stress the importance of some of the more recent concepts of central nervous system histology and biochemistry. I have particular reference to the neurological system.* These cells (the astrocytes, oligodendrocytes and the ependymal cells) account for some 40 percent of the bulk of the brain. There are an estimated ten neurological cells for each neurone. In contrast to neurones they can regenerate and take part in repair processes.

Since they perform many essential biochemical functions (e.g. oligodendrocytes—oxidative phosphorylation and maintenance of myelin sheaths; astrocytes—sodium, glucose, water and metabolite exchange between blood vessels and neurones; ependymal cells—secretion, spinal fluid) it would be logical to suppose that the biochemical lesions from CNS toxins would in many cases be found in these cells as well as in neurones.

Kudsk: Timm† has developed a histochemical sulfide-silver method for the demonstration of heavy metals based upon the principal of physical development known from photography. In experimental mercury intoxication the sulfide-silver method showed an accumulation of mercury within cytoplasmic granules which probably were identical to the lysosomes in the kidney, liver and brain.‡ A similar localization of heavy metals has been demonstrated in normal hepatic and brain cells by means of the method of Brun and Brunk.§ This method has been applied to electron microscopy but the sensitivity seems to suffer because of the preparatory techniques used hitherto.

I would like to add that in my opinion it would be of interest to investigate if methylene blue and the related phenothiazine compounds have any influence upon the fate and deposition of mercury in the body, including the cellular and subcellular levels, especially in conjunction with exposure to metallic mercury.

Gage: Concerning your suggestion that hydroxyl free radicals are involved, have you tried any other model hydroxylating systems such as that described by Udenfriend?**

Kudsk: No, but I have been thinking of trying.

Gage: Do you know that ethanol inhibits by mopping up free radicals?

Kudsk: It's known that ethanol has an high affinity for the primary hydrogen peroxide catalase complex so I think that it competes on the enzyme with mercury but I really don't know yet. The inhibition of mercury oxidation caused by aminotriazole which is competitive with ethanol for the reaction with the catalase complex supports this view.

* Philip Handler (Ed.): *Biology and the Future of Man.* London, Oxford Press, 1970, Chap. 9.
† *Dtsch Z f gerichtl Med* 46:706, 1958.
‡ *Int Arch Gewebepath Gewebehyg,* 22:236, 1966.
§ *J Histochem Cytochem,* 18:820, 1970.
** *J Biol Chem,* 208:731, 1954.

AUTHOR INDEX

(Italicized page numbers indicate discussion.)

Åberg, et al., 241
Albanus, L., 294
Altman, P.L., 273, 274
Ashbel, S.I., 255
Ashe, W., et al., 8, 9, 15

Bakulina, A.V., 32
Barber, J., *92, 95, 324, 343, 344, 345*
Benesch, R., et al., 149, 158
Berglund, F., 294, 295
Berlin, M., *20, 21, 48, 54, 94, 157, 158, 160, 175, 206, 207, 208, 241, 242, 275, 294, 295, 311, 370, 371*
Berlin, M., et al., 6, 7, 172, 297
Beutler, E., et al., 361
Bidstrup, P., et al., 13
Blanck, S., 99
Bolanowska, W., 292
Bornmann, G., et al., 6
Brendeford, M., 276
Brown, I.A., 32
Butler, W.H., 181

Cafruny, E.J., 125, 137
Cafruny, E.J., et al., 125
Carpenter, *291, 292*
Carroll, R., et al., 169
Carter, N.W., 350
Cerletti, U., 27
Cernik, A.A., 211, 212, 348
Christie, A.A., 327
Clarkson, T.W., *20, 21, 25, 48, 49, 110, 122, 175, 178, 210, 265, 266, 269, 275, 323, 352*
Clarkson, T.W., et al., 7, 144, 149, 174, 175, 358, 359, 369
Cotter, L.H., 60, 61
Creamer, B., 275

Cross, J.M., 144
Cross, R.J., 126, 127

Dahhan, S.S., 32
Daniel, J.W., 64
Daniel, J.W., et al., 175
Dutkiewicz, T., 255

Edwards, G., 26
Evans, D., 333

Falk, R., 238, 241
Fang, S.C., *108, 261, 291, 292, 293, 353*
Farah, A., 134
Fassett, D., *20, 110, 137, 138, 156, 208, 242, 261, 262, 275, 276, 293, 343, 344, 371*
Fiserova-Bergerova, V., 6
Fitzhugh, O.G., et al., 60, 62, 63, 278
Foulkes, E., *19, 49, 93, 108, 109, 110, 122, 137, 156, 241, 242, 262, 369*
Frankenberg, L., 294
Friberg, L., 8, 13, *19, 21, 25, 47, 48, 49,* 174
Fries, N., 318

Gage, J., 6, 48, 64, *93, 108, 110, 123, 156, 206, 212, 274, 292, 293, 343, 344, 352, 353, 354, 371*
Goldwater, L., *19, 21, 49, 108, 158, 184, 185, 231, 242, 260, 261, 310*
Grant, C.A., *50,* 188, *207, 241, 310, 311, 312*
Greco, A., 26
Greenland, L., 99
Greenwood, M., 175
Gussin, R.Z., 125

von Haartman, U., 294, 295
Haeusler, V., 13
Hammer, D., 92, 242, 312
Handler, P., 371
Harada, Y., 33, 34
Havill, et al., 185
Hayes, A., 6, 288
Hellberg, J., 294
Hepp, H., 27
Herdman, R.C., 42, 54
Herman, S., 48, 242
Hook, J.B., 127, 137, 138, 273
Horwitz, L., 358, 359
Hughes, W.L., 356
Hunter, D., 31, 306, 308
Hunter, D., et al., 30, 31, 303, 304, 306, 308

Ivy, A.C., 274

Jablonska, J., 247
Jacob, 369
Jakubowski, et al., 262, 276, 278
Jandl, 369
Jellum, 266
Jernelöv, A., 207, 208, 274, 294, 311, 312, 316, 323, 324, 325
Jones, V.D., et al., 114, 117
Joselow, M., 55, 311
Juliusberg, F., 6

Kägi, J.H.R., 250
Katsunuma, H., 209
Katzantzis, G., 20, 21, 54, 109, 206, 241, 242, 272, 275, 292, 311, 312, 353
Kesic, B., 13
Kessler, R.H., et al., 140, 152
Khera, K.S., 39
Kitamura, S., et al., 211, 316
Klein, D., 50, 108, 260
Koizumi, 209
Kolbye, A., 54
Kournossov, V., 11
Kuchinskas, et al., 157
Kudsk, N., 6, 369, 370, 371
Kurland, L.T., 50, 54, 55, 207

Landner, L., 315, 316, 318, 319
Lenz, et al, 157

Lewis, A.E., 127
Lucier, G., 276

Magos, L., 6, 95, 110, 137, 157, 185, 186, 207, 208, 211, 212, 231, 262, 263, 273, 348, 358, 370
McDuffie, B.R., 20, 41, 50, 53, 54, 95, 240, 242, 345
McMaster, P.D., 274
Merville, R., et al., 64
Miettinen, J., 25, 53, 206, 207, 208, 232, 240, 241, 242, 311, 370
Miller, T.B., et al., 142, 156, 210, 344
Miyakawa, T., et al., 31, 303
Moriyama, H., 32
Morrow, P., et al., 7
Mulder, D.W., 32
Munro, J.R., 127

Neal, P., et al., 13
Nechay, B.R., 54, 94, 109, 122, 123, 156, 242, 344
Nelson, J.A., 111
Nordberg, G., 10, 294
Norseth, T., 22, 32, 95, 207, 210, 265, 266, 269, 272, 273, 274, 275, 276, 324, 352
Nose, K., 210

Oginski, M., 255
Öhman, H., 9
Oliver, J., 263
Orfaly, H., 32
Östlund, K., 38, 208

Petrovic, C., 60
Piechoka, J., 60
Pinajian, J.J., 144
Piotrowski, J.K., 185, 260, 261, 262, 263, 276, 292, 293, 353
Piotrowski, J., et al., 262
Piscator, M., 260, 263, 274, 292
Poleshajev, N., 11, 13
Prickett, C., et al., 7

Rather, J.L., 263
Rentos, P., 13
Ross, C.R., 134

Author Index

Rothstein, A., 6, 82, *92, 93, 94, 95, 109, 110, 122, 137*, 288, *292, 293, 343, 344*
Russell, D.S., 31, 306, 308

Sadakane, T., 210
Saito, M., et al., 303
Sapota, A., 247
Schamberg, J., et al., 6
Schütz, A., 188
Seligman, E., 13
Shieh, Y.J., 325
Silverman, M., et al., 100
Skerfving, S., 294, 295
Skerfving, S., et al., 40
Small, H., 266
Smith, J., 53
Smith, C.R., 274
Smith, H.W., et al., 17
Smith, R., et al., 9, 13, 15, 16
Sodee, D., 9
Soldatovic, D., 60
Stannard, J.N., *94*
Stock, A., 358
Sundvall, A., 294
Suzuki, T., *21, 47, 49, 231, 310, 353*
Suzuki, T., et al., 32
Swensson, Å., 166, 169

Taggart, J.V., 126, 127
Takeuchi, T., 29, 31, 165, 303
Takeda, Y., et al., 210, 211, 289
Takeshita, M., 81
Taylor, W., et al., 32

Teisinger, J., 6
Tejning, S., 9
Trachtenberg, I., 10, 12, 13, 14
Trojanowska, B., et al., 182

Ulfvarson, U., 8, 166, 169, 348
Ullberg, S., 175, 189

Vallee, B.L., 250
Vallee, B.L., et al., 248
Vostal, J.J., 75, *94, 137, 156, 157, 158, 160, 241, 242, 261, 262, 263, 273, 274, 275, 292, 312, 353, 354*
Vostal, J.J., et al., 157, 158,

Wahlberg, J., 6, 7
Warren, M.M., 212
Webb, L.L., 69, 341
Weiner, I.M., et al., 114, 142, 143, 148, 149, 156, 350, 351
Weiss, B., *20*
Went, F.W., 326
Westöö, G., 234, 318
Whitaker, J.R., 279
White, J., 82
Whitmore, F.C., 56
Wineiwska, J.M., et al., 278, 288
Wood, R. *208*
Wright, G.F., 57

Xintaras, C. *206, 207*

Yamaguchi, S., et al., 60

SUBJECT INDEX

A

Abavit, 63
Absorption, 6, 23, 37
Absorption
  inorganic mercury, 238
  skin, 58
Accumulation
  methylmercury, 299
  squirrel monkey, 299
Active transport, 88
Acute exposure, 35
Aerosols, 7, 13
Agano River, 29
Agricultural use, 27, 28
Alamogordo, 30, 32, 36
Albumin globulin ratio, 60
Algal systems
  biotransformation, 325–345
  mercury uptake, 325–345
  volatilization, 325–345
Alkoxy and miscellaneous mercurials (list), 58
Alkoxyalkymercurials, 56–67
Alkymercury, 23–55
Allowable daily intake (ADI), 24, 40, 41
Alpha amino isobutyric acid (AIB), 102, 103, 106
Aminoaciduria, 99–110
Amino triazole, 359, 371
Amylase, 292
Amyotrophic lateral sclerosis (ALS), 32
Analysis, 60
Aneuploidy, 40
Anion channel, 93
Anion pathway, 80–81
Anion permeability, 88
Anionic mercurials
  meralluride, 124–138
  mercaptomerin, 124–138
Aquatic biotransformation, 29
Arsenic, 50
Arylmercurials, 56–67
Ascorbic acid
  mercury retention, 170, 171
  mercury uptake, 167, 169
  protective action, 184
Asthenic vegetative syndrome, 14
Astrocytes, 371
Astrocytosis, 300
Ataxia, 23, 33, 296, 306
Athetosis, 33
ATPase, 88
Autoradiography, 296
Autoradiography
  plants, 328

B

Babinski's sign, 31
Bacterial systems
  biotransformation, 325–345
  mercury uptake, 325–345
  volatilization, 325–345
BAL, 29
BAL
  bile Hg excretion, 275
  brain mercury, 177
  Hg Cl$_2$ excretion, 176, 177
  maleate effect, 180
Behavior, 10, 40
Betaine, 318
Bile
  flow, 273
  hydroxy phenylmercury, 354
  methylmercury binding, 270
  phenylmercury, 353
  rat, 268, 273, 274
Biliary excretion, 158

376

Subject Index

Biological half life, 38, 51, 52
Biotransformation, 7
Blindness, 26, 30
Blood, 21, 34, 36, 37, 41, 48, 51, 52, 59, 60, 64, 68, 70
Blood
  metallic mercury, 174
  methylmercury dicyandiamide, 174
Blood brain barrier, 26, 27, 28, 35
Blood brain barrier
  elemental mercury, 356
  ethylmercury, 229
Blood brain ratio
  species difference, 310
Blood mercury
  monkey, 187
BMHP — 1-bromomercuri-2-hydroxypropane, 82
Body burden, 38, 49
Brain, 7, 8, 9, 10, 20, 21, 24, 32, 34, 35, 38, 39, 50, 60, 69, 78
Brain
  inorganic mercury, 172
  mercury removal, 165
  mercury vapor, 172
  methylmercury dicyandiamide, 174
Brain development, 40

C

Cadmium, 110
Cadmium
  metallothionein induction, 261
Cadmium sulfide, 327, 330
Calcarine sulci, 300
Cammermeyer technique, 296
Canada, 54
Carbon mercury bond
  stability, 148
Carvone, 326
Cation channel, 93
Cation permeability, 88
Cats, 10
Cats
  blood methylmercury, 306
  brain methylmercury, 306
  methylmercury, 285
CCCP, 338

CCCP
  oxidative phosphorylation, 338
  photosynthetic phosphorylation, 338
C-mitosis, 6, 24, 39, 40
Cell membrane, 23, 27, 28, 78, 79
Cell membrane permeability, 39
Central nervous system, 5, 13, 15, 16, 27, 34, 69, 72, 78
Cerebellar cortical atrophy, 31
Cerebellum, 42
Cerebral palsy, 33, 40, 43
Cerebrospinal fluid, 21, 34
Ceresan (*see* ethylmercuric chloride)
"Chemical specificity," 80
Chick, 59
Chlorella pyrenoidosa, 325
Chlorella pyrenoidosa
  inorganic mercury uptake, 334
Chloromerodrin, 74–95, 111, 112, 143
Chloromerodrin
  bile excretion, 157
Choline, 318
Chromosome breakage, 24
Chromosome breaks, 40
Chronic exposure, 30
Clorometholxy propylmercuric acetate, 64
CoA, 69
Colloidal osmotic pressure, 88
Complexing agents, 176–178
Conditioned reflexes, 15, 20
Congenital disease, 32
Cortex
  atrophy, 299
Cortical damage
  atrophy, 31
  individual differences, 307
Crab, 54
Creatinine clearance, 101
Cyanide
  mercury uptake, 370
Cysteine, 69, 71, 89, 111, 144, 149, 152, 153, 319, 350
Cysteine
  complex injection, 176
  dissociation constants, 157
  mercury uptake, 167
Cytochrome C., 292

D

Deafness, 26
Desquamation
  intestinal cells 275
  tubular cells, 181, 182
Determinations, 13
Determinations
  inorganic mercury, 145
  metallothionein, 249
  polarography, 158
Dettol, 331
Diethylaminoethyl diphenyl propylacetate, 354
Diethylmercury, 26
Diffusible mercury, 73, 75, 89
Diffusible sulfhydryl groups, 73
Diffusion, 80
Dimercaptopropanol, 157, 158, 319
Dimethylmercury, 38, 92, 324
Dimethylmercury
  sewage, 345
Dinitro fluorobenzene (DNFB), 84
Dissociation constants, 71
Distribution, 25, 34, 38
Distribution
  methylmercury, 299, 304
  squirrel monkey, 299
Dithizone, 352
Diuresis
  lock and key theory, 141, 151, 152
Diuretic effects
  in acidosis, 112, 146
  in alkalosis, 112
Diuretics, 56, 57, 60, 64, 78, 94, 110
Diuretics
  chloromerodrin, 124–138
  kidney levels, 146, 147
  meralluride, 110–123
  mercaptomerin, 110–123
  mercuric chloride, 110–123
  mersalyl, 110–123
  PCMB, 101–105, 110–123, 124–138
  PCMBS, 111–123
  WIN 8518, 115
DNP
  kidney metabolism, 167
  mercury uptake, 179
Dogs, 8, 9, 74, 111, 124, 143

Dorsal roots, 31, 42, 303
Dorsal tracts, 31
Drosophila melanogaster, 39
Dysarthria, 30

E

Ecology, 25
Edema
  mercury ascorbic acid, 171–172
Effective renal plasma flow (ERPF), 101
Elemental mercury, 5, 6, 7, 19
Elemental mercury
  airoxidation, 356
  biological oxidation, 355–371
  microorganisms, 323, 324
  ozone, 356
  partition coefficient, 356
  peroxisomal oxidation, 367
  solubility, 356
Embryotoxicity, 39, 40
Endoplasmic reticulum
  mercury binding, 276
Enterohepatic circulation, 38
Environmental mercury, 28
Enzyme activity, 69
Ependymal cells, 371
Equilibration of mercury
  in blood, 76–77
erethismus mercurialis, 5
Ethyl alcohol
  brain mercury uptake, 370
Ethylmercuric chloride, 32
Ethylmercury, 26, 32
Ethylmercury
  bile, 229
  biological half time, 219, 228, 229
  biotransformation, 210
  blood, 215
  blood brain barrier, 229
  brain, 210, 212, 217, 218, 219, 222, 224, 226, 227, 228
  cerebellar cortex, 217
  feces, 212
  hair, 211, 215, 217, 232
  HCl extraction, 214
  human intoxication, 210

Subject Index

human plasma, 214–219
inorganics, 224
kidney, 210, 211, 212, 217, 219, 223, 224, 226, 227, 228
kidney binding, 280–291
LD_{50}, 210
liver, 210, 211, 219, 223, 224, 226, 227, 228
MAC, 209–210
metabolism, 211, 229
mice, 210, 212, 213, 219
nervous system, 217
neurotoxicity, 210, 218, 230
plasma, 211
protein profiles, 283
rats, 210
red cell plasma ratio, 210
red cells, 211, 212, 219
retention, 221–228
skin, 217
spleen, 217
thimerosal, 211
urine, 211, 212, 215, 229
Ethylmercury chloride, 60
Ethylmercury p-toluene sulphanilide, 32
Excretion, 8, 9, 15, 16, 34, 38
Excretion
  ethylmercuric acetate, 164
  mercuric acetate, 164
  phenylmercuric acetate, 164

F

F.D.A., 60
F.D.A. guidelines, 41, 54
Feces, 8, 9, 38
Fetal cells, 40
Fish, 23, 28, 29, 33, 34, 36, 37, 40, 43, 47, 48
Fish
  methylmercury, 295
Fish consumption, 47
Food chain, 28
Fungicides, 23, 27, 61

G

Gallbladder
  mercury reabsorption, 274
Galveston Bay, 54

Gastrointestinal tract, 5, 6, 7, 23, 37
"Geographical specificity," 80
Germicides, 61
Glia cells
  methylmercury, 299
  proliferation, 300
Glomerular filtration rate (GFR), 101
Glucose transport, 125
Glutathione, 29, 69, 79, 319, 359, 360
Glutathione
  in bile, 271
Glutathione peroxidase, 360, 367
Granosan-M (see ethylmercury p-toluene sulphanilide)
Guinea pig, 7, 59

H

Hair, 33, 34, 35, 36, 37, 41, 47, 48, 49, 51, 52
Hair blood ratio, 36, 48
Halibut, 52, 54
Hearing, 23
Heart, 63
Hemoglobin, 23, 28, 32, 87, 90
Hemoglobin-SH, 73
Homocysteine, 319
Human, 7, 12, 16, 37, 56, 58, 62, 64, 74
Human poisonings, 28
Hydrogen peroxide catalase, 359
Hydrothermal orebodies, 332
Hydroxy phenylmercury, 353
Hydroxy phenylmercury acetate, 351, 352
Hyperkinesia, 31
Hypoplasia, 33
Hypoxydosis
  convulsions, 308

I

Individual sensitivity, 43
Individual variation, 15, 16, 37
Industrial exposure, 21, 23, 28
Industrial use, 28
Inhalation, 6
Inorganic mercury, 5, 6, 7, 8, 25, 53, 78
Inorganic mercury
  absorption, 238

biological half time, 238,
brain, 172
cation transport system, 333
excretion, 238
kidney binding, 280–291
leaves, 328
membranes, 341
peroxisomes, 367
protein profiles, 283
rats, 256
red cell plasma ratio, 238
retention, 241
roots, 328
unithiol, 256
xylem sap, 328, 329
Intellectual deterioration, 23
Intellectual retardation, 33
Intellectual symptoms, 29, 30
International Conference of Occupational Medicine, 187
Intestinal bacterial flora, 344
Intestinal cells
  desquamation, 275
  turnover rates, 275
Intestinal epithelial turnover, 38
Iraq, 32
Iron, 295
Itai Itai disease, 49

J

Japan, 23, 48

K

Kidney, 5, 8, 9, 10, 16, 20. 21, 28. 37, 60, 61, 62, 63, 69
Kidney
  adenosine triphosphatase activity, 111–123
  argenine transport, 109
  aspartate transport, 105, 108, 109
  binding sites, 178
  binding time course, 286, 287, 288, 289
  ethylmercurials, 174
  glumatate carrier system, 101, 106
  half times, 287–290
  Hg-metallothionein complex, 258
  innulin, 100, 107, 109
  lumenal membrane effects, 99–110
  lysine transport, 109
  mercury accumulation, 280, 281
  mercury binding, 277–293
  mercury kinetics, 167–186
  mercury release, 259
  mercury removal, 165
  mercury removal rates, 287–290
  metabolism, 167
  metallothionein, 288
  methionine transport, 101, 106
  methoxyethylmercury, 174, 175
  peritubular membrane effects, 99–110
  phenylalanine transport, 101, 106
  protective role, 167, 168
  protein profiles, 282, 283
  repeated doses Hg, 262
  retention, 258
  single dose Hg, 262
  target organ, 168
  thioacetamide, 254
  toxic levels, 258
  tubular cell desquamation, 167
  tubular necrosis, 167, 181
  uranium, 103, 107, 108
Kidney brain ratio, 35
Kidney levels
  in acidosis, 146, 147, 151, 152
  in alkalosis, 146, 147, 151, 152

L

LD^{50}
  alkoxyalkyl, 64
  ascorbic acid effect, 169
  mercuric chloride, 169
  mercury vapor, 75
  phenylmercurials, 59
Leptomeninges
  lymphocytes, 299
  eosinophils, 299
Ligands, 5, 79
Lipid pathway, 80–82
Lipoate, 69
Liver, 37, 60, 61, 63
Liver
  inorganic mercury, 169
  mercury removal, 165

Subject Index

methylmercury dicyandiamide, 174, 175
Liver brain ratio, 35
Liver catalase
  elemental mercury, 361
Lobster, 54
Long ashton nutrient, 327
Long Island, 54

M

MAC, 58
Manganese, 50
Meiotic chromosomal disjunction, 39
Membrane sulfhydrils, 80–95
Membrane
  aqueous channels, 87
  leakage, 137
Membrane SH populations, 82
Menadione
  red cell mercury uptake, 360
Mentha spicata, 325, 326
Meralluride, 110–123
Mercaptans, 68–95
Mercaptides, 71–72
Mercaptoacetic acid, 319
Mercaptomerin, 110–123
Mercurial diuretics
  acid base balance, 142
  cysteine potentiation, 139
  glutathione potentiation, 139
  mercurial ion hypothesis, 147, 156
  N-acetylpenicillamine effects, 139
  penicillamine effects, 139
  prevention of action, 142
  steric requirements, 152
  structural requirements, 140, 141
Mercuric chloride, 29, 110–123
Mercuric chloride, 144
Mercuric ion hypothesis, 147
Mercurous mercuric ratio, 358
Mercury
  edema treatment, 140
Mercury complex
  blood serum, 260
Mercury cysteine, 110
Mercury cysteine
  formation constant, 157

Mercury distribution
  kidney, 278
  liver, 278
  rats, 278
Mercury nitrate, 21
Mercury penicillamine
  formation constant, 157
Mercury vapor, 9, 12, 13, 16, 23–55, 76, 77, 78
Mercury vapor
  blood, 358
  brain, 172, 213, 241
  ethyl alcohol, 359, 363, 364
  glutathione, 363, 364
  hair, 213
  HCl extraction, 214
  hydrogen peroxide, 363, 364
  mice, 213
  uptake, 356
Mercury vapor exposure
  inorganic estimations, 212, 213
Mersalyl, 110–123, 143
Metabolism, 15, 24, 28, 34, 38
Metabolism
  depression, 137
  methoxyethylmercury, 346–354
  phenylmercury, 346–354
  PMA (methoxymercuric Cl), 64
Metallothionein, 178, 185, 247–263, 288
Metallothionein
  alkylmercury, 260
  binding capacity, 258, 259
  biosynthesis, 249, 250
  cadmium, 258
  cadmium effect, 261
  cadmium stimulation, 262, 263
  cysteine incorporation, 252
  cytoplasm, 258
  determination, 249
  detoxification, 259
  inorganic mercury, 260
  kidney, 250
  kidney binding, 261
  kidney location, 261
  liver, 248, 249, 250, 258
  molecular weight, 248
  phenylmercury, 260
  purity, 260
  rats, 248, 250

selectivity, 258
SH stability, 292
urine, 255
Metallothionein biosynthesis
thioacetamide, 255
Methanobacterium, 315
Methanogenic bacteria, 315
Methemoglobin
peroxidase, 370
Methemoglobin reductase, 360
Methionine
biosynthesis and methylation, 316
Methoxyethylmercury
bile, 348
biological half life, 174
excretion half time, 348
kidney, 350
liver, 350
metabolism, 346–354
Methoxyethylmercury nitrate, 26
Methylating enzyme
feedback, 320
Methylation, 23, 28, 315–324
Methylation
anaerobic, 315
F4-OR8-1, 318
hydrogen sulfide, 315, 316
Methylene blue
red cell mercury uptake, 359
Methylmercury, 23, 24, 25, 26, 27, 28–55, 69, 74, 75, 77, 78, 82, 106
Methylmercury
absorption, 187, 204
ansiform lobes, 306
astrocytosis, 300
autoradiograms, 195–198
axon fragmentation, 304
axon swelling, 304
bacterial degradation, 265
behavior, 187–208
bile duct ligation, 265–267
biliary complexes, 264–276
biological half life, 38, 190, 206
biological half time, 234, 235, 236, 237, 238, 239
biotransformation, 207, 265
blood levels, 190, 191
brachial plexus, 304
brain, 187–208, 273, 347

brain distribution, 194–196
calcarine cortex, 195
cats, 295
cerebellar lesion, 206
cerebellum, 306, 308
cerebral cortex, 308
cortical atrophy, 300
cortical lesion, 206
degradation, 274
distribution, 304
dorsal roots, 303
EEG, 206
elimination, 235
enterohepatic circulation, 265–269
excretion, 204
feces, 235, 238
fish, 208, 242, 295, 310, 311
fish protein, 234
foveal vision, 205
fur, 207, 208
gastrointestinal tract, 265
gill inflammation, 311
growth dilution, 324
hair, 242
head, 238
humans, 233–243
kidney, 273
lingula, 306
liver, 273
maximum acceptable blood concentration, 311
metabolism, 207
mice, 303
microgliosis, 300
morphology changes, 195
motor coordination, 198, 205
myelin sheath disintegration, 304
neocortex, 307
neurolemmal sheath, 304
nodulus, 306
"no effect level," 311, 312
occipital region, 195
organ distribution, 265
pathology, 294–312
periferal nerves, 207, 303, 308
periferal vision, 205
plasma, 234
posterior column lesion, 206
rats, 303, 304

Subject Index

red cell plasma ratio, 238, 267
red cells, 234
redistribution, 308
Saimiri sciureus, 187–208
Schwann cell proliferation, 304
sciatic nerve, 304
sensory nerves, 308
sewage, 345
solubility, 275, 276
species differences, 310
spleen, 311
squirrel monkey, 187–208, 295
urine, 235, 238
uvula, 306
visual cortex, 205, 207, 308
visual discrimination, 198, 202
visual discrimination test, 302, 303
visual impairment, 196, 197
Methyl Mercury in Fish, 24, 25
Methyl Mercury in Fish. A Toxicologic-Epidemiologic Evaluation of Risks, 50
Mice, 39, 59
Microbial transformation, 25
Microgliosis, 300
Microsomal enzyme system
    phenylmercury, 352
Milk, 32
Minamata, 29, 32, 35, 36, 40, 49, 50, 54
Minamata Bay, 25, 33
Minamata disease, 24, 30, 35, 55, 165
Monkey, 74
Muramidase, 292
Mutagenesis, 40

N

N-acetylpenicillamine, 152, 153
NADPH dependent diaphorase, 360
Natriuretic action, 112
Natriuretic effect
    in acidosis, 134
    in carbon mercury bond, 134
    pH effect, 134
Nephrotic syndrome, 6
Nerve cells, 9, 33, 92
Nerve conduction velocity, 31
Nervous function, 10, 11
Nervous system, 23, 28, 29, 42, 43
Nervous system
    fixation, 296, 297
    pathology, 296
Neuronal changes
    Nissl's "ischaomic" type, 307
Neuronal damage, 28
Neurons
    methylmercury, 299
    nutrition, 312
Neurospora
    detoxification, 318
    mercury tolerance, 318
Neurospora crassa, 315, 316
New Mexico, 23, 55
Niigata, 24, 29, 35, 36, 47, 48, 49, 50, 242
Niigata Report, 206
NMN (N-methylnicotinamide)
    uptake inhibition, 124
Nutrition, 24

O

Occupational exposure, 13–16, 20, 58
Oligodendrocytes, 371
Oral administration, 37
Organ concentrations, 34
Organ specificities, 77–78
Organic Compounds of Mercury, 56
Organic mercury, 26
Ossious abnormalities, 40
Ouabain, 117
Oysters, 54

P

PAH, 124–138
PAH
    transport inhibition, 125
    uptake inhibition, 124
PAH clearance, 100, 101
PAH S/M ratio, 128, 129, 130, 131, 132
PAH uptake
    diuretic inhibitory potency, 134
Pancreatic secretions, 38
Paresis, 31
Paresthesia, 23
Pathology, 31
Pathology
    methylmercury, 294–312

PCMB, 74–95, 101, 102, 103, 104, 105, 110–123, 143
PCMBS, 74–95, 111–123
Penicillamine, 29, 152, 153, 154, 156, 178
Penicillamine
  dissociation constants, 157
  maleate effect, 179–180
  metabolism, 157
Pentose shunt
  methylene blue, 360
  menadione, 360
Periferal nerves, 23, 31, 303
Permeability, 78–80, 81
Pharmacokinetics, 162–243
Phenothiazine, 371
Phenylmercuric salts (lists), 57
  of inorganic acids
  of aromatic acids
  of aliphatic acids
Phenylmercury, 22, 23
Phenylmercury
  brain, 347
  excretion, 350, 351
  kidney binding, 280–291
  metabolism, 346–354
  protein profiles, 283
  stability, 149
Placental barrier, 23, 28, 32
Plant systems
  biotransformation, 325–345
  mercury uptake, 325–345
  volatilization, 325–345
Plasma, 76
Plasma red cell ratio, 157, 158
PMA (phenylmercuric acetate), 26, 32, 56
Polyploidy, 40
Posterior root ganglion, 31
Potassium exchange
  temperature, 339
Potassium exchange system, 334, 337
Potassium exchange system
  cysteine effect, 337
Potassium leakage, 87, 88
Potassium permeability, 87, 88
Protein configuration, 72
Proteins, 5
Proteinuria, 6

Proximal tubule cells
  anion transport, 125
  cation transport, 125
Pseudomonas fluorescens, 325
Psychologic deterioration, 30
Purkinje cells
  methylmercury, 307

R

Rabbit, 8, 9, 10, 20, 59, 60, 62, 74, 99
Radio-chemical Centre, 327
Rat, 6, 8, 9, 11, 12, 20, 39, 56, 59, 60, 61, 62, 74, 75, 117, 124, 169, 174
Red blood cell model, 78–95
Red blood cells (erythrocytes), 8, 9, 28, 34, 37, 40, 70, 76
Red cell membrane, 71
Red cell plasma ratio
  bile derivative, 273
  inorganic mercury, 238
  methylmercury, 238
  methylmercury cysteine, 273
Red cells, 47, 49
Reflexes, 30
Renal diuresis
  mercuric ion, 139–160
  organomercurials, 139–160
Renal electrolyte transport, 111–123
Renal failure
  inorganic mercury, 168
Renal function, 10
Renal organic ion transport, 124–138
Renal sodium transport, 111–123
Retardation, 30
Retention, 28
Ribonuclease, 292

S

Safety factors, 311
Salyrgan, 60
Scatchard Plot, 82, 84
Sediment, 28
Sediment
  aerobic, 316, 317
  anaerobic, 316, 317
Seed dressings
  methylmercury dicyandiamide, 346

Subject Index

Seed grain, 30
Seizures, 33
Selectivity, 23
Selenium, 50
Serum, 36, 70
Sex differences, 15
SH enzymes, 39
SH groups, 38
Shellfish, 28, 33, 54
Short carbon chain alkylmercuries, 27
Shrimp, 54
SITS, 81–95
SITS-insensitive channel, 93
SITS-sensitive channel, 93
Sixteenth International Congress on Occupational Health, 21
SKF 525A, 354
Skin, 6, 7, 30
Sodium azide, 369
Sodium flouride
  mercury excretion, 181
Sodium maleate
  diffusable thiols, 167
  kidney mercury, 179
  mercury binding, 186
  renal mercury, 167
Sodium permeability, 87, 88
Sodium potassium exchange, 340
Sodium transport
  gull salt gland, 112
  toad bladder, 112
Soil air mercury concentrations, 333
Spasticity, 33
Species differences
  bile secretion, 269
  blood binding, 269
  fectal excretion, 270
  liver, 271
  methylmercury, 310
  red cell plasma ratio, 270
Specificity, 80
Spleen, 60
Spleen
  inorganic mercury, 169
Spongy encephalopathies, 50
Squid axons, 92
Squirrel monkey, 21

Squirrel monkey
  cerebral cortex, 297
  methylmercury, 295
Staphylococcus, 316
Stockholm, 50
Stockholm Meeting (Nov '68), 48
Stockholm Symposium, 5
Sweden, 48, 49
Swedish Commission, 52
Swedish Commission on Evaluating the Toxicity of Mercury in Fish, 40
Swordfish, 41, 42, 51, 52
Syphilis, 56

T

Teratogenesis, 24, 54
Teratogenicity, 39, 40
Terpenes, 326
Thallium, 50
Thiamine, 295
Thioacetamide, 254
Thioacetamide
  Hg-metallothionein complex, 255
  membrane permeability, 255
  metallothionein biosynthesis, 255
Thioglycolate, 69
Thiol groups, 5
Threshold
  cortical damage, 303
Threshold limit value, 15, 20
Thyroid, 12, 14, 20
Thyroxin, 20
Tissue distribution, 76
TLV (methylmercury), 48
TLV (phenylmercury), 58
Tokyo, 187
Toluene
  oxyhemoglobin, 369
Torry Research Institute, 328
Transaminase inhibition, 39
Transmethylation
  methionine, 316
Transpiration, 326–332
Transpiration
  organic volatiles, 326
Tremor, 16, 31
Tubular reabsorption, 99
Tuna, 51, 52

Turbot, 52
Two point discrimination, 31

U

Unithiol, 255
Unithiol
  kidney, 257
  kidney damage, 259
  mercury distribution, 257
  rats, 256
Urine, 8, 9, 15, 16, 19, 20, 51, 52, 58, 59

V

Vision, 23, 30

Visual discrimination test, 187
Visual field testing, 48
Visual impairment
  squirrel monkey, 297
Vitamin E, 295

W

Weight reduction, 40
Weight-Watchers, Inc., 41
Whale meat, 54
Wilson's disease, 153
Wisconsin Alumni Research Foundation (WARF), 62